Tourism Geography,
Second edition

Tourism is an intensely geographic phenomenon. It stimulates large-scale movement of people on an increasingly globalised scale and forges distinctive relationships between people – as tourists – and the spaces and places that they visit. It has significant implications for processes of physical development and resource exploitation, whilst the presence of visitors will frequently exert a range of economic, social, cultural and environmental impacts that have important implications for local geographies.

This second edition of *Tourism Geography: A New Synthesis* aims to develop a critical understanding of how many different geographies of tourism are created and maintained. Drawing on a range of social science disciplines, and adopting both an historical and a contemporary perspective, the discussion strives to connect tourism to key geographical concepts relating to globalisation, mobility, new geographies of production and consumption, and post-industrial change. The work is arranged in three main parts. In Part I the discussion is focused on how spatial patterns of tourism are formed and how they evolve through time. Part II offers an extended discussion of how tourism relates to places that are toured through an examination of physical and economic development, socio-cultural and environmental relations and the role of planning for tourism as a mechanism for the spatial regulation of the activity. Part III develops a range of material that is new for this second edition and which considers some of the newer influences upon geographies of tourism, including place promotion, the development of new forms of urban tourism, heritage, identity and embodied forms of tourism.

Drawing on an extended set of new case studies from across the world and supported by up-to-date statistical information, the text offers a concise yet comprehensive review of geographies of tourism and the ways in which geographers can interpret this important contemporary process. Written primarily as an introductory text for students, the book includes guidance for further study in each chapter, summary bibliographies and useful Internet sites that can form the basis for independent work.

Stephen Williams is Professor of Human Geography and Head of Applied Sciences at Staffordshire University, UK. He is also the Director of the University's Institute for Environment, Sustainability and Regeneration. His publications include *Outdoor Recreation and the Urban Environment* (Routledge), and *Tourism and Recreation* (Prentice Hall).

'This revised edition of Stephen Williams's book contributes significantly to an understanding of the importance of geographical issues in the theory and practice of tourism. Leavened with useful case studies and questions this new edition links some conventional and basic issues to the intricacies of the geographical perspectives to provide some new insights into practices of tourism.'

Dr Michael Fagance, *The University of Queensland, Australia*

'This new edition provides a comprehensive introductory text based on an excellent range of case studies.'

Professor Gareth Shaw, *University of Exeter, UK*

Tourism Geography
A new synthesis

Second edition

Stephen Williams

Routledge
Taylor & Francis Group

LONDON AND NEW YORK

First published 2009
by Routledge
2 Park Square, Milton Park, Abingdon, Oxon, OX14 4RN

Simultaneously published in the USA and Canada
by Routledge
270 Madison Avenue, New York, NY 10016

Routledge is an imprint of the Taylor & Francis Group, an informa business

Typeset in Times New Roman
by Keystroke, 28 High Street, Tettenhall, Wolverhampton
Printed and bound in Great Britain
by TJ International Ltd, Padstow, Cornwall

British Library Cataloguing in Publication Data
A catalogue record for this book is available from the British Library

Library of Congress Cataloguing in Publication Data
Williams, Stephen, 1951 May 30–
    Tourism geography : a new synthesis / Stephen Williams. — 2nd ed.
        p. cm.
    Includes bibliographical references and index.
    1. Tourism. I. Title.
    G155.A1W49 2009
    338.4′791—dc22   2008048783

ISBN 10: 0–415–39425–2 (hbk)
ISBN 10: 0–415–39426–0 (pbk)
ISBN 10: 0–203–87755–1 (ebk)

ISBN 13: 978–0–415–39425–3 (hbk)
ISBN 13: 978–0–415–39426–0 (pbk)
ISBN 13: 978–0–203–87755–5 (ebk)

Contents

Plates

Figures

Tables

Boxed Case Studies

Acknowledgements

Once again I am pleased to acknowledge the assistance and support of a number of individuals who have helped bring this book to completion. Andrew Mould and Michael P. Jones at Routledge have provided support and encouragement throughout the duration of the project and shown great patience with the many delays that the work has suffered. I would also like to thank the four anonymous reviewers of the draft manuscript for their constructive comments and suggestions for improvement to the text.

Amongst my colleagues at Staffordshire University I would like to thank Louise Bonner and Fiona Tweed for the use of their photographs of the Taj Mahal (Louise) and Snowdonia (Fiona). However, particular thanks go to Rosie Duncan (Cartographer in the Department of Geography) who devoted many hours to the production of the maps and diagrams that are such an important part of the book as a whole.

I am grateful to the following publishers for permission to reproduce material under copyright: CAB International Ltd (Figure 5.4 reprinted from Brennan, F. and Allen, G. (2001) 'Community-based ecotourism, social exclusion and the changing political economy of KwaZulu-Natal, South Africa', in Harrison, D. (ed.) *Tourism and the Less-developed World: Issues and Case Studies*, p. 208); Elsevier Science Ltd (Figure 4.2 reprinted from Mbaiwa, J. (2005) 'Enclave tourism and its socio-economic impacts in the Okavango Delta, Botswana', *Tourism Management* Vol. 26 (2) p. 158; Figures 4.4 and 4.5 reprinted from Smith, R.A. (1991) 'Beach resorts: a model of development evolution', *Landscape and Planning*, Vol. 21, pp. 189–210 ; Figure 7.5 reprinted from Mitchell, R.E. and Reid, D.G. (2001) 'Community integration: island tourism in Peru', *Annals of Tourism Research*, Vol. 28 (1) p. 116 and Figure 7.7 reprinted from Bramwell, B. and Sharman, A. (1999) 'Collaboration in local tourism policy making', *Annals of Tourism Research*, Vol. 26 (2) p. 405); Pearson Education Ltd (Figure 9.6 reprinted from Page, S.J. and Hall, C.M. (2003) *Managing Urban Tourism*, p. 50); Routledge Ltd (Figure 1.1 reprinted from Hall, C.M. and Page, S.J. (1999) *The Geography of Tourism and Recreation: Environment, Place and Space*, p. 4; Figure 6.1 reprinted from Murphy, P.E. (1985) *Tourism: A Community Approach*, p. 124; Figure 11.4 reprinted from Carmichael, B.A. (2005) 'Understanding the wine tourism experience for winery visitors in the Niagara region, Ontario, Canada', *Tourism Geographies*, Vol. 7 (2) p. 191); Sage Publications Ltd (Figure 1.3 reprinted from Shaw, G. and Williams, A.M. (2004) *Tourism and Tourism Space*, p. 148 and Figure 11.1 reprinted from du Gay, P. (ed.) (1997) *Production of Culture/Cultures of Production*, opposite p. 1); John Wiley & Sons Ltd (Figures 2.1 and 2.2 reprinted from Towner, J. (1996) *An Historical Geography of Recreation and Tourism in the Western World: 1540–1940*, pp. 179 and 196; Figure 7.2 reprinted from Inskeep, E. (1991) *Tourism Planning: An Integrated and Sustainable Development Approach*, p. 39); and to the National Railway Museum Science and Society Picture Library for permission to reproduce Plate 8.2.

Except where acknowledged above, all other photographs were taken by the author.

Introduction

1 ▸ Tourism, geography and geographies of tourism

Tourism is an intensely geographic phenomenon. If we view human geography as being concerned essentially with the patterns and consequences of the economic, social, cultural and political relationships between people, and between people and the spaces and places that comprise their environment, then the annual migration of millions of travellers worldwide within the activity that we label as 'tourism' is a process that human geographers should not ignore.

Part of the contemporary significance of tourism arises from the sheer scale of the activity and the rapidity with which it has developed. From a position at the end of the Second World War when less than 25 million people worldwide travelled outside their home countries for the purposes by which we define tourism, the number of international travellers reached 900 million people annually in 2007 (WTO, 2008). The gross receipts from the activities of these tourists amounted to US$620 billion in 2004 (WTO, 2005a), accounting for almost 9 per cent of world trade. To these foreign travellers and their expenditure must be added the domestic tourists who do not cross international boundaries but who, in most developed nations at least, are many times more numerous than their international counterparts. In the UK holiday tourism sector, for example, for each foreign visitor there are around three domestic holidaymakers and over 100 day trippers (SCPR, 1997; BTA, 2001).

The absolute number of tourists, however, accounts for only a part of the significance of tourism. Tourism acquires added importance, first, through the range of impacts that the movement of people on this scale inevitably produces at local, regional, national and, increasingly, at international level. Second, and less obviously, tourism is acquiring a new level of relevance through its emblematic nature, both as a mirror of contemporary lifestyles, tastes and preferences and, more fundamentally, an embedded facet of (post)modern life. The sociologist John Urry has argued that mobility – in its various guises – has become central to the structuring of social life and cultural identity in the twenty-first century (Urry, 2000) and tourism is an essential component in modern mobilities.

Tourism impacts are felt across the range of economic, social, cultural and environmental contexts. Globally, an estimated 200 million people derive direct employment from the tourism business: from travel and transportation, accommodation, promotion, entertainment, visitor attractions and tourist retailing (Milne and Ateljevic, 2001). Tourism is highly implicated in processes of globalisation (Shaw and Williams, 2004) and has been variously recognised: as a means of advancing wider international integration within areas such as the European Union (EU); as a catalyst for modernisation, economic development and prosperity in emerging nations in the Third World (Britton, 1989); or a pathway for regenerating post-industrial economies of the First World (Robinson, 1999). It may contribute to the preservation of local cultures in the face of the homogenising effects of globalisation; it can encourage and enable the restoration and conservation of special environments; and promote international peace and understanding (Higgins-Desbiolles, 2006).

Yet tourism has its negative dimensions too. Whilst it brings development, tourism may also be responsible for a range of detrimental impacts on the physical environment: pollution of air and water, traffic congestion, physical erosion of sites, disruption of habitats and the species that occupy places that visitors use, and the unsightly visual blight that results from poorly planned or poorly designed buildings. The exposure of local societies and their customs to tourists can be a means of sustaining traditions and rituals, but it may also be a potent agency for cultural change, a key element in the erosion of distinctive beliefs, values and practices and a producer of nondescript, globalised forms of culture. Likewise, in the field of economic impacts, although tourism has shown itself to be capable of generating significant volumes of employment at national, regional and local levels, the uncertainties that surround a market that is more prone than most to the whims of fashion can make tourism an insecure foundation on which to build national economic growth, and the quality of jobs created within this sector (as defined by their permanence, reward and remuneration levels) often leaves much to be desired. More critically, perhaps, tourism can be a vehicle for perpetuating economic inequalities, maintaining dependency and hence, neo-colonial relationships between developed and developing nations (Higgins-Desbiolles, 2006).

The study of tourism impacts has become a traditional means of understanding the significance of tourism and dates from seminal work conducted in the 1970s by writers such as Mathieson and Wall (1982). More recently, work in cultural geography and related fields such as anthropology have drawn out new areas of significance for tourism. Thus tourism and tourist experience are now seen as an influential arena for social differentiation (Ateljevic and Doorne, 2003); as a means by which we develop and reinforce our identities and locate ourselves in the modern world (Franklin, 2004); as a prominent source for the acquisition of what Bourdieu (1984) defines as 'cultural capital'; and a key context within which people engage with the fluid and changing nature of modernity (Franklin, 2004). Franklin and Crang (2001: 19) summarise this new reading of tourism and emphasise its new-found relevance thus:

> The tourist and styles of tourist consumption are not only emblematic of many features of contemporary life, such as mobility, restlessness, the search for authenticity and escape, but they are increasingly central to economic restructuring, globalization, the consumption of place and the aestheticization of everyday life.

Readers will detect within this medley of themes and issues much that is of direct interest to the geographer and to disregard what has become a primary area of physical, social, cultural and economic development would be to deny a pervasive and powerful force for change in the world in which we live. Modern tourism creates a very broad agenda for enquiry to which geographers can contribute, especially because the nature of tourism's effects is so often contingent upon the geographical circumstances in which the activity is developed and practised. The spaces and places in which tourism occurs are usually fundamental to the tourist experience – and space and place are core interests for human geographers. Furthermore, realisation of the contingent nature of tourism has encouraged a shift in critical thinking around the subject, away from traditional binary views of tourism and towards more relational perspectives. Thus, for example, rather than perpetuating a conventional view of tourism impacts as being necessarily either positive or negative in effect, or the relationship between so-called hosts and guests as being shaped around the dependency of the former on the latter, recent work in tourism geography has promoted more nuanced, equivocal understandings that have provided new insight into the ways in which tourists relate to the world around them.

This book is essentially concerned with developing an understanding of how tourism geographies are formed and maintained through the diverse and increasingly flexible relationships between people and the places that are toured and how those relationships become manifest across geographical space. However, it takes as its point of departure a key assumption – namely that to *understand tourism geography one must also understand tourism*. Hence, for example, in the following sections important issues are introduced relating to:

● an understanding of what tourism is and some of the inherent problems associated with the study of tourism;
● some of the ways in which tourists may be differentiated (since such a vast body of people is clearly far from homogeneous);
● how tourist motivation and experience may be understood.

This material is included, not because it is inherently geographical per se, but because the differentiation of tourists, their motivations and the experiences that they seek are often intimately bound to resultant geographical patterns and behaviours. It is probably a fair criticism that geographers have not made a particularly significant contribution to the development of any of these core concepts (especially the differentiation of tourists or the development of tourism motivation theory and concepts of tourism experience), but the understandings that other disciplines have developed are still essential to comprehending tourism geography.

What is tourism?

What is tourism and how does it relate to associated concepts of recreation and leisure? The word 'tourism', although accepted and recognised in common parlance, is nevertheless a term that is subject to a diversity of meanings and interpretations (Leiper, 1993). For the student this is a potential difficulty since consensus in the understanding of the term and, hence, the scope for investigation that such agreement opens up, is a desirable starting point to any structured form of enquiry and interpretation. Definitional problems arise because the word 'tourism' is typically used not only as a single term to designate a variety of concepts (Gilbert, 1990), but also as an area of study in a range of disciplines that includes geography, economics, business and marketing, sociology, anthropology, history and psychology. The differing conceptual structures and epistemologies within these disciplines lead inevitably to contrasts in perspective and emphasis. Furthermore, whilst there has been some convergence in 'official' definitions (i.e. those used by tourism organisations, governments and international forums such as the United Nations (UN)), public perception of what constitutes a tourist and the activity of tourism may differ quite markedly. More fundamentally, perhaps, recent critical analysis has begun to offer significant challenges to traditional concepts of tourism as a bounded and separate entity, and therefore one that is open to meaningful definition (see, e.g., Franklin, 2004; Shaw and Williams, 2004).

Traditional definitions of tourists and tourism – as found, for example, within dictionaries – commonly explain a 'tourist' as a person undertaking a tour – a circular trip that is usually made for business, pleasure or education, at the end of which one returns to the starting point, normally the home. 'Tourism' is habitually viewed as a composite concept involving not just the temporary movement of people to destinations that are removed from their normal place of residence but, in addition, the organisation and conduct of their activities and of the facilities and services that are necessary for meeting their needs.

Simple statements of this character are actually quite effective in drawing attention to the core elements that may be held to distinguish tourism as an area of activity:

- They give primacy to the notion that tourism involves travel but that the relocation of people is a temporary one.
- They make explicit the idea that motivations for tourism may come from one (or more than one) of a variety of sources. We tend to think of tourism as being associated with pleasure motives, but it can also embrace business, education, social contact, health or religion as a basis for travelling.
- They draw attention to the fact that the activity of tourism requires an accessible supporting infrastructure of transport, accommodation, marketing systems, entertainment and attractions that together form the basis for the tourism industries.

Official definitions of tourism have tended to be similarly broad in scope. For example, the World Tourism Organization (WTO) definition published in 1994 saw tourism as comprising:

> the activities of persons travelling to and staying in places outside their usual environment for not more than one consecutive year for leisure, business or other purposes.

> (WTO, 1994)

This approach acknowledges that tourism occurs both between and within countries (i.e. international and domestic tourism) and that it covers visitors who stay as well as those who visit for part of a day (Lickorish and Jenkins, 1997). The recognition of forms of day visiting as constituting a part of tourism is important, primarily because the actions, impacts and, indeed, the local geographies of day visitors and excursionists are often indistinguishable in cause and effect from those of staying visitors, so to confine the study of tourism to those who stay, omits an important component from the overall concept of tourism (see Williams, 2003).

Recently, however, traditional conceptions of tourism of the kind set out in the preceding definitions have come under a sustained attack as developments in critical analysis of tourism have raised fundamental challenges to many previous assumptions concerning its distinctive nature. As the discussion of motivation and experience later in this chapter will explain, the development of tourism was generally held to be a form of escape, a quest to experience difference and, in some readings, to find an authenticity that could not be obtained in normal routines (MacCannell, 1973, 1989). However, since the 1980s, post-industrial restructuring of economy, society and culture has been progressively linked to what has been termed a process of 'de-differentiation', whereby formerly clear distinctions (e.g., between work and leisure; home and away; or public and private) have been blurred and eroded (Lash and Urry, 1994; Urry, 1994a; Rojek and Urry 1997). In globalising societies what was once different is now familiar and the necessity to travel to encounter difference is greatly diminished as the experience of foreign cultures, practices, tastes and fashions become routinely embedded in everyone's daily lives. Franklin (2004: 24) asserts that 'it is difficult (and pointless) to define tourism in spatial terms: it is simply not behaviour that only takes place away from home' – a thesis that is reinforced by Urry's (2000) articulation of modern mobilities where he argues that in the excessively mobile societies of the twenty-first century, much of life is lived in a touristic manner. Hence, concepts of home and away (and their associated experiences) become less meaningful and sometimes meaningless in situations where, for example, people possess multiple homes.

Consequently, Shaw and Williams (2004: 9) confidently describe the quest for definitions of tourism as an 'arid debate' given the progressive blurring of boundaries between tourism and daily life, whilst Franklin (2004: 27) is openly hostile to what he perceives as the limiting effects of conventional definitions that place the travel and accommodation industry and the associated provision and purchase of commodities at the heart of tourism, rather than tourist behaviour and culture. This tendency, he argues, 'denudes tourism of some of its most interesting and important characteristics'. Franklin's thesis places tourism at the core of individual engagement with the fluid and changing conditions of modernity and he is content to reflect both this belief and his resistance to industry-focused definitions through radically different descriptions of the subject, such that, for example, tourism is described as 'the nomadic manner in which we all attempt to make sense of modernity (and enjoy it) from the varied and multiple positions that we hold' (Franklin, 2004: 64).

These recent attempts to ground tourism as constitutive of daily experience rather than a distinct and separate entity that expresses resistance to the everyday (e.g., through notions of escape and a quest for difference), raises the wider issue of the relationship between tourism, recreation and leisure. As areas of academic study (and not least within the discipline of geography), a tradition of separate modes of investigation has emerged within these three fields, with particular emphasis upon the separation of tourism. Unfortunately, the terms 'leisure' and 'recreation' are themselves contested (see, e.g., Rojek, 1993a, 1997), but if we take a traditional view of 'leisure' as being related either to free time and/or to a frame of mind in which people believe themselves to be 'at leisure' (Patmore, 1983) and of 'recreation' as being 'activity voluntarily undertaken primarily for pleasure and satisfaction during leisure time' (Pigram, 1983: 3), then some significant areas of tourism are clearly congruent with major areas of recreation and leisure. Not only does a great deal of tourism activity take place in the leisure time/space framework, but much of it also centres upon recreational activities and experiences (e.g., sightseeing, travelling for pleasure, leisure shopping, eating and drinking, socialising) that may occur with equal ease within leisurely contexts that exist outside the framework of tourism. Similarly, as has been argued above, tourism permeates day-to-day lifestyles, in both leisure and work. We read about tourism in newspapers or magazines and view television travel shows; we spend leisure time reviewing home videos or photo albums of previous trips and actively planning future ones; and we import experiences of travel into our home and working lives; for example, by eating at foreign-food restaurants, or by including foreign clothing styles within our wardrobe. Thus, Carr (2002) argues that many forms of tourist behaviour are extensions of established behaviours in the leisure environment of our daily lives and hence rather than conceiving of leisure and tourism as polar opposites, it is more meaningful to visualise the different forms of engagement with leisure and tourism as being arranged along a continuum. This raises interesting questions relating to *where* tourism takes place.

In approaching the study of tourism, therefore, we need to understand that the relationships between leisure, recreation and tourism are much closer and more intimate than the disparate manner in which they are treated in textbooks might suggest. There is considerable common ground in the major motivations for participation (attractions of destinations, events and experiences; social contacts; exploration), in the factors that facilitate engagement with activity (discretionary income; mobility; knowledge of opportunity) and the rewards (pleasure; experience; knowledge or memories) that we gain from tourism, recreation and leisure. Figure 1.1 provides a familiar representation of these relationships as overlapping spheres of experience and draws attention both to areas of coincidence and to areas of potential separation. However, rather than viewing each sphere as a discrete and clearly delineated zone of practice and experience, it is more meaningful to emphasise the permeability of boundaries (as indicated by the use of broken lines) and hence a fluidity in the relationship between the different elements.

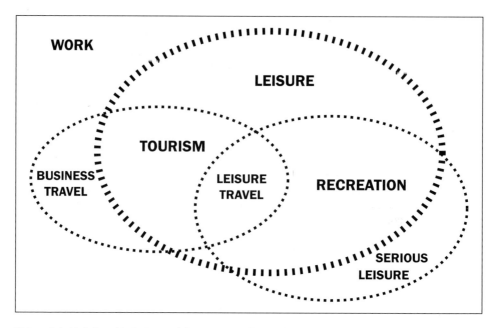

Figure 1.1 Relationship between leisure, recreation and tourism

Problems in the study of tourism

The definitional complexities of tourism and the uncertain linkages with the allied fields of recreation and leisure are basic problems that confront the student of tourism geography, though they are not the sole difficulties. Three further problems merit brief attention at this introductory stage.

First, in later chapters a range of statistics is used to map out the basic dimensions and patterns of tourism. This is a common starting point in understanding the geography of tourism since statistical enumeration of arrivals and departures at differing geographical scales (e.g., continental, national, regional or at the level of individual destinations) is a primary means of isolating and then describing the movements and concentrations of tourists. But it is important to appreciate that in many situations, comparability across space and time is made difficult or sometimes impossible by variation in official practice in distinguishing and recording the levels of tourist activity. At a global scale, for example, there are some critical differences of approach between – on the one hand – the UN World Tourism Organization (WTO), and – on the other – the World Travel and Tourism Council (WTTC). The latter organisation, with its strong focus on business, has developed what it terms a 'tourism satellite accounting' process (TSA) as a means of measuring tourism. The TSA derives indicators of tourism activity that are primarily reflective of economic performance (such as Gross Domestic Product, employment, demand and investment) and which are measured in terms of their financial value in US dollars. In contrast, the WTO bases much of its measurement of tourism on data that enumerate arrivals and departures as tourist headcounts. But because these two primary sources of global scale data adopt different approaches, the picture that each paints of the state of world tourism may also be rather different.

Moreover, the enumeration of tourists at a national level may also be problematic. Some countries, for example, do not even count the arrivals of foreign nationals at their borders.

The relaxation of border controls between the fifteen European states that are now signatories to the Schengen Agreement (first signed by five states in 1995) permits largely unrestricted (and hence undocumented) movement of tourists between these countries. Elsewhere, the presence of foreign nationals may be recorded at points of entry, although local definitions of tourist status or a failure to identify precise motives for visiting can also lead to an inability to enumerate tourists exactly. Some states count business travellers as tourists whilst others may not. Within states, tourism statistics may also be compiled through sample surveys of visitors or by reference to hotel registrations, although these will naturally be selective and prone to imprecision. Hotel-based figures, for example, will exclude those visitors who lodge with friends or relatives. Data, therefore, are seldom directly comparable and always need to be treated with some caution.

Second, there are problems inherent in the definition of tourism as an industry, even though there are some practical advantages in delineating tourism as a coherent and bounded area of economic activity. It has been argued that designating tourism as an 'industry' establishes a framework within which activity and associated impacts may be mapped, measured and recorded, and, more critically, provides a form of legitimisation for an activity that has often struggled to gain the strategic recognition of political and economic analysts and hence a place within policy agendas. However, tourism, in practice, is a nebulous area and the notion that it may be conceived as a distinctive industry with a definable product and measurable geographic flows of associated goods, labour and capital has in itself been a problem. Conventionally, an industry is defined as a group of firms engaged in the manufacture or production of a given product or service. In tourism, though, there are many products and services, some tangible (provision of accommodation, entertainment and the production of gifts and souvenirs), others less so (creation of experience, memories or social contact). Many of the firms that service tourists also provide the same service to local people who do not fall into the category of tourists, however it may be cast. Tourism is not, therefore, an industry in any conventional sense. It is really a collection of industries which experience varying levels of dependence upon visitors, a dependence that alters through both space and time.

A third practical problem is the lack of a unified conceptual grounding for the study of tourism (Williams, 2004a). Meethan (2001: 2), for example, describes the study of tourism as 'under-theorised, eclectic and disparate'. Such criticisms are important because, in the absence of theoretical underpinning, adopted methodologies tend to regress towards a broadly empirical/descriptive approach and the insights that can arise from the more structured forms of analysis that a sound conceptual framework permits are harder to realise. This is not to imply that there has been no conceptualisation within the study of tourism, for (as several of the following chapters will demonstrate) the understanding of many aspects of tourism has benefited from varying degrees of theoretical thought within particular disciplines. But what is largely absent is the broader synthesis of diverse (though still related) issues and perspectives (Llewellyn Watson and Kopachevsky, 1994). As an intrinsically eclectic discipline, geography is better placed than many to provide the type of holistic perspective that a multi-dimensional phenomenon such as tourism evidently merits and this is very much central to the approach adopted in this book. But there are still limits to the level and extent of understanding that any one discipline, in isolation, can afford. The student of tourism geography must, therefore, be predisposed towards adopting multi-disciplinary perspectives in seeking to understand this most contradictory and, at times, enigmatic phenomenon.

Tourist motivation

The question of why people travel is both obvious and fundamental to any understanding of the practice of tourism and is often directly influential on tourism geographies. The spatial patterns of movement and the concentrations of people – as tourists – at preferred destinations is not an accidental process but is shaped by individual or collective motives and related expectations that by travelling to particular places, those motives may be realised. Of course, other elements – such as the supply of tourist facilities or the promotion of places as tourist destinations are also central to the process – but understanding motivation is a key part of understanding the geography of different forms of tourism.

As Shaw and Williams (2004) note, many motivational theories are grounded in the concept of 'need', as originally conceived by Maslow (1954). This is evident in some of the early work on motivation (e.g., Compton, 1979; Dann, 1981) which placed at the heart of the understanding of tourist motivation notions of a *need* to escape temporarily from the routine situations of the home, the workplace and the familiarity of their physical and social environments. Such needs arise, it is argued, because individuals strive to maintain stability in their lives (what is termed 'homeostasis') which is disrupted when needs become evident and is restored – in theory – once those needs have been met. Hence an extended period of work might create a perceived need for rest and relaxation that might be met through a holiday. Embedded within these core motives are a range of related motivational components. Compton (1979) for example, proposed that tourists might seek opportunities to relax; to enhance kinship or other social relations; to experience novelty and be entertained; to indulge regressive forms of behaviour; and to engage in forms of self-discovery. In a not dissimilar vein, Beard and Ragheb (1983) emphasised four motivational components: an intellectual component (in which tourists acquired knowledge); a social component (through which social networks were maintained or extended); a competence component (in which skills were developed); and a stimulus-avoidance component (which reflects the desire for release from pressured situations – such as work – and attain rest or relaxation).

Implicit in these conceptualisations are two important propositions. First, tourist motivations are formed around combinations of stimuli that, on the one hand, encourage tourist behaviours (push factors) and, on the other, attract tourists to particular destinations or forms of activity (pull factors). Second, tourists expect to derive benefit (or reward) from activities undertaken. These two assumptions are brought together in Iso-Ahola's (1982) model of the social psychology of tourism. Here elements of escape from routine environments are juxtaposed with a parallel quest for intrinsic rewards in the environments to be visited. By envisaging these key elements as the axes on a matrix (Figure 1.2) it is possible to construct a set of theoretical 'cells' in which elements of escape and reward are

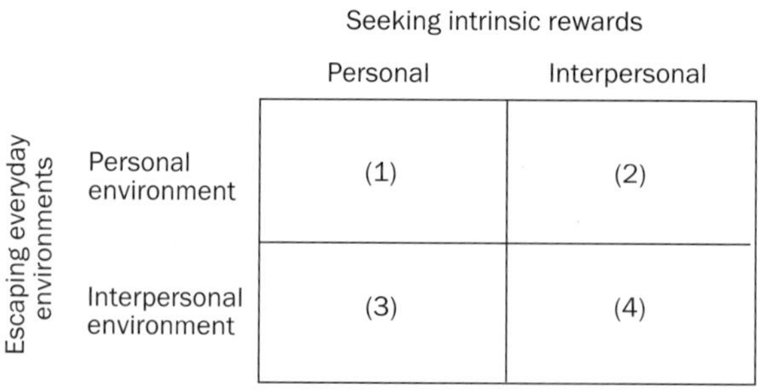

Figure 1.2 Iso-Ahola's model of the social psychology of tourism

combined in differing ways and within which tourist motives may be located, depending upon their particular circumstances and objectives at any one time.

It is also implicit in conceptions of tourism as a form of escape that behavioural patterns will be adjusted to reflect motivation. One of the most interesting expositions of this idea is Graburn's (1983a) explanation of tourist 'inversions' – shifts in behaviour patterns away from a norm and towards a temporary opposite. This might be shown in extended periods of relaxation (as opposed to work); increased consumption of food, and increased purchases of drinks and consumer goods; relaxation in dress codes through varying states of nudity; and, most importantly from a geographical perspective, relocation to contrasting places, climates or environments. Graburn proposes several different headings or 'dimensions' under which tourist behavioural inversions occur, including environment, lifestyle, formality and health (Table 1.1). Graburn emphasises that within the context of any one visit, only some dimensions will normally be subject to a reversal, and this allows us to explain how the same people may take different types of holiday at different times and to different locations. It is also the case that actual behaviour patterns will usually exhibit degrees of departure from a norm, rather than automatically switching to a polar opposite. This accommodates those people whose behavioural patterns as tourists show minimal differences from most of the normal dimensions of their lives, whilst still emphasising the notions of escape and contrast as being central to most forms of tourism experience.

It is also understood that the motivations that shape the patterns of tourism of an individual will alter through time as well as across situations. This idea has been articulated by Pearce (1993) in his concept of the travel career ladder (Figure 1.3). The travel career ladder builds directly on Maslow's (1954) ideas of a hierarchy of needs by proposing five levels of motivation that ascend from the comparatively simple matter of relaxation and the meeting of bodily needs, to an existential quest for self-esteem and fulfilment (see Cohen, 1979 – below). Lower order needs are satisfied first and as the tourist gains experience, so higher order motives are accessed. However, whilst the model has value in emphasising the

Table 1.1 *Examples of 'inversions' in tourism*

Dimension	Continua	Tourist behavioural pattern
Environment	Winter *vs* summer Cold *vs* warmth Crowds *vs* isolation Modern *vs* ancient Home *vs* foreign	Tourist escapes cooler latitudes in favour of warmer places. Urban people may seek the solitude of rural or remote places. Historic sites attract tourists who live in modern environments. Familiarity of the home is replaced by the difference of the foreign.
Lifestyle	Thrift *vs* indulgence Affluence *vs* simplicity Work *vs* leisure	Expenditure increased on special events or purchases. Experiences selected to contrast routines of work with rewards of leisure.
Formality	Rigid *vs* flexible Formal *vs* informal Restriction *vs* licence	Routines of normal time-keeping, dress codes and social behaviours replaced by contrasting patterns and practices based on flexibility and informality.
Health	Diet *vs* gluttony Stress *vs* tranquillity Sloth *vs* exercise Age *vs* rejuvenation	Tourists indulge through increases in consumption. Relaxation sought as relief from routine stresses. Active holidays chosen as alternative to sedentary patterns in daily life. Health spas and exercise used to counteract process of ageing.

Source: Adapted from Graburn (1983a)

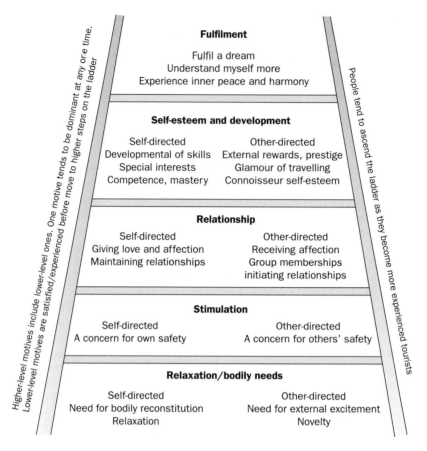

Figure 1.3 The travel career 'ladder'

importance of experience in shaping tourist motivation and behaviours, the notion of a progressive development of experience through a travel 'career' is confounded by the observable tendency for contemporary tourists to seek different kinds of experience at the same stages of their 'career'. In particular, the trend towards multiple holiday-taking (see Chapters 2 and 3) allows tourists to indulge a range of motives, more or less simultaneously, rather than sequentially as the model implies.

These different concepts and models offer what we might consider as 'traditional' readings of tourist motivation. Since perhaps the early 1990s, work in fields such as cultural studies has brought new perspectives to bear on the question of why people travel and how they choose between alternative destinations, some of which offer significant challenges to received wisdom. In particular writers such as Crouch (1999), Franklin and Crang (2001) and Franklin (2004) have developed persuasive lines of argument that emphasise the progressive embedding of tourism into daily life, in which – as a consequence – tourism practice becomes not just a means of relaxation, entertainment, social development or bodily reconstitution, but also an expression of identity and of social positioning through patterns of consumption. Thus tourism is not only a vehicle for accessing the world through travel, but increasingly a way of locating ourselves within it. Whether people (as tourists) consciously recognise such motives in shaping the choices they make is a moot point, but if we accept Franklin's (2004) assertion that tourism is a way of *connecting* to the (post)modern world rather than escaping from it, many of the established theories of motivation need to be reappraised. What is equally important from the geographical perspective is that such processes clearly encourage alternative spatial patterns of tourism and new ways of engagement of people (as tourists) with place and space, as we will see in greater detail in Part III of this book.

Tourism typologies

Murphy (1985: 5) is probably correct when he writes that 'there are as many types of tourist as there are motives for travel'. The early realisation that tourism was an area characterised by complexity has stimulated repeated attempts at creating typologies of the contrasting forms of tourism and different categories of tourist, essentially as a means of bringing some semblance of order – and hence, understanding – to the subject. Recently, the creation of typologies as a means for comprehending tourism has attracted some critical comment, largely because authors such as Franklin (2004) see such frameworks as a limitation. If tourism is truly an integral feature of postmodern life, then structures that compartmentalise or which infer boundaries to experience that have limited meaning in reality, become barriers rather than pathways to developing understanding. That said, the fact remains that comprehension of the diversity of tourism requires some means of differentiating one form of activity from another and so some consideration of typological approaches is probably merited.

The benefits of recognising typologies of tourists and tourism are that they allow us to identify key dimensions of the activity and its participants. In particular, typological analyses help us to:

● differentiate types of tourism (e.g., recreational or business tourism);
● differentiate types of tourist (e.g., mass tourists or independent travellers);
● anticipate contrasting motives for travel;
● expect variations in impacts within host areas according to motives and forms of travel;
● expect differences in the structural elements within tourism (e.g., accommodation, travel and entertainment) that different categories of tourists will generate.

From a geographical perspective, these key dimensions are also central to the processes that demarcate the different forms of geographical space in which tourism may occur and the contrasting ways in which tourism relates to those spaces. For example, in the sophisticated city destinations of the business tourist; or the highly developed resorts or the accessible countryside that attract the mass tourists; or the remoter, undeveloped places that attract independent travellers and tourists on existential journeys of 'discovery'.

Attempts at the categorisation of tourism normally use the activity that is central to the trip as a criterion around which to construct a subdivision. Thus we may draw basic distinctions between, say, recreational tourism (where activities focus upon the pursuit of pleasure, whether through passive enjoyment of places as sightseers or through more active engagement with sports and pastimes), or business tourism (where the primary focus will be the development or maintenance of commercial interests or professional contacts). However, it is also recognised that people may travel to secure treatment for medical conditions, for educational reasons, for social purposes or, in some cultures, as pilgrims for religious purposes. Furthermore, most of these categories may themselves be subdivided. It is, though, risky to push such distinctions too far or to assume that tourists travel for a narrow range of reasons. Most tourists choose destinations for a diversity of purposes and will combine more than one form of experience within a visit. One of the intractable problems of isolating generalities within patterns is that the real-world complexity of tourism admits a whole spectrum of motives and behaviours that in many cases will co-exist within visits. So, for example, the business traveller may visit friends, take in a show or tour a museum, alongside the business meetings that provide the primary motive for the trip.

One of the earliest and most influential attempts to classify tourists was proposed by Cohen (1972). Cohen developed a four-fold categorisation of tourists, differentiated according to whether they were institutionalised (i.e. effectively managed through the travel industry) or non-institutionalised (i.e. very loosely attached – or independent of the tourist establishment). The two institutional categories are described by Cohen as organised mass tourists and individual mass tourists, whilst the non-institutional categories embrace people that Cohen labels as explorers and drifters.

Organised mass tourists characteristically travel to destinations that are essentially familiar rather than novel – familiarity commonly having been gained through previous experience, through reported experiences of others or through media exposure. The sense of familiarity is reinforced by the nature of goods and services that are available at the destination, which are often tailored to meet the tastes of dominant tourist groups. The mass tourist is highly dependent upon travel industry infrastructure to deliver a packaged trip at a competitive price and with minimal organisational requirements on the part of the tourist. Incipient tourists, feeling their way into foreign travel and new destinations for the first time, may typically operate in this sector, at least until experience is acquired. This sector is dominated by recreational tourists.

Individual or small-group mass tourists are partly dependent upon the infrastructure of mass tourism to deliver some elements of the tourist package, especially travel and accommodation, but will structure more of the trip to suit themselves. The experiences sought are still likely to be familiar but with some elements of exploration or novelty. The sector will contain business tourists alongside recreational travellers and is also more likely to accommodate activities such as cultural or educational forms of tourism.

Explorers generally arrange their own trips and seek novelty and experiences that are not embodied in concepts of mass tourism or the places that mass tourists visit. Hence, for example, contact with host societies will often be a strong motivation amongst explorers. It is possible, too, that people with very specific objectives in travelling (e.g., some business tourists or religious or health tourists) would be more prominent here. There may be a residual dependence upon elements in the tourism industry – travel and accommodation bookings being the most likely point of contact, but these are minimal.

The people that Cohen labels as 'drifters' probably do not consider themselves to be tourists in any conventional sense. They plan trips alone, shun other tourist groups (except perhaps fellow drifters) and seek immersion in host cultures and systems. People engaged in this form of tourism may sometimes be considered as pioneers, constituting the first travellers to previously untouched areas, but in the process, of course, may initiate new spatial patterns of travel that through time become embedded in changed geographical patterns of tourism.

To some extent these typological subdivisions of tourists may be linked to contrasting patterns of tourist motivation. The actions of organised, mass tourists, for example, have been widely interpreted as essentially a quest for diversionary forms of pleasure through an escape from the repetitive routine of daily life and a desire for restorative benefits through rest, relaxation and entertainment. The individual or small-group traveller may retain all or some of these motives but might equally replace or supplement them with an experiential motive, a desire to learn about or engage with alternative customs or cultures – MacCannell's (1973) quest for authenticity or meaning in life. This tendency becomes most clearly embodied in the motives of the explorers and the drifters whom, it is argued, seek active immersion in alternative lifestyles in a search for a particular form of self-fulfilment. However, we should exercise caution in making too many assumptions concerning the links between motivation and forms of travel since as Urierly (2005: 205) reminds us, 'the inclination to couple external practice with internal meaning needs to be resisted'.

The patterns of activity and behaviour that might be associated with different types of tourism may also lead to a range of particular impacts upon the local geography of host areas. Organised mass tourism, for instance, generally requires infrastructure development such as extensive provision of hotels and apartments, entertainment facilities, transportation systems and public utilities, the development of which inevitably alters the physical nature of places and will probably affect environments and ecosystems too. Furthermore, the actions of tourists en masse will usually have an impact upon local lifestyles. In contrast, the much smaller numbers of explorers make fewer demands for infrastructure provision and through different attitudes and expectations towards host communities, exert a much reduced impact upon local life, although even these forms of tourism are not impact-free.

These ideas are summarised in Figure 1.4, which offers a typological framework of tourism and tourists that builds upon Cohen's classification. In interpreting this summary, however, it is important to reiterate that differing forms of tourism may be combined within a single trip and as individuals we can and will move within the framework, especially as we progress through the life cycle. For example, people who were strongly independent travellers in their youth may gravitate towards mass forms of tourism in later life, perhaps when acquiring a family or with the onset of old age when the capacities to travel independently may diminish.

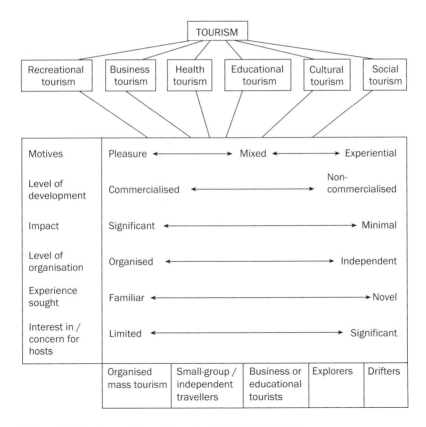

Figure 1.4 Tourism and tourists: a typological framework

Cohen's work provides a useful summary of the forms of tourism that are broadly reflective of the modernist tradition that developed under the so-called 'Fordist' pattern of mass production and consumption, with its emphasis upon packaging and standardisation of tourism products. But tourism is seldom a static entity. For example, Poon (1989) noted a shift that became evident from perhaps the mid-1980s onwards, towards new patterns of tourism that are characterised by high degrees of segmentation within tourism markets, with highly flexible patterns of provision that are customised to meet the diverse demands of niche markets (see also Urry, 1994a). The diversity of these 'post-Fordist' forms of tourism is not so effectively captured in typologies of the style developed by Cohen (although the distinctions between mass and independent forms of travel remain relevant, if less-clearly etched) and recent work on tourist typologies has tended to focus on how the segmented markets that are characteristic of post-Fordist patterns are formed. Shaw and Williams (2004) provide a range of examples based around the emerging popularity of eco-tourism (see Chapter 5) that illustrate how typologies have been constructed around variables such as visitors' levels of interest in, or knowledge of, the natural environment; their degree of dedication; levels of physical effort entailed in undertaking visits; as well as more conventional criteria relating to levels of organisation (or otherwise) of tours. The different perspectives offered in these typologies is revealing – not just of the changing nature of tourism – but also, the more flexible ways in which the study of tourism needs to be approached.

The nature of the tourist experience

Alongside interest in motivation and the typological structure of tourism, the need to understand the nature of tourist experience has also been a recurring theme in the development of tourism studies (see, e.g., Cohen, 1979; Urierly, 2005). However, it is important to recognise some significant changes in the way in which tourist experience has been viewed. Urierly (2005) suggests that this shift is reflective of transitions from a modernist perspective in which tourism was seen as essentially distinct from everyday life, to a postmodern perspective in which tourism becomes an embedded facet of life and where the meanings attached to the act of touring are negotiated at an individual level and are contingent on the context.

The seminal work in the modernist tradition is probably Cohen's (1979) essay on the phenomenology of tourist experience. Here Cohen proposed five 'modes' of experience (recreational, diversionary, experiential, experimental and existential) that he envisaged as spanning a spectrum from the outwardly simple pursuit of pleasure acquired through the experience of difference (the recreational mode) to a quest to establish a meaning to life through exposure to the lives of others (the existential mode). In articulating the experimental and existential modes in particular, Cohen suggests a number of parallels with MacCannell's (1973: 591) interpretation of tourist experience as a quest for authenticity – a desire 'to see life as it is really lived' and which is pursued as an antidote to the perceived inauthenticity of the modern lives that are lived out by tourists in their home settings.

The conceptual understandings developed by writers such as Cohen and MacCannell typically frame the tourist experience as a bounded event that stands apart – and is therefore distinct – from the routines and the geographical spaces of day-to-day life. However, even a simple analysis of how a tourist event is constructed and experienced challenges this assumption. Figure 1.5 illustrates a theoretical summary which proposes a tourism event as comprising a series of key phases and related processes:

● An initial phase of planning the trip in which destinations, modes of travel, preferred styles and levels of accommodation are evaluated and a destination selected. The

Figure 1.5
Structure of the tourist experience

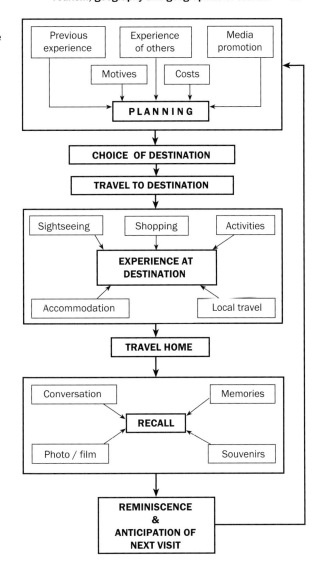

planning phase is informed by a number of potential inputs (including previous experience, images and perceptions of places and suggestions made by others) and will be reflective of motives and intentions for travelling.

● Outward travel. All tourism involves travel, and it is important to realise that travelling is often more than just a means to an end. In many tourism contexts, getting there is half the fun, and in some forms of tourism – most conspicuously in sea cruising – the act of travelling rather than visiting places often becomes the central element within the tourism experience as a whole.

● Experience of the destination. This element is normally the main component within the visit and most clearly reflects the category or categories of tourism in which the trip is located and the motivations of the visitors. In general forms of tourism, experience of the destination will typically include elements of sightseeing, leisure shopping and the collection of souvenirs and memorabilia. It may also include varying levels of contact with host populations, society and culture, the extent and significance of which will vary. Other forms – for example, adventure tourism – may reveal a more overt focus on activity and engagement with pursuits.

- Return travel, which, as with the outward journey, may be an integral part of the tourism experience, although it may not realise the same degree of pleasure, anticipation and excitement, as the trip is nearing its end and fatigue may have begun to affect the tourist.
- Recall. The trip will be relived subsequently and probably repeatedly, in conversation with friends and relatives, in holiday photographs and/or videos, or in response to the visual prompts offered by souvenirs that may now be arranged around the home. The recall phase will also inform the preliminary planning of the next visit and may be a positive, mixed, or negative stimulus, depending upon the perceived levels of success or failure of the trip.

This approach to understanding the structuring of experience around a specific tourism event makes three fundamental points. First, by emphasising how the actual visit is prefigured by a planning phase and then subsequently relived through memory, the model demonstrates the holistic nature of experience and the fact that the total experience of tourism is much more than the visit itself. Second, the model shows how experience is strongly grounded in geography since the places in which the experience is located and the geographic transitions between those places are seen as central to the overall process. Finally, the model shows how important aspects of the tourist experience occur in the home environment and thus become enmeshed in aspects of daily life, rather than being confined to the trip itself.

The tendency to question conventional wisdoms regarding the separate nature of tourist experience and its grounding in concepts such as authenticity has become much more pronounced within recent and contemporary tourism research. Passing reference has already been made to some of the ways in which globalisation (especially in areas such as the media or in wider patterns of material consumption) infuse tourist experiences of difference into daily life. In turn, daily life directly shapes much of tourism experience. Ritzer and Liska (1997: 99) note that 'people increasingly travel to other locales in order to experience much of what they experience in their day-to-day lives', or as Franklin (2004: 10) observes, we travel 'within the realm of the familiar'. Moreover, in societies that are increasingly formed around mobilities (Urry, 2000), tourism becomes an expression of that way of life rather than a form of resistance to it.

Tourism has also become a more overtly embodied and sensual form of experience (Macnaghten and Urry, 2000; Crouch and Desforges, 2003) (e.g., through the development of adventuresome forms of tourism) and has acquired a diversity (e.g., in visiting friends and relatives, in beach holidays, in nature tourism, in activity holidays or visiting theme parks) that defies adequate explanation through conventional concepts of authenticity. Indeed, in a world in which tens of millions of people base their tourist experiences in what Eco (1986) describes as the 'hyper-reality' of Disney-style theme parks or the artificial environments of resorts such as Las Vegas, the notion of authentic experience seems, at one level at least, to be very out-dated. However, as Wang (1999) explains, such forms of tourism may acquire a different form of authenticity that is no less important (see Chapter 6).

But although we have clearly moved into an era in which tourism has acquired an embedded place in (post)modern lifestyles and the boundaries between tourism and other aspects of life have become blurred, we should not assume that the activity of touring has surrendered all meaning or all its claims to distinction. Tourism – especially in the form of holidays – remains a prominent component in the ordering of individual and family life and a very significant area of personal expenditure. Despite the outward familiarity of many forms of contemporary tourism, most tourist trips still deliver experiences of varying degrees of difference and of change from routine. Consequently, tourism still endows most people with experiences that are sufficiently distinct to form memories that survive long after other (routine) events are forgotten.

Geography and the study of tourism

Although tourism (with its focus upon travelling and the transfer of people, goods and services through time and space) is essentially a geographical phenomenon, it has occupied what Coles (2004: 137) has described as a 'curiously estranged' position within human geography. Initially the issue was one of the credibility and legitimacy of serious academic investigation of a fun-related activity, but even when acceptance was generally forthcoming, the treatment of tourism within the literature of human geography has remained extremely uneven. For example, a recent top-selling UK text on human geography (Daniels et al., 2005) indexes just one page on which tourism is discussed. Fortunately others have been more willing to recognise the significance of tourism within human geography (e.g., Aitchison et al., 2001; Crouch, 1999), both as a valid subject in its own right and, equally important, as a 'lens' through which a range of contemporary issues may be examined.

Geographical approaches to the study of tourism have moved through a number of evolutionary phases. Butler (2004) suggests that three distinct eras of development may be discerned: pre-1950; 1950 to *circa* 1980; and *circa* 1980–present. The pre-1950 period is labelled by Butler as 'the descriptive era'. Here the study of tourism was uncommon within human geography and an activity of marginal interest or relevance. Where work was conducted it was characteristically descriptive and related to traditional, existing interests within the wider field. Gilbert's (1939) study of the growth of seaside resorts as a form of urban geography is an example.

Second, between 1950 (when the first reliable data on tourism began to emerge) and the early 1980s, Butler (2004) argues that the geographical study of tourism entered 'the thematic era' as connections between tourism and some of the wider agenda of the discipline became more evident. As Ateljevic (2000) notes, the geographic approach at this time was strongly spatial in focus, deploying largely positivist perspectives to describe and record geographies of tourism. Issues such as the effect of scale, spatial distributions of tourism phenomena and of tourist movement, people–land relationships and tourism impact, and the spatial modelling of tourism development were typical foci for geographical work which established a basic approach to tourism geography that remained influential into the 1990s (see, e.g., Boniface and Cooper, 1987; Burton, 1991; Mathieson and Wall, 1982; Pearce, 1987; 1989; Williams, 1998). Within such analyses geographical approaches centred on some now-familiar questions:

● Under what conditions (physical, economic, social) does tourism develop, in the sense of generating both demand for travel and a supply of tourist facilities?
● Where does tourism develop and in what form? (The question of location may be addressed at a range of geographical scales whilst the question of what is developed focuses particularly upon provision of infrastructure.)
● How is tourism developed? (This question addresses not just the rate and character of tourism development but also the question of who are the developers.)
● Who are the tourists (defined in terms of their number, characteristics, travel patterns, etc.) and what are their motives?
● What is the impact of tourism upon the physical, economic and socio-cultural environments of host areas?

Third, Butler (2004) describes the period since the mid-1980s as being 'the era of diversity'. As the scale of tourism has grown and become more diverse in its composition (e.g., through the emergence of niche markets in areas such as adventure and eco-tourism or the widening popularity of heritage tourism), so the approach to the study of tourism has, in itself, tended to become more diverse. So the focus of work has extended beyond the issues that characterised Butler's 'thematic' era and added new areas of interest. These include important areas of work relating to, amongst others: tourism and communities (Murphy, 1985); tourism and capitalist political economies (Britton, 1991); tourism, production, consumption and the 'new' economic geography (Shaw and Williams, 1994); cultural change and new cultural readings of tourism (Crouch, 1999); tourism as an agent of urban regeneration and place promotion (Gold and Ward, 1994; Law, 1992; 2000); and tourism as a sustainable form of development (Mowforth and Munt, 2003).

Underlying Butler's 'era of diversity' are several important shifts in the nature of geographical approaches to the study of tourism which reflect wider change in the epistemology of human geography. Three areas of change are worth noting.

First, and perhaps most influential, has been the impact on tourism studies of the so-called 'cultural turn' and associated rise of postmodern critical perspectives in human geography. Arguably the most significant aspect of the new cultural geography is that it offers a sustained challenge to conventional views of the pre-eminence of political and economic understandings of the world in which we live and, instead, emphasises a different set of perspectives on the way that we think about human geography. Issues of how places and their people are represented (and the subjective nature of those representations), how identities are constructed (especially in relation to difference, or – in the language of cultural studies – Others), or how patterns of consumption become embedded in cultural rather than economic processes, have defined a new agenda for the subject (Crang, 1998). These shifts in critical thinking have directly affected approaches to tourism, not just because tourism entails as central components the representation of places and people or that it is a primary area of consumption that is shaped around culture and identity, but because, in some readings, tourism is now seen as a *practice* (rather than a product) that is actively made and re-made through complex human and social engagements, relations and negotiations (Crouch, 1999).

Second, arising from these new cultural perspectives, approaches to understanding tourism have become more relational in character. Thus, traditional binary readings of tourism's impacts and effects have been – and are being – replaced by more nuanced interpretations that recognise the negotiated and contingent nature of how tourism relates to the places, peoples, societies and cultures that are toured. For example, until quite recently, social science perspectives (including human geography) tended to present the relationship between tourists and the people they visited in terms of 'hosts and guests' (Smith, 1977) or – in economic terms – producers and consumers. However, the notion of hosts and guests has been challenged through application of the work of writers such as Castells (1997) and Shurmer-Smith and Hannam (1994) on the construction of power relations. Consequently there is now a better understanding of how outcomes, rather than being predicated on fixed relationships (in which there is a presumed dominance of the tourist), are often negotiated in variable and quite surprising ways. Cheong and Miller (2000) argue that whilst tourism outcomes are often regarded as being driven by the tourist, in practice a tripartite power structure of tourists, locals and brokers creates a dynamic flexibility in which there is no fixed, one-sided relationship between the power of one group over another. Indeed, in many situations the tourist operates from positions of insecurity rather than influence. They may be located in unfamiliar political, cultural or geographical areas; they may be subject to new social norms and expectations; and they may be required to communicate from a distinct linguistic disadvantage. Similarly, Ateljevic (2000) (drawing on work by Johnson (1986) and du Gay et al. (1997)) shows how assumed notions of tourists as passive consumers of 'products' within a uni-directional relationship between producer and consumer, are being replaced by new understandings of the producer-consumer relationship as a circular process. Hence the nature of the product is seen as a negotiated outcome in which the product is continuously reproduced in light of shifting tastes, preferences and even meanings that are expressed by consumers through the process of consumption. The making and remaking of Las Vegas that is discussed in Chapter 9 is a good example, expressed on a spectacular scale, but the key point is that many areas of tourism production are subject to the same essential process.

Finally, the understanding of tourism has been enriched by closer association with a number of new critical and conceptual positions within human geography and the wider social sciences. In some respects this has become a reciprocal relationship in so far as whilst the understanding of tourism has benefited from the application of new critical positions, so tourism has become widely recognised as a 'lens' through which those same positions may be studied to advantage. Five areas of conceptual thinking that are relevant to tourism geography are worth highlighting:

Modernity and mobility Urry's (2000) thesis is that under modernity, mobility – as both a metaphor and a process – is at the heart of social life and that travelling has become a means through which social life and cultural identity is recursively formed and reformed. Mobility encompasses goods, information, images, ideas, services, finance and, of course, people (Shaw and Williams, 2004) within complex patterns that Urry (2000) suggests are structured around systems of what he terms 'scapes' and 'flows'. (The 'scapes' comprise the networks of machines, technologies and infrastructures that enable mobility – such as airports, motorways and computer networks, whilst the 'flows' are the movements of people, goods, ideas or images.) Tourism and tourist spaces, it is argued, are directly structured by the patterns of scapes and flows and as one of the most significant areas of modern mobility, tourism is deeply implicated in a key area of change.

Globalisation An important aspect of Urry's concept of mobility is that the scapes and flows transcend (and in many situations dissolve) national boundaries. Mobility therefore connects directly to processes of globalisation which may be seen as one of the primary consequences of the time–space compressions that are associated with enhanced and accelerated mobilities (Harvey, 1989). However, globalisation is not simply about greater levels of physical connectivity, more importantly it is both an economic and a cultural phenomenon that is shaped by progressively more complex and extended networks of interchange within transnational systems of production and consumption (Robins, 1997). Tourism is a prominent component of the process of globalisation and is a primary channel for economic and cultural exchange, but it is also shaped by globalisation through the evolving system of scapes and flows.

New geographies of production and consumption Globalisation connects closely to new geographies of production and consumption. Change in patterns of production is complex and involves both spatial and sectoral shifts. Thus, for example, there has been significant migration of manufacturing capacity from old centres of production such as western Europe to new spaces of production in regions such as south-east Asia. As a linked process, we have also seen sectoral shifts from declining manufacturing industries to expanding service industries, especially in older industrial economies. At the same time we have seen a move from Fordist patterns of mass production of standardised products, to post-Fordist patterns of flexible production in which goods and services are outwardly matched to the needs of different market segments. But there are countervailing tensions here since there is extensive evidence that in the realms of consumption, the spread of global capital and its associated consumer cultures provides a profound source of erosion of cultural traditions or difference that is quite capable of overwhelming local and regional experience (Crang, 2005). Ritzer's (1998) well-known thesis of the 'McDonaldisation' of society articulates these concerns with particular clarity, but there is much within the contemporary literature on tourism that explores the same themes.

Consumption and identity Although consumption is evidently an economic process, it is not exclusively so. Indeed, advocates of cultural readings of contemporary society will be quick to point to the many ways in which consumption is socially and culturally produced (Crang, 2005). Thus although many of the goods and services that we consume are still mass produced, in the process of consumption people will impart individual meanings and significance to products by the way in which they are utilised. Crang (2005) provides the example of the motor scooter – developed in Italy as a fashionable means of urban travel for a largely female market, it was adopted (and adapted) as an iconic component of a largely male 'Mod' culture in 1960s Britain. In these ways consumption becomes one of the primary mechanisms through which people form and then project their identity and tourism – as an arena of conspicuous consumption and a recurrent focus of popular social discourse – has assumed an increasingly central role in this process.

Sustainability As Sharpley (2000: 1) notes, the period since the publication of the report of the World Commission on Environment and Development (1987) on 'Our Common Future', the concept of sustainable development has become the focus of increasing attention amongst tourism theorists and practitioners. This has served several purposes. First, by connecting tourism with the wider agenda of sustainability, the debate has helped to emphasise the relevance of tourism as a significant arena for interaction between people and their environments. In this way, the political significance of tourism has been reinforced. Second, although – as authors such as Clarke (1997) have illustrated – there are significant difficulties in capturing the essence of what sustainability actually means, the discourse surrounding the sustainable nature of tourism has helped to refocus traditional debates on tourism impacts. Hence, in place of approaches that sought to isolate tourism impacts as positive or negative in their effect, the perspective of sustainability casts a more revealing light on the *processes* by which tourists might affect the places that they visit. As we shall see in subsequent sections of the book (especially Part II), a focus on process rather than outcome is often a more revealing way of understanding tourism's effects.

The structure of the book

Coles (2004: 140) writes that 'knowledge (of tourism) is shaped not only by recent influences . . . but also by the embedded nature of certain ideas, approaches, perspectives and traditions that have persisted over a longer time'. This is the approach that has been consciously adopted in redrafting this book in its second edition. The work proposes that tourism geography is essentially concerned with understanding how people – as tourists – relate to the places that are toured. However, this outwardly simple statement conceals some much more complex and detailed issues that tourism geographers need to address, particularly if we want to progress beyond the basic description of spatial patterns of tourism (which constitutes a very simple notion of what tourism geography is about) and instead access the much more interesting territory that relates to the explanation of those patterns and the meanings and values that might be embedded within them.

In developing that understanding, part of the discussion in this book draws on a traditional geographic interest in the processes of tourism development through which the modern geography of tourist space is shaped. In most situations, such spatial patterns are a combination of a legacy of historical–geographic processes, modified by contemporary change and part of the discussions that follow aim to demonstrate that contemporary tourism geography often has an important history that needs to be recognised and understood.

This theme forms the focus for Part I of the book – 'Tourism development and spatial change'. Here we examine some of the primary ways through which the spatial patterns of tourism and associated development of tourist places, especially in the form of resorts, have been shaped. The discussion is organised around the familiar geographic concern for scale and examines evolving patterns in both domestic and international tourism. The approach is broadly a historical–geographic one, but parts of the discussion also connect to issues of mobility as well as contemporary concerns for the impacts of globalisation.

Embedded within the types of spatial development that are discussed in Part I are complex and highly variable relationships between tourism and the spaces and places in which it is developed. These relationships embrace the physical settings of tourism, the environment, the economies, the societies and the cultures that tourist visit. Most importantly, these relationships are neither fixed nor consistent through time and over space so that an essential challenge to tourism geographers is to isolate and explain the processes by which these relations become manifest and to understand why the effects that tourism creates are often so variable from place to place.

The chapters grouped in Part II – 'Tourism relations' – explore the broad theme of the interaction between tourism and host environments, economies and societies, together with the role of planning as a mechanism for managing tourism development. The concept of tourism 'relations' (rather than the more traditional concern for impacts) is deployed as a more reflective way of thinking about how tourism relates to the places that are toured and the discussion also connects to wider issues of sustainability, circuits of production and consumption, commodification and power relations.

To some extent the thematic content of Parts I and II mirror many of the established traditions of tourism geography to which Coles (2004) refers. However, the final Part of the book – 'Understanding the spaces of tourism' – aims to take the discussion into some of the newer readings of tourism as a geographical phenomenon. In particular, this section of the book attempts to examine some of the differing ways in which tourist places are created, experienced and understood by the tourist (e.g., through the rise of heritage tourism or the remaking of cities as tourist places) as well as making some important connections to several of the newer foci of interest in human geography, such as the changing relations between production, consumption and identity. Here themes such as the embodied nature of tourism, its role in identity formation, its infusion into daily experience, and the role of tourism as a means of enabling people to make sense of the world in which they live, come to the fore.

Summary

Tourism has become an activity of global significance and, as an inherently geographical phenomenon that centres upon the movement of people, goods and services through time and space, it merits the serious consideration of geographers. Our understanding of tourism is, however, complicated by problems of definition, by the diversity of forms that the activity takes, by the contrasting categories of tourists, and by the different disciplines in which tourism may be studied. Geography, as an intrinsically eclectic subject with a tradition in the synthesis of alternative perspectives, is better placed than many to make sense of the patterns and practices of tourists, although the development of geographical perspectives on tourism has been characteristically uneven. However, as a significant contemporary phenomenon, tourism provides a valuable lens through which a number of contemporary themes in geography may be studied. These include new relationships between modernity and mobility; globalisation; new patterns of production and consumption; the links between consumption and identity; and finally, sustainability.

Discussion questions

1   Why is the definition of 'tourism' problematic?
2   What can geographers bring to the study of tourism?
3   Explain why it is important in tourism geography to distinguish between different categories of tourist and forms of tourism.
4   How does an understanding of tourist motivations help us to interpret geographical patterns of tourism?

Further reading

A number of recent texts provide useful overviews of geographical approaches to the study of tourism, including:
Hall, C.M. (2005) *Tourism: Rethinking the Social Science of Mobility*, Harlow: Prentice Hall.
Hall, C.M. and Page, S.J. (2005) *The Geography of Tourism and Recreation: Environment, Place and Space*, London: Routledge.
Lew, A.A., Hall, C.M. and Williams, A.M. (eds) (2004) *A Companion to Tourism Geography*, Blackwell: Oxford.
Shaw, G. and Williams, A.M. (2004) *Tourism and Tourism Spaces*, London: Sage.
Williams, S. (2003) *Tourism and Recreation*, Harlow: Prentice Hall.

Issues of tourist motivation have been explored in some classic papers that include:
Cohen, E. (1972) 'Towards a sociology of international tourism', *Social Research* Vol. 39: 164–82.
Compton, J.L. (1979) 'Motivations for pleasure vacation', *Annals of Tourism Research* Vol. 6 (4): 408–24.

More recent perspectives are set out in:
Crouch, D. (ed.) (1999) *Leisure/Tourism Geographies*, London: Routledge.
Franklin, A. (2004) *Tourism: An Introduction*, London: Sage.

Classic papers on tourist experience include:
Cohen, E. (1979) 'A phenomenology of tourist experiences', *Sociology* Vol. 13: 179–201.
MacCannell, D. (1973) 'Staged authenticity: arrangements of social space in tourist settings', *American Journal of Sociology* Vol. 79 (3): 589–603.

The growing significance of mobilities in contemporary life are explored in:
Urry, J. (2000) *Sociology Beyond Societies: Mobilities for the 21st Century*, London: Routledge.

For an excellent review of the changing approaches in geography to the study of tourism, see:
Butler, R. (2004) 'Geographical research on tourism, recreation and leisure: origins, eras and directions', *Tourism Geographies*, Vol. 6 (2): 143–62.

Part I
Tourism development and spatial change

In this first substantive part of the book we address a question that is essential to any geographical understanding of tourism, namely, 'how do basic geographies of tourism develop?' This is a question that has been a central concern in tourism geography from the inception of this sub-set of the discipline and whilst part of this attention has focused upon the development of descriptive approaches to the spatial patterns of tourist activity, a far more important line of investigation has centred upon the development of *explanations* of the geographical patterns that tourism reveals. This is the approach that has been consciously adopted in the following two chapters, both of which offer outline geographic descriptions of change in, respectively, domestic and international forms of tourism, but which concentrate upon developing an understanding of the factors that have shaped those changes.

Both chapters reveal a strong historic dimension to the narratives, but readers are encouraged to look beyond the basic historic narratives and to appreciate how the processes that underpin these narratives are quintessentially geographic in nature. The explanations of processes of change connect to some established and very significant areas of geographic interest: spatial diffusion; environmental perception; the nature of place; time–space compression; mobility and, ultimately, globalisation.

Part of the argument that is implicit in this discussion is that the historic processes that established early places of tourism created a basic spatial framework of activity that remains both evident and – in many contexts influential – in contemporary geographic patterns. Those patterns have often been modified by subsequent development (in which the spatial diffusion and geographic relocation of activity to new tourism centres is a recurring theme that is well-illustrated in the following chapters), but in many situations there remain evident connections between past and present tourism geographies. In this sense the narrative of tourism development is more than just an intrinsically interesting story of – say – socio-economic evolution; it is a fundamental part of understanding the contemporary pattern of tourism in many parts of the world.

The examination of the development of coastal resorts and rural tourism that is pursued in Chapter 2 provides an excellent illustration of how change in the way in which people view and evaluate their environments produces new geographic patterns of activity. The historical geography of resort development and the adoption of the countryside – especially wilder countryside – as a place of tourism, illustrates directly how changes in environmental perceptions and values influences spatial patterns and triggers change. It also illustrates the enduring importance of place because whilst there are many commonalities that underpin processes of development, there is often significant variation in the outcomes of those processes across geographic space which is due, in no small measure, to the distinctive and localised character of geographic places.

These themes inform our understanding of the more extended spaces of international tourism too, but Chapter 3 also brings into focus the impacts of processes such as time–space compression (through transport innovation); the significance of mobility in extending geographical patterns of activity; and the onset of global systems of connection and exchange. Of course, time–space compression and changed levels of mobility are an important aspect of domestic tourism development, in both a historic and a contemporary context, but it is perhaps the enhanced geographic scale associated with international travel that emphasises the role that these processes exert upon tourism geography.

2 Tourism places and the place of tourism: resort development and the popularisation of tourism

One of the fundamental questions that interests tourism geographers is 'how do places develop as centres of tourism?' This chapter aims to explore this question by using an historical–geographic perspective on the spatial, social and structural development of domestic forms of tourism – that is, tourism that occurs within the boundaries of a single country. Two contrasting types of tourist place – urban coastal resorts and the picturesque countryside – are examined, using the case of Britain as a primary example, although with some comparative reference to other parts of Europe and North America. The British experience offers particularly good case studies of the development of tourist places since Britain was one of the first nations to develop modern practices of tourism and clearly exemplifies many of the factors that have shaped the subsequent geography of domestic tourist activity. However, Britain is by no means unique and other industrialised nations such as France, Germany and the USA have a rich tradition in local tourism from which many lessons may be learned.

The socio-geographic development of tourism has been influenced by numerous elements but there are four factors that, it is argued here, are especially important in understanding how and why the patterns and character of tourist places have evolved in the way that they have.

First, we should acknowledge the significance of change through time in attitudes and motivations. In the modern world, tourist travel is becoming a seemingly natural and incidental part of life and in most Western nations – and in a growing number of countries in the developing world – millions of people now harbour expectations of becoming tourists at least on an annual basis, if not more frequently. However, this was not always the case. For most of recorded history, travel was difficult, expensive, uncomfortable and often dangerous, so the desire to travel must initially have been prompted by powerful and very basic motives. It is not surprising, therefore, that amongst early 'tourists' we find religious pilgrims motivated by a strong sense of spiritual purpose, or travellers who journeyed in the quest for health (one of the most fundamental human concerns). As travel became less difficult and more affordable, it became easier to admit other motives as a basis to tourism, especially the pursuit of pleasure. However, whenever differences in priorities emerge, changes in the needs, expectations and attitudes of visitors usually alter the geography of tourism and re-work the character of the tourism experience.

Second, the social and economic emancipation of the urban middle classes and (particularly) the proletariat is also important. For ordinary people to bring tourism into their lifestyles, extensive and fundamental change was required in the way in which lives were lived. Central to this transformation is the liberation of blocks of time that are free from work and which are sufficiently extended to permit tourism trips and, equally significantly, the ability to accumulate reserves of disposable income that can be expended on a discretionary purchase such as a holiday.

Third, mass forms of tourism became possible only with the development of efficient and affordable systems of transportation. The railway, in particular, made mass travel a reality in the second half of the nineteenth century, extending the range over which people could travel for pleasure and prompting development of new tourist regions, much in the same way that developments in civil aviation following the end of the Second World War underpinned the more recent emergence of popular forms of international tourism (see Chapter 3).

Finally, modern tourism also requires organisational systems and the enterprise that is associated with such systems to provide the supporting infrastructure of facilities and personnel able to run the tourism business and to promote tourism places to potential visitors. With the exception, perhaps, of the more solitary and explorative forms of tourism practised by the lone travellers and drifters discussed in Chapter 1, most forms of travel will not develop in the absence of either the basic facilities of support, or active promotion designed to raise public awareness. Essential facilities include accommodation, transport, entertainment, retail services and (increasingly) forms of packaged tourism in which all these elements may be purchased within a single transaction.

The development of tourist places: a conceptual perspective

The narrative of tourism development is both an extended and a fascinating story, as seminal work by writers such as Pimlott (1947), Soane (1993), Towner (1996), Turner and Ash (1975), Walton (1983a, 2000) and Walvin (1978) reveals. However, in order to distil that narrative into a more concise format, it is helpful to conceive the process of establishing tourism places as passing through a successive series of development phases. This approach has been most successfully – and influentially – captured by Butler (1980) in his well-known model of the cycle of evolution of tourist areas. This adapts the product life-cycle model – as developed in marketing theory to chart the normal pattern of uptake of a new product in the market place – and applies the same broad principles to the development of tourist destinations.

In summary, the model proposes that development processes are initiated in an *exploratory* stage in which small numbers of tourists, sometimes acting in the mode of Cohen's (1972) 'explorers', pioneer the discovery of the new destination area. If such activity persists, some local residents will react to the new economic opportunities by providing basic facilities for tourists (such as accommodation), triggering an *involvement* stage in which characteristics such as a tourist season might also be expected to emerge. As the reputation of the destination area becomes established (whether through word of mouth or through advertising and promotion), it begins to attract inward investment (and associated increases in the level of external control), acquires more sophisticated infrastructure and marketable attractions, and develops a rapidly expanding clientele that is drawn from a wider market – the *development* stage. Eventually the rate of increase in visitors will slow (although tourist numbers continue to increase) as the *consolidation* stage is reached. At this point tourism is established as a significant part of the local economy and usually fundamental to the wider prosperity of the local area. Tourists will outnumber local residents for much of the main season, with an increased potential for conflict. Investment and promotional activity is strongly aligned to maintaining the position of the resort in a competitive market. The penultimate phase is described by Butler as the *stagnation* stage in which the capacity levels of the area have been attained and the demand is no longer growing. Continued prosperity depends upon attracting repeat visitors and containing the potentially negative impacts of a widening range of potential problems; for example, physical deterioration in older infrastructure, over-development or congestion, leading to a loss of image. In the final phase of the model (now widely referred to as the *post-stagnation* phase, although Butler does not actually use that term), several alternative trajectories based upon rejuvenation or decline are proposed. The natural tendency may be towards some form of decline, as virtually all tourist destinations will lose their attraction unless there is intervention (e.g., through planned investment) to arrest that process. However, where a destination is subjected to a process of restructuring that is designed to reinvent the area to appeal to new markets (see Agarwal, 2002), then theoretically that process of rejuvenation will trigger a further cycle of development.

Butler's model has subsequently been subjected to a number of detailed criticisms and refinements, especially in relation to the post-stagnation phase (see, amongst many, Agarwal, 1994, 1997, 2002; Cooper and Jackson, 1989; Haywood, 1986; Priestley and Mundet, 1998). The common criticisms that are derived from these papers draw attention to several perceived deficiencies:

- As a universal evolutionary model it fails to capture the uniqueness of place and the capacity for local economies to resist broader national or international processes. In particular, it does not reflect with any clarity the articulation of the internal–external relationships that affect resort development in differing ways, dependent upon a range of contextual attributes.
- It downplays the role of human agency in mediating processes of development and change in ways that mean that outcomes are not always inevitable and predictable.
- It implies a seamless and continuous transition from one phase to the next, whereas in practice phases are likely to overlap and may also be subject to periodic reversal.
- Phases are difficult to identify without the experience of hindsight since the longer-term trends only become established and recognisable with the passage of time. (In this regard it is particularly difficult to separate the stagnation from the post-stagnation phases, since features of decline – and local response to that decline – will in practice be evident in both stages.)
- It fails to separate causes and consequences – especially at the decline stage. Is decline caused by changing patterns of production and consumption, or is change in production and consumption of resorts a consequence of decline that is prompted by other processes?
- It presents development as a uni-directional process with, according to some critics (e.g., Priestley and Mundet, 1998), an excessively pessimistic and seemingly inevitable progression towards decline.

However, despite these reservations, the model survives as a useful conceptual framework in which to explore the dynamics of how tourist places develop, and so in the sections that follow, it is used to shape the discussion of the development of seaside resorts and of tourism to the picturesque countryside.

Exploration and involvement: the formation of resorts and the emergence of rural tourism

Although tourism today is found widely across cities, countryside and coast, historically its development was most evident in the formation of resorts and there remains a strong, visible legacy of resort-based tourism within contemporary geographical patterns, both in Britain and elsewhere. In Britain, the first resorts were the inland health spas – towns and villages that possessed local mineral waters that were believed to have curative qualities and which attracted people who were seeking a remedy for particular conditions. Mineral water therapies were not an innovation – as the Roman remains at Bath testify   and the intermittent and usually localised popularity of spas was a feature of sixteenth- and seventeenth-century life in both Britain and Europe. However, in the mid-eighteenth century, watering places such as Buxton, Scarborough, Tunbridge Wells and Bath itself all enjoyed a significant increase in fortune, as mineral cures became widely popular amongst people of wealth. In Europe, spa towns such as Baden (in Germany) and Vichy (in France) flourished under similar conditions.

Initially, of course, spas were predominantly the resort of the sick, but because they were often promoted by shrewd entrepreneurs as exclusive places – many of which also benefited from royal patronage – the spas also developed as places of fashion that attracted leisure-seekers who had no need for a cure, but who were drawn by the social life that developed at the resort. In order that visitors might be entertained, facilities were provided not just for taking the waters, but for concerts and theatre, dances, walks and promenades, and at the best spas (Bath or Tunbridge Wells, for example) there soon emerged a microcosm of fashionable metropolitan life.

The geographic extension in this early form of tourism from inland spas to coastal resorts came about through an almost incidental shift in medical thinking that suggested that sea bathing and, in some cases, drinking of sea water was a more effective treatment than many of the cures offered at inland spas. Sea bathing was not, however, a new practice and both Corbin (1995) and Towner (1996) draw attention to much older cultures of sea bathing that were widely established across Europe and in parts of Britain and which were quite independent of the elite cultures that were soon to emerge in fledgling resorts. Towner (1996: 171) writes that 'along the coasts of the Baltic, North Sea and Mediterranean there was a tradition in sea bathing that lay beyond the codified practices of the leisured classes' and that 'on the shores of the Mediterranean it was the peasant classes who bathed for pleasure well before it was adopted by the ruling classes'. In the British context, Walton (1983a: 11) makes a similar observation, noting that prior to the local expansion of resorts after 1800, a popular sea-bathing culture existed on the Lancashire coast, not emulating the rich but having a 'prior and independent existence'. Comparable patterns have also been identified at Santander and San Sebastian on the northern coast of Spain (Walton, 1997a). Critically, though, such practices were seldom seen as engaging with a medicinal purpose and therefore attracted neither the advocacy of the professions nor the patronage of the wealthy that was crucial to the subsequent organised development of coastal resorts.

In Britain the first recorded uses of sea water treatments were noted in the Yorkshire coastal town of Scarborough as early as 1667 (where a local doctor – Robert Whittie – enjoyed considerable local success with a combination of mineral and sea water cures) and at nearby Whitby where, according to Travis (1997), a fashionable style of sea bathing was noted in 1718. However, credit for the wider promotion of sea water treatments is usually given to a Dr Richard Russell who practised near Brighton and who published an influential text on the subject in 1750 (see Case Study 2.1). Russell's book rapidly caught the imagination of the upper classes (who were the only social group that could afford the time and the expense to travel to the seaside) and in a process that almost precisely mirrored the development of inland health spas, a string of fashionable and exclusive new sea bathing resorts sprang up, especially along those parts of the coasts of Kent and Sussex that were relatively accessible from London (Figure 2.1).

The new fashion for sea bathing (and, in due course, the seaside holiday) soon became what Walton (1997a: 37) describes as a 'prominent cultural export' that diffused from England into France, Germany and the Low Countries, and then later to Spain and Italy. Corbin (1995) shows how a formal sea bathing season had been established in Dieppe, Boulogne and Ostende by 1785; the first German resort at Doberan was developed from 1794 (which is broadly contemporary with the development of Scheveningen as the leading Dutch resort); whilst the origins as a resort of San Sebastian in northern Spain can be traced to visits by the Spanish royal family around 1828 (Walton and Smith, 1996). This pattern of diffusion

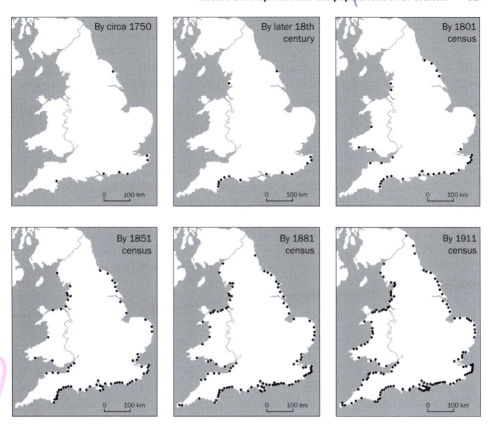

Figure 2.1 Expansion of sea bathing resorts in England and Wales, 1750–1911

was not, however, simply confined to Europe. The colonial influence of Britain in North America (where a network of inland spas was already established) led to the early development of coastal resorts on the eastern seaboard of the USA. Nahant (north of Boston), Long Branch (south of New York) and Cape May (south of Philadelphia) were all established as fashionable sea bathing resorts before 1800 (Towner, 1996).

These early coastal resorts reflected most, if not all, of the criteria set down in Butler's model. They were small in scale and because they were exclusive, they depended upon a comparatively small group of visitors who were 'early adopters' of the fashion for sea bathing. The provision of basic infrastructure, such as lodging houses, was largely local in organisation and limited in scale, not least because the 'season' for seaside visiting – whilst an established feature – would have represented a comparatively brief sojourn within the extended patterns of visiting to places of fashion that were practised by elite groups at this time.

The extension of elite leisure practices from inland spas to sea bathing resorts was remarkable not only as a geographic process but also because it reflected quite profound changes in public attitudes towards the coastline. From a twenty-first century perspective, the attraction of the sea seems entirely natural, but historically the sea and its coastlines were viewed quite differently. As Corbin (1995) explains, the coast was often a place of fear and repulsion. It was a zone of tension, associated with pirates and smugglers, shipwrecks and places of invasion, whilst the sea itself was an unfathomable mystery, a home to monstrous creatures and a chaotic remnant of the Great Flood that was capable of unleashing awesome, destructive powers upon the coastline. As if to reinforce the point, the incidence of seasickness amongst early tourists who did venture onto the oceans must have confirmed for many that this was not a natural and proper place for people. Similar reservations were also evident in the internal arrangement of coastal settlements which were generally shaped by a need for protection from the elements rather than to take advantage of what would be later considered as a pleasing view (Urry, 1990).

Yet by the beginning of the nineteenth century, the sea and the coast had become central to the popular imagination. This attitudinal change may be broadly viewed as emanating from the Enlightenment in Europe and included the new popularity and influence of natural theology (in which the enjoyment of natural spectacles such as the sea was now seen as a celebration of God's work); interest in 'new' sciences such as geology and natural history that focused attention upon coastlines as field laboratories; and the emergence of a public taste for the picturesque in the latter half of the eighteenth century and then the influence of the romantic movement of the early nineteenth century (see Corbin, 1995 for an extended discussion of this change in taste).

This realignment in public sensibilities did not only affect coastal areas, it was also fundamental to the early emergence of rural areas as new objects of the tourist gaze. Prior to the middle of the eighteenth century, the countryside – and especially wild regions of moorland and mountain – had, like the coast, been generally disregarded as places to be visited for pleasure. The pervasive preferences in rural landscape were for what Towner (1996: 139) describes as 'the humanised scene of cultivation . . . as evidence of the successful mastery of nature', and these sentiments were reflected in early tours of rural areas reported by writers such as Defoe (1724) and, later, Cobbett (1830). In contrast, mountain areas were perceived as untouched by the organising hand of civilised people and these places and their rough inhabitants were widely shunned and actively avoided (Plate 2.1). However, the eighteenth century was a critical period in the advancement of understanding of natural

Plate 2.1 Mountain landscapes in Snowdonia, UK: environments that were shunned before the Romantic Movement but are now revered objects of the tourist gaze

systems and relationships between people and nature, and in affluent society in Britain, Europe and North America, the wilder landscapes acquired a new level of attraction for people of taste (Bunce, 1994; Williams, 2003).

The new tastes for wilder landscapes introduced tourism to places that would eventually become highly valued destinations. In Britain, these included the English Lake District, upland Wales and highland Scotland (Andrews, 1989; Urry, 1994c), whilst in the USA, the Catskill Mountains were quickly adopted as a regular place of visiting for fashionable New York and Boston society (Demars, 1990). However, unlike early resort-based seaside tourism which was essentially place-specific, rural tourism was more typically focused upon touring as a *practice* and because of the difficulties of travel before the advent of the railway, it was generally conducted in very small, independent groups. The development of rural resorts (such as Windermere in the English Lakes) as places of popular visiting was generally delayed until the second half of the nineteenth century.

Development: the popularisation of the seaside and the country in the nineteenth century

Although by the turn of the nineteenth century both the seaside and the picturesque countryside had become focal points for a nascent popular interest, each remained relatively inaccessible in both a spatial and a social sense. In an era when roads were badly maintained and travel (mainly by stagecoach) was slow and expensive, the numbers who journeyed to the seaside – whether for a cure or simply for pleasure – remained small. Touring within picturesque rural districts was an even more selective preoccupation. Yet within a very few years of the turn of the nineteenth century, several key developments began a process of transformation in the nature of seaside tourism and would eventually impact upon rural tourism too.

The first of these changes came in transportation. In Britain, the invention of the steamship in the early years of the nineteenth century initially prompted the growth of new resorts on the Thames estuary (such as Margate) and also encouraged the development of small resorts on the estuaries of the Forth and, especially, the Clyde in Scotland. By the end of the nineteenth century, these Scottish resorts had developed – as Figure 2.2 shows – into a complex network of leisure places and associated steamer services (Durie, 1994). In America, the early development of Coney Island was also dependent on steamer services from New York. But more important changes generally followed the development of passenger railways after 1830.

The railways transformed the nature of tourism by shortening journey times whilst increasing dramatically the numbers that might be moved on any one journey. Some writers have argued that the primary effects of the railway were to trigger growth by bringing existing resorts within range of the growing urban populations and (eventually) to develop new markets for popular forms of tourism through a reduction in the cost of travel (Urry, 1990; Towner, 1996). For example, in the USA, Atlantic City developed very rapidly as a popular destination, once the railroad link to Philadelphia had been completed in 1860 (Towner, 1996). However, in some situations the development of national railway networks also helped to open up new areas to both resort development and rural tourism. In France, for example, Brittany had been 'discovered' by romantic artists, writers and travellers by 1820 (Plate 2.2), but it was not until after 1850 when the railways began to connect the region to major urban areas such as Paris, that a resort system developed around the Breton coast (Towner, 1996). The spread of the French railway network also triggered the extension of resorts into western Normandy. Similarly, the creation of new rail links between Madrid and the northern Spanish towns of San Sebastian and Santander aided the geographical expansion of tourism along that coastline (Walton, 1997a).

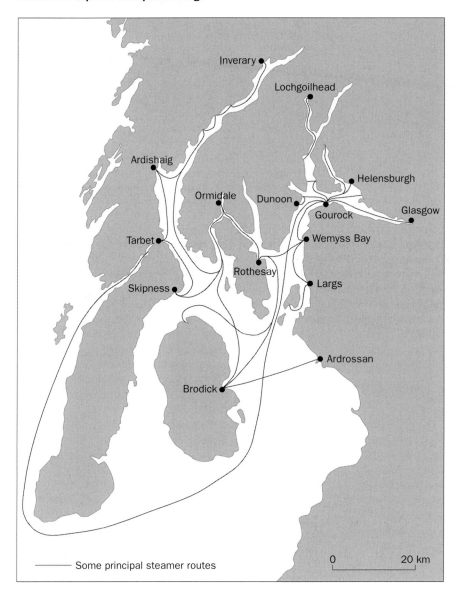

Figure 2.2 Principal pleasure steamer routes on the Firth of Clyde, Scotland, *circa* 1890

Allied with changes in mobility came equally significant changes in social access to travel and tourism. Although popular travel for working-class families remained inhibited by a number of obvious constraints (especially lack of time and shortage of money), the expansion of the international economy from the middle of the nineteenth century spawned a new and prosperous professional middle class who were not so constrained and clearly possessed the inclination to imitate the habits of the aristocracy in resorting to the coast for day trips and holidays (Soane, 1993). The effect of this influx of new tourists on the resorts was often to displace (both spatially and temporally) the elite groups that had pioneered the resort development to places that were further removed from the urban conurbations that were the primary source of new tourists. In some situations, therefore, distance became a key factor in maintaining the social tone of fashionable resorts.

Plate 2.2 Part of the picturesque landscapes of Brittany that were discovered by tourists in the second half of the nineteenth century: the river-front at Auray

By the last quarter of the nineteenth century, the holiday habits of the middle classes had generally begun to filter down to working people. In Britain, reductions in the length of the working week and the first statutory holidays that followed Lubbock's Bank Holidays Act of 1871 had made more time available for seaside excursions, whilst by the 1880s and 1890s gradual improvements in levels of pay, when combined with the Victorian virtue of thrift – which had often been essential to basic survival in the early phases of urban industrialisation – were paying dividends in terms of the abilities of many working families to save money for excursions and holidays. Walton (1981) observes how rising living standards in the last quarter of the nineteenth century released a flood of new seaside visitors drawn from skilled working families. In industrial communities, especially in the north of England for example, saving through co-operative or friendly societies was actively encouraged and the benefits became manifest in a number of ways, including the taking of holidays. The development of working-class holidays in this area also drew impetus from the older tradition of wakes weeks. Wakes had originated in religious festivals in the eighteenth century but, by the 1870s, had evolved into large-scale industrial holidays in which significant portions of the working populations in towns such as Oldham (Poole, 1983) would decamp *en masse* for short holidays to the seaside or on excursions to the coast or the countryside (Freethy and Freethy, 1997; Hudson, 1992).

A third key set of changes were essentially structural in character and reflect three related processes.

- the physical developments of facilities in resorts;
- the role of local municipal control in resort development;
- the early organisation of a tourism industry.

As the demand for seaside holidays grew in mid-Victorian Britain, resorts witnessed significant developments of hotels and boarding houses, places of entertainment (which signalled most clearly that the motives for visiting the seaside were now as much pleasurable as it was health-related), as well as civic facilities and service industries that supported or developed around tourism. This is the era in which iconic landmarks of the seaside, such as the pier and the promenade, became established features of resort landscapes, supplemented – in the more fashionable resorts – by pavilions, public gardens, theatres and concert halls and in the more popular seaside towns by funfairs, music halls, amusements and cheap outlets for food and drink. In Britain during the middle years of the nineteenth century, seaside resorts recorded the fastest growth rates amongst all categories of urban centre, including London and the major industrial cities of the north.

An important regulatory influence upon physical development of resorts was the level of control exerted by local municipalities. This was a key factor that shaped not only the style of resort development but also the social tone. Walton (1983b) explains that municipal government commonly adopted a dual role – as a regulator of development and activity through the application of local bye-laws, and as a provider of public amenities such as parks, libraries, museums and other potential attractions. In time, municipal government would also become active in the promotion of resorts to potential visitors. However, critically, both regulatory and investment policies often reflected local desires to maintain a particular style of resort and an associated tone. So, for example, the refined character of the English resort of Bournemouth was effectively sustained by regulation and investment by the local authority in a particular range of facilities that reflected the tastes and preferences of a middle-class clientele (Roberts, 1983; Soane, 1993). Similarly, the fashionable qualities of the resort of San Sebastian in northern Spain owed much to municipal intervention that applied strict planning regulations, effective law and order enforcement, active engagement with sanitary improvements and investment in quality facilities (Walton and Smith, 1996). In contrast, municipal authorities in Blackpool adopted a very different approach by actively promoting popular forms of seaside visiting and investing in facilities that matched the preferences of those markets (Walton, 1983b, 1994).

The early development of an organised tourist industry is perhaps reflected most clearly in the emergence of the first tourist excursions. These are commonly attributed as the invention of Thomas Cook, although Towner (1996) observes that travellers from London to parts of France and Switzerland had been able to purchase inclusive tours in the 1820s. Cook was a bookseller and a Baptist preacher who, on the way to a temperance meeting in 1841, had the inspired idea of chartering special trains to transport supporters of temperance to their meetings at low cost. However, almost immediately, the far more profitable idea of organised excursions for pleasure occurred to him, and by 1845 'Cook's Tours' had become a recognised phrase and the beginnings of a travel industry that would serve a popular rather than an elite market. By the middle decades of the nineteenth century, popular excursions were an established feature of leisure patterns in Britain and were quickly adopted elsewhere. As well as serving to widen social access to tourism, excursions were also instrumental in helping to shape a spatial extension of resorts into new areas of development (such as the north Wales coast in Britain) during the second half of the nineteenth century (see Figure 2.1).

Trips to the seaside were a key part of Cook's business, but thanks to the Victorian taste for the picturesque and the romantic, new tourist areas in north Wales, the Lake District, the Isle of Man, Scotland and Ireland were soon added to his itineraries, representing the first significant move away from a pattern of tourism centred in coastal towns. Excursions to both urban and rural destinations in Europe were soon on offer too. These trends reflected the widening popular interest in rural forms of tourism that became more prominent as the century progressed and which often built upon initial patterns of visiting established by elite tourists in search of the picturesque. In northern England, for example, the arrival of the railway on the fringes of the Lake District in the late 1840s introduced a new social class and new forms of tourism to towns such as Windermere (Towner, 1996; Urry, 1994c). In comparison with seaside resorts the level of visiting was low, but was sufficiently sustained to enable the development of resort functions that centred on genteel enjoyment of scenery and outdoor activity (O'Neill, 1994). Similarly, Green (1990) notes rising levels of both middle- and working-class tourist visits from Paris to the Seine Valley from the 1840s and Sears (1989) charts a developing tourist industry in rural areas of New York state, based upon Albany, the Hudson Valley and the Catskill and Adirondack Mountains. This diffusion of activity into progressively more distant rural areas was an essential component in the development of rural tourism and provided many of the foundations on which more consolidated patterns of activity were to develop, particularly after 1920.

The preceding narrative provides a broad illustration of the essential features of the exploration, involvement and development phases of Butler's model as applied to tourism development in resorts and in rural areas. To complete these sections of the chapter, Case Study 2.1 offers a specific example of the early development of the resort of Brighton.

CASE STUDY 2.1

The early development of Brighton as a sea bathing resort

The development of Brighton illustrates very well the key factors that often shaped the process of early resort development, in particular: the role of patronage; the impacts of transport on accessibility; the importance of investment in resort infrastructure and the shaping of an emergent tourism industry; and the progressive widening of social accessibility.

Patronage was an essential part of the early establishment of Brighton as a sea bathing resort. In the early 1750s, a Dr Richard Russell – a local man who had gained a national reputation as a specialist in the use of sea water treatments for common illnesses – opened a practice in Brighton. Russell's reputation attracted a small but growing clientele of wealthy patrons to the town. Initially the majority of visitors came for treatment but, as was the case with inland spas, sea bathing resorts also acquired a reputation as fashionable places of leisure which, through time, became the dominant attraction. The most influential patron of Brighton was the Prince of Wales (later King George IV) who first visited the town in 1783 and who continued to make annual and extended visits until 1827. The presence of the Prince and his entourage helped to position the resort as an exclusive and fashionable place to which, through time, a widening range of visitors would aspire.

However, for Brighton to develop from the small and exclusive place that was patronised by elite Regency society to the popular resort that it became, required improved accessibility, new infrastructure and local reorganisation of the resort as a place of tourism. Initially, access to Brighton was dependent upon stage coach services, most of which were slow and of very limited capacity. However, the opening of the railway from London in 1841 exerted something of a transformative effect: by reducing significantly journey times to the resort; lowering the cost of travel; and increasing dramatically the number of visitors that could be transported by a single journey. Hence, by 1850, the total annual number of visitors to Brighton was put in excess of 250,000 people, most of whom came by train, often travelling on the new, cheap excursions.

As visitor levels rose, the provision of new infrastructure to meet their needs quickly became evident. The Regency phase of Brighton's development had seen some physical development of the town: through the construction of hotels; through the laying out of public

continued

spaces along the seafront; and through the building of new areas of exclusive housing for the visitors, as well as cheaper housing for the growing population of permanent residents that were necessary to sustain the resort and which had risen to over 65,000 people by 1851.

By the middle of the ninteenth century the level of visiting to Brighton and the rising demand by middle-class tourists for staying holidays stimulated concerted programmes of development and reorganisation. Central to this process was a major programme of large hotel construction that included Brighton's famous Grand Hotel (1864), but which also included the West Pier (1868) and the Brighton Aquarium (1871). In keeping with the new taste for entertainment at the seaside, facilities such as the piers as well as many of the older, established theatres began to display many of the attractions that are now widely associated with the Victorian seaside but which at the time bore the key attribute of novelty. These included military band concerts, music hall acts, black and white minstrel shows, *pierrots* and, later, amusements and fairground sideshows. Many of these new attractions reflected the tastes and preferences of the working classes who, by 1900, formed a particularly prominent group of visitors to the resort.

This widening of social accessibility that is characteristic of early phases of resort development had important implications for reworking both the temporal and the spatial patterns of resort-based tourism. After the death of King George IV, both William IV and then Queen Victoria continued for a while the royal tradition of visiting Brighton. However, once the railway began to open the resort to large-scale visiting by middle- and then working-class tourists, the elite groups soon adapted their patterns, either in favour of visits outside the popular summer season and/or to places (such as neighbouring Hove) that still retained an air of exclusivity. This process of social displacement has been integral to the wider diffusion of resort-based tourism and is a recurring theme in the development of both domestic and international geographies of resorts.

Sources: Gilbert (1975); Pimlott (1947); Walvin (1978)

Consolidation: domestic tourism between 1920 and 1970

The First World War marked a watershed in many aspects of life, none more so than the incidence and practice of popular forms of leisure. Whatever vestiges of exclusivity in the traditional resorts that may have survived the onslaught of middle- and working-class tourists in the nineteenth century were largely swept away by the collapse across Europe of the old orders and the migration of social elites away from the older resorts to new and exclusive places, as yet untouched by popular demands. The processes of social emancipation in access to tourism and a number of related areas of change that had been gaining momentum at the end of the nineteenth and in the early years of the twentieth centuries, exerted a number of significant effects after 1918.

These developments served to provide the *consolidation* that Butler (1980) identifies and which is reflected in:

- further increases in the numbers of tourists being received in destinations;
- consolidation of resorts as urban places with a diverse range of functions;
- further enhancement of the significance of tourism in local economies;
- the extension of tourism into contiguous zones and the consequent creation of resort (or tourism) regions, often based around new, low-cost forms of accommodation such as holiday camps, camping and (later) caravans.

To illustrate these ideas, the following discussion focuses specifically on Britain which, at the start of the twentieth century, is described by Walton (2000: 27) as possessing 'a system of coastal resorts whose scale and complexity was unmatched anywhere else in the world'.

Prior to the 1950s there is no reliable data by which to chart the expansion in the numbers of domestic tourists in Britain, even though there is ample evidence from other sources (e.g., contemporary photographs of holiday crowds or summer passenger figures from railway companies) that show that mass forms of tourism were clearly established by the late 1930s. On the eve of the Second World War, it is estimated that some 11 million holidays and an uncounted number of day excursions were being taken annually within Britain (Walvin, 1978), the vast majority of which were directed at the seaside resorts. In 1938, for example, Blackpool alone received an estimated 7 million visitors. But these are 'snapshot' data that do not establish clear trends.

However, by 1951, when the first of what was to become an annual survey of holiday-making was conducted by the British Travel Association (later to be redesignated the British Tourist Authority, both hereafter referred to as the BTA), an estimated 26.5 million holidays were taken by the British, including 1.5 million abroad. Figure 2.3 charts the expansion in total holidays taken between 1951 and 1970 and reveals a distinct pattern with pronounced growth throughout the 1950s, followed by a period of relative stability in the 1960s.

Explanation for the increased level of holidaymaking needs to take account of several factors:

- The latent demand that had built up in the latter part of the 1930s and during the war years was finally released as the Holidays with Pay Act 1938 came fully into force, creating popular expectations of an annual holiday.
- Real wages increased bringing improved living conditions and more widespread household purchases of luxury items, including holidays. The increased purchasing power of the lower-middle-class and the working-class populations was especially important in this respect (Walton, 2000).
- Holidays were actively promoted within the media, by transport operators and a rapidly developing travel industry, raising public awareness and fuelling demand.

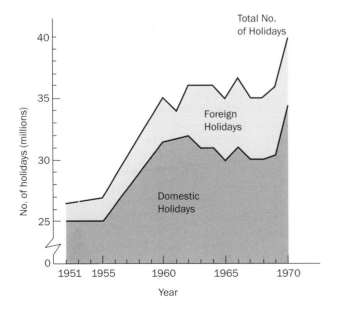

Figure 2.3
Growth of British holidaymaking, 1950–1970

- Programmes of modernisation and investment in new amenities in the modernist style had endowed many of the larger seaside resorts with enhanced levels of appeal, particularly to ordinary people (Hassan, 2003).
- Rising levels of mobility and reductions in the relative cost of travel, first through the development of public bus services and, subsequently, private car ownership, made resort areas more accessible.

The popularity of resort areas during the late 1930s and, especially, the 1950s – an era that has been interpreted as a 'golden age' for the traditional seaside (Demetriadi, 1997) – helped to draw a wider range of amenity-related activities that reinforced (and hence consolidated) the position of resorts within the wider urban framework. In some instances – for example at Bournemouth, the use of resorts as places of retirement had been noted during the nineteenth century (Soane, 1993) but this became a much more established trend in the inter-war years. By 1951 the proportion of the population that was of pensionable age in popular retirement resorts such as Worthing was almost 30 per cent (Walton, 1997b). At the same time, commuting populations began to develop in coastal towns that enjoyed rapid rail links to London and other major centres. Hence resorts such as Clacton, Southend and Walton on the Essex coast more than doubled their populations between 1911 and 1951 and the share of the total population for England and Wales that lived in resorts rose from 4.5 per cent in 1911 to over 5.7 per cent in 1951 (Walton, 1997b). This population growth had very positive impacts on local service economies that supported the increased numbers of permanent residents.

Part of the appeal of resorts to permanent populations – especially the elderly – lay in a combination of a nostalgic construction of the seaside as places of youthful enjoyment, married to a residual confidence in the therepeutic values of the seaside (Hassan, 2003). However, of greater importance was the active promotion of resorts and the investment in civic amenities that many municipal authorities pursued. Walton (1997b) observes that the 1920s and 1930s marked a high point in local investment in seaside amenities – for example, promenades, parks, gardens, lido, tennis courts and golf courses – that were designed to appeal to new popular tastes and establish resorts as exciting and fashionable places (Hassan, 2003).

A key element in the consolidation of seaside resorts in this period was the related development of large-scale provision of affordable accommodation in holiday camps and caravanning and camping grounds. On popular coastlines, hierarchies of primary-, secondary- and even tertiary-level urban resorts had generally developed before 1920 (Figure 2.1), through natural processes of urban expansion and diffusion of activity into neighbouring places. After 1920, however, the spatial extension of tourism was reinforced by these new developments. In many instances, holiday camps were located in close proximity to resorts in order to take advantage of the resort amenities. However, the rising levels of demand for the affordable holidays that camps provided also encouraged encroachment onto previously undeveloped stretches of coastline, often in haphazard and poorly planned developments (Ward and Hardy, 1986). Figure 2.4 illustrates the distribution of camps in England and Wales in 1939 and emphasises the extent of development but, more importantly, also shows how in some areas extended zones of coastal tourism within what we might label 'resort regions' were being formed through the combination of urban resorts and camps. (As an example, Case Study 4.3 in Chapter 4 describes the resort region that has developed on the coast of north Wales.)

Within the overall patterns of growth in domestic tourism in Britain during the post-1945 period, there have been some significant changes in the geography of tourism. Long-term analyses of regional shifts are frustrated by the periodic redefinition of BTA regions and data areas. However, even allowing for the uncertainties that this practice creates, we may be

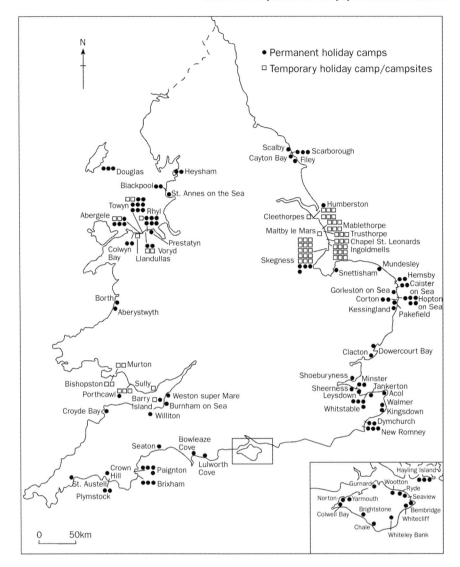

Figure 2.4 Distribution of holiday camps in England and Wales, 1939

confident that there has been a pronounced development of tourism in the South West of England (and to a lesser extent in Wales) and relative stagnation and even decline in the older holiday regions such as the North West and the South East of England (which include traditional resorts such as Blackpool, Brighton and Eastbourne). Table 2.1 illustrates estimated regional shares of the domestic market for a selection of regions that are broadly consistent in definition at a range of dates and shows the extent to which the South West now dominates the British market.

These regional shifts in domestic tourism reveal, once again, the impact of transport technology. The comparative remoteness (especially of Cornwall) had ensured that in the nineteenth century, the South West had not been extensively developed as a tourist destination, although Devon did possess some established resorts of regional importance such as Torquay. However, tourism to Devon and Cornwall developed substantially from the turn of the twentieth century onwards, especially in response to the active promotion of the Great Western Railway (Thomas, 1997; Thornton, 1997), which invented the image of an 'English Riviera' for this region (see Chapter 8). From about 1960 onwards, rapid increases in car ownership and the spatial flexibility that the car permits have allowed widespread diffusion of tourism, not only across the South West, but into other peripheral localities too. The shift in holiday transport from the public modes of travel by train and bus to the private car has been one of the most persistent changes in the structure of tourism in Britain (Table 2.2) and has directly promoted many new tourist localities, as well as one of the most popular tourist pastimes: recreational motoring.

Table 2.1 *Changes in the regional share of domestic tourism markets in the UK, 1958–98*

	Percentage share of market			
Region	*1958*	*1968*	*1978*	*1998*
South West England	14	23	20	26
North West England	13	11	7	5
South East England	9	10	8	7
Wales	10	13	14	9

Sources: BTA (1969); BTA (1995); ONS (2000)

Table 2.2 *Change in the share of holiday transport markets in the UK amongst primary modes, 1951–98*

	Percentage share of market					
Transport Mode	*1951*	*1961*	*1971*	*1981*	*1991*	*1998*
Car	28	49	63	72	76	71
Train	48	28	10	12	10	8
Bus or Coach	24	23	17	12	8	12

Sources: BTA (1969); BTA (1995); BTA (2001)

Stagnation and new directions: domestic tourism since 1970

In the final phases of the destination area model, Butler (1980) proposes that tourism places eventually stagnate and unless active steps are taken to reinvent or rejuvenate the resort, processes of decline are likely to set in as alternative and more appealing destinations emerge. Each of these scenarios is evident in established domestic tourism markets such as Britain, although the pattern is far from clear.

At an aggregate level, the traditional British domestic tourism markets (based on holidays away from home of at least four nights) show clear evidence of stagnation and decline. Data suggest (Figure 2.5) that from a peak level in 1974 when around 40 million holidays (of at least four nights duration) were taken in Britain by the British, persistent decline becomes an established feature thereafter. At the same time, the numbers of foreign holidays increase significantly, for reasons that will be explored more fully in Chapter 3. However, it is important to emphasise that these data represent growth in holidays, not

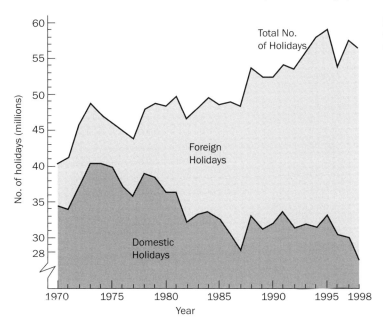

Figure 2.5
Growth of British holidaymaking, 1970–1998

necessarily growth in the numbers of holidaymakers. Indeed, after the initial expansion in numbers during the 1950s, which certainly did reflect a situation in which more people were taking a holiday, there is clear evidence that from the mid-1960s onwards much of the apparent growth in domestic tourism was solely accounted for by the increased incidence of people taking more than one holiday. Table 2.3 provides selected data from 1975 to 1998 and shows that the proportion taking a holiday scarcely changed but the numbers who took multiple holidays (and who would therefore be counted more than once in survey statistics) rose noticeably. This trend is also reflected in a 'flattening' of the holiday season, with a less pronounced peak in the traditional holiday months of July and August and greater activity in early and late summer (Table 2.4). This reflects the growing habit of taking foreign holidays in high summer and shorter breaks closer to home at other times.

At a detailed level, however, the pattern is more complex. Up to perhaps 1960, patterns of domestic tourism in countries such as Britain essentially sustained the emphasis upon coastal resorts, but by the 1970s, several significant changes had emerged that

Table 2.3 *Level and frequency of holidaymaking by UK residents, 1975–98*

	Percentages					
	1975	*1980*	*1985*	*1990*	*1995*	*1998*
Taking 1 holiday	44	43	37	36	35	34
Taking 2 holidays	14	14	14	15	17	16
Taking 3 or more holidays	4	5	6	7	10	9
Total taking a holiday	62	62	57	58	62	59
Total taking no holiday	38	38	43	42	38	41

Sources: BTA (1969); BTA (1995); BTA (2001)

Table 2.4 *Change in seasonal patterns of domestic holidays by UK residents, 1951–99*

Month of main holiday	Percentages					
	1951	*1961*	*1971*	*1981*	*1991*	*1999*
May	4	3	5	9	10	9
June	17	15	14	13	12	8
July	32	37	33	19	18	16
August	32	28	31	25	23	23
September	11	11	12	14	14	9
Other month	4	5	4	18	23	35

Sources: BTA (1969); BTA (1995); BTA (2001)

have conspired to redraw the map of post-war tourism and which contain elements of both decline and expansion. Three related areas of change invite comment:

- the stagnation and, in many circumstances, decline that has begun to affect many – although not all – traditional resorts;
- the recent growth in rural tourism;
- the rediscovery of urban areas as tourist places.

The decline of traditional resorts

There is a broad consensus that most British resorts (and, for that matter, many traditional resorts in Europe and North America) now find themselves in a critical phase of their development. Having consolidated their position in the boom years of the 1950s and 1960s, they now have to confront the uncertainties of the alternative pathways that Butler envisages as following the 'stagnation' phase.

The decline of the traditional domestic holiday market (which in Britain has seen levels at the start of the twenty-first century no higher than those of the 1960s) has had a range of impacts upon the conventional seaside resorts, but in many places a number of distinctive and often serious problems have arisen. These include:

- A loss of traditional markets – during the 1980s, British resorts lost over 39 million visitor nights – or one-fifth of their market (Cooper, 1997). Long-stay visitors have migrated to new destination areas at home and – especially abroad – and have been widely replaced by day trippers and touring visitors. In 1973, 21 per cent of all holidays taken in England were located at the seaside, by 1998 that share had reduced to just 14 per cent (English Tourism Council cited in Shaw and Williams, 2004). This trend has resulted in extensive closures of (particularly) smaller hotels and guest houses.
- A movement down-market – 70 per cent of visitors to British seaside resorts are now believed to be drawn from the elderly and/or the less affluent C, D and E socio-economic groups.
- Low spending patterns by these visitors have set in motion a downward spiral of loss of income, reduced investment, diminished attraction and loss of image.
- A failure to adapt and compete with new destinations, both within the UK and overseas has had adverse effects upon competitive position. Alongside the newer attractions in rural and urban tourism discussed below, new concept holiday centres (such as Center Parcs) are also providing an additional source of competition. Although an estimated

179 million day trips were still being made to the British seaside in the late-1990s (Shaw and Williams, 2004), more than six times as many visits were made to the countryside (SCPR, 1997).

Such difficulties are not unique to Britain. Agarwal and Brunt (2006) identify some comparable difficulties in established Mediterranean resorts, including a downturn in visitor numbers, over-dependence upon particular market sectors, an excess of low-quality accommodation and deterioration in both the built and the natural environments.

The net effect of these changes has commonly been to impose a process of restructuring on resorts and whilst some of the more vigorously competitive places (e.g., Brighton or Torquay in Britain, or Atlantic City in the USA) have been able to attract new investment for refurbishment and redevelopment (e.g., in stylish new shopping malls, leisure centres, conference facilities, marinas and, in Atlantic City, casinos), lesser places have often had to adapt to situations in which tourism plays only a minor role.

Agarwal (2002) identifies two primary approaches to resort restructuring: product reorganisation (which essentially addresses ways in which tourism is provided to visitors) and product transformation (which is concerned with modifying the nature of the product itself). Within the latter category, key areas of adjustment include: enhancement in the quality of services; enhancement in the quality of the resort environment; repositioning of products to appeal to new markets and diversification into new sectors (such as conference tourism); and wider use of public/private partnerships to encourage development and support new initiatives.

However, where resorts have been unable to pursue such restructuring, other processes of adjustment in resorts have commonly been evident. These include:

- contraction in long-stay holiday provision and increased emphasis upon short-break/off-season markets;
- promotion of business and conference tourism;
- movement into 'pseudo-resort' functions, that is, roles that benefit from traditional perceptions of the attractiveness of resorts and their therapeutic advantages, but are not actually tourism-related; for example, the conversion of small hotels and guest houses into retirement and nursing homes or office space.

The historic English resort of Scarborough, for example, saw a 55 per cent reduction in its tourist bed spaces between 1978 and 1994, with many establishments that originally provided tourist accommodation (especially in the small hotel and guest house sector) being converted into nursing and retirement homes, flats, offices and hostels for people on state benefits. Day trips and short breaks to Scarborough District have increased, but in 1994 nearly half of these visits were directed at rural parts of the district rather than the town itself. Over the same 1978–94 period, business tourism increased significantly, with a 185 per cent growth in the number of conferences hosted in the town, but it is highly unlikely that the growth of the conference trade will be capable of compensating for the removal of other forms of tourism to alternative destinations, either within the domestic travel area or, increasingly, outside the UK (Scarborough Borough Council, 1994).

Explanations for the processes of change that have triggered the decline of seaside resorts have pursued several pathways. In some assessments, resorts are seen – in part – as authoring their own demise – especially through lack of investment in key facilities or the exercise of appropriate control over development. Some of these problems are deeply embedded. Hassan (2003), for example, shows how – during the periods of resort expansion between 1880 and 1960 – many British resorts failed to provide effective management of sewage disposal other than through short-range outfall of untreated waste into the sea. Beach and seawater pollution thus became an accepted part of the Victorian seaside. However, the persistence of such practice into the modern era (in which the European Union now requires that member states conform to environmental directives that set demanding levels for the cleanliness of their beaches and sea water) has endowed many resorts that fail to reach these standards with an unwelcome – although often deserved – reputation as dirty and unhealthy places. In a similar vein, the failure of Spanish authorities to regulate the rapid development of Spanish Mediterranean resorts in the 1960s created poor-quality urban environments that were soon widely perceived as projecting a negative image, with predictable impacts on visitor levels (Pollard and Rodriguez, 1993).

Much attention – especially in the 1990s – has also been focused on structural explanations for resort decline and related, managed solutions (see, e.g., Cooper, 1990b, 1997; English Tourist Board, 1991). In particular, concerns were widely expressed about the impact of over-dependence upon a traditional model of resort development that is often characteristically inflexible or slow to adapt and which thus becomes vulnerable to changing tastes and preferences. The effects of structural problems are readily conceived as a spiral of decline in which problems such as contraction in visitor numbers affects (in turn): profits and associated levels of investment, maintenance of infrastructure, retention of image and hence the ability to attract return visits, let alone new patrons (Figure 2.6). Walton (2000: 42) alludes to one of the consequences of this process in noting that many resorts 'captivated a rising inter-war generation and failed to speak to succeeding ones'.

However, Walton's comment also signals a more fundamental point. A number of writers (e.g., Agarwal, 2002 and Gale, 2005) have observed that structural problems such as loss of market share and an inability to maintain resort infrastructure are *symptoms* of decline rather than causes. Instead, it is argued, the causes should be sought in more deep-seated processes of cultural change. Urry (1990) and Shields (1990) have each argued that resorts developed as places of distinction that attracted a clientele through the contrasts that they offered to routine life in industrial cities and the opportunities for ritualised enjoyment of the carnavalesque pleasures that most resorts provided. Franklin (2004: 158) describes how 'pure pleasure places such as these [resorts], spangled and noisy and designed to deliver you into a state of elation, contrasted dramatically with the industrial landscape of work', and

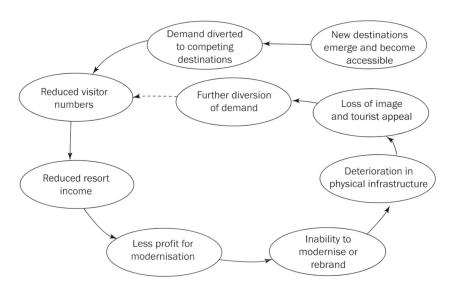

Figure 2.6 Processes of decline in the attraction of traditional resorts

from the 1930s onwards, exerted a powerful appeal to working-class communities in particular. But as Gale (2005) explains, the cultural transition from modernism to postmodernism has unpicked much of the rationale on which traditional resorts were based. The deindustrialisation of economies such as Britain's has diluted the *communitas* that once surrounded working industry and its leisure practices and dissolved much of the demand for the traditional seaside. At the same time, the spectacle and fantasy that defined the seaside has become routinised in the themed, spectacular landscapes of post-industrial cities (Franklin, 2004), whilst the virtual and corporeal mobilities that Urry (2000) marks out as a defining feature of contemporary life means that, for many people, there is little that is exceptional about the seaside – to the contrary, it is over-familiar. In a cultural milieu in which consumption is becoming more susceptible to aesthetics, traditional resorts are now seldom seen as places of distinction.

Recent growth in rural tourism

Ironically, perhaps, although tourists have begun to desert traditional seaside destinations in favour of new leisure places, one of the functions that helps to sustain coastal resorts in many tourism regions is their evolving role as bases from which a wider hinterland of rural and other urban spaces may be explored. Rural tourism, in particular, has not experienced the stagnation and decline that has affected many traditional seaside resorts and the period since 1920 (and especially after 1970 and the on-set of mass ownership of private cars) has seen significant extension in public enjoyment of the countryside.

As we have already seen, such enjoyment is not a recent development and it was noted earlier how some of the first excursions of Thomas Cook took trippers to country areas rather than the seaside. In Britain, as soon as the railways penetrated areas of attractive countryside, such as the Lake District or Snowdonia and areas of mid-Wales, the Victorian excursionist rapidly followed. Even before the First World War, bicycles (which appeared in number from the 1890s) opened up extensive areas of both coast and countryside in Britain, France and the USA to affordable forms of exploration (Tobin, 1974; Lowerson, 1995; Williams, 2003). After the First World War, bus travel began to make an impact too. In the inter-war years, coach (or charabanc) trips, as a distinctly working-class form of holiday, enjoyed widespread popularity as large numbers of people made excursions to coast and countryside, often in organised groups from churches, factories or neighbourhoods. Rambling, camping and youth hostelling developed too, as rural tourism began to emerge strongly in an era in which outdoor pursuits acquired new levels of interest and support (Walker, 1985).

The subsequent development of rural tourism from 1945 has built upon this inherited set of practices but has been further shaped by five key themes:

- Growth in the levels and frequency of participation – in Britain, for example, some 9 million short-break holidays are taken in the countryside each year (Beioley, 1999) and the number of day (or part-day) visits to rural sites exceeds 1 billion, although many of these are of a recreational rather than a touristic character (SCPR, 1997).
- Diversification of activities and the spaces that form the basis for rural tourism – theme parks, heritage sites, integrated holiday villages and a widening range of outdoor pursuits and new sports have added significantly to the attraction of rural space.
- Increased commercialisation and commodification of experience which has helped to make many rural sites into marketable attractions.
- Progressive integration of tourism into the wider framework of rural production and consumption, especially in agriculture and forestry – in the USA, for example, the national forest system absorbs more than 340 million visitor days annually (US Dept. of Commerce, 2002).

- Progressive extension in the designation of rural land into zones of conservation that are often aligned with meeting a range of tourist demands – visits to the US national parks and other national level designations, such as national monuments, parkways, recreational areas and seashores, exceeded 285 million in 2000 (US Dept. of Commerce, 2002).

These changes have been stimulated by a number of factors. Enhanced levels of personal mobility through car ownership enable people to visit sites that formerly were less accessible when travel depended upon the fixed networks of rail and bus services. Rural tourism also feeds off the accentuated sense of difference between town and countryside in ways that are comparable with some of the contrasts that shaped previous relationships between coastal resorts and the industrial cities. Commodification of rural experience (in ventures such as integrated holiday villages, working farms or craft centres) has also been influential in making rural sites accessible to urban visitors who may possess an instilled interest in the countryside that is often acquired through media sources such as television programmes, but who lack the expertise to explore the countryside on their own terms. Such sites commonly draw on a range of popular images of rural life that, it has been argued, both promote and depend upon a rising level of nostalgia for traditional ways of living. This has been detected by a number of authors as a response to many of the stresses attached to life under (late) modernity (Urry, 1990; Hopkins, 1999). (Similar explanations have been rehearsed in the wider discussion of heritage attractions and will be explored more fully in Chapter 10.) The net effect of these changes, however, has been to establish rural tourism as a very prominent component in domestic patterns of leisure across the developed world.

The redevelopment of urban areas as tourist places

Traditional resorts have also lost parts of their market to competition from major urban areas. Historically, tourism was generally about escape from the confines of towns and cities but in one of the many reversals in convention that accompanied the onset of a post-industrial pattern of life in the late twentieth century, cities themselves have now become major tourist attractions. Of course, international cities (e.g., London, Paris, Rome and Venice) have enjoyed a flourishing tourist industry for many decades, as have major provincial cities such as Edinburgh and York in the UK, particularly where there is a historical basis to their appeal. What is new is the manner in which cities where there was no strong tradition of tourism (such as Leeds or Liverpool in the UK; Baltimore in the USA; or Victoria in Canada) have, through shrewd promotion, active development of attractions and associated re-imaging of place, been able to develop tourist industries of their own.

As will be revealed in much greater detail in Chapter 9, the impetus behind these transformations is closely shaped by several powerful processes of change in the contemporary urban environment. These include:

- the impacts of globalisation which have altered fundamentally the relationships between cities and heightened the levels of competition between urban places;
- economic restructuring that has seen a progressive reduction in the role of production and an associated rise in the significance of consumption in urban economies;
- active remaking of cities and their identities (through urban regeneration and place promotion).

Tourism has emerged as a central component in each of these processes, helping to develop and project positive, attractive images of regenerated urban places and contributing in both direct and indirect ways to the development of consumption-based urban economies. Cities have become marketable 'products' (Law, 2002) in which place promotion and the active production of new and exciting sites of consumption centred around themed shopping malls, regenerated waterfronts, state-of-the-art museums, galleries or concert halls and fashionable zones of retailing, cafe-life and entertainment, become essential elements in enhancing the appeal of urban places to both their citizens and outsiders.

In the process, these urban destinations have emerged as formidable competitors for many of the older, traditional urban seaside resorts, establishing strong counter-attractions and deflecting significant levels of demand (especially in the short-break market). As we will see in Chapter 9, precise enumeration of levels of urban tourism is not easy, but, as an example, Table 2.5 lists recent visitor levels at a range of urban attractions in British cities and indicates something of the scale of activity in this emerging tourism sector.

Table 2.5 *Annual visitor levels at selected urban tourist attractions in the UK, 2002*

Attraction/Location	Visitor numbers
Tate Modern, London	4,618,632
British Museum, London	4,607,311
National Gallery, London	4,130,973
The London Eye *	4,090,000
Natural History Museum, London	2,957,501
Victoria and Albert Museum, London	2,661,338
Tower of London	1,940,856
York Minster	1,570,500
Legoland, Windsor	1,453,000
Canterbury Cathedral	1,110,529
Edinburgh Castle	1,153,317
Westminster Abbey, London	1,058,856
Kelvingrove Gallery, Glasgow	955,671
Windsor Castle	931,042
The Lowry, Salford *	810,200
National Museum of Photography, Film & Television, Bradford	795,371
National Railway Museum, York	742,515
Cadbury World, Birmingham	534,766
New Lanark World Heritage Village, nr Glasgow	409,500

Source: Visit Britain (2003)
* estimate

Summary

This chapter has presented a highly condensed account of an extensive and often complex process. However, in summary, the development of geographies of domestic tourism may be seen as having been shaped by a range of key processes. These include: patterns of patronage by elite groups; changes in public attitudes and associated tastes that favoured the development of new destination areas. Changing levels of physical access through innovations in transport and widening social access through the progressive emancipation of middle- and working-class groups were also influential, as was the application of both private and municipal enterprise to the physical development of destinations and their tourism infrastructure. The spatial diffusion of tourism from initial sites of development to a widening range of destination areas, many of which ultimately become competitors to original destinations, prompted a re-alignment in the form, character and function of these older tourism destinations.

Discussion questions

1   With reference to a country of your choice, examine the primary effects of changes in transport technology on the geography of domestic tourism.
2   To what extent may successive geographies of domestic tourism be seen as responses to changes in social attitudes and expectations?
3   How well does the Butler model of resort development describe the evolution of British seaside resorts since 1750?
4   Taking as an example a seaside resort with which you are familiar, what evidence do you find of actions or policies directed at meeting competition from new tourism places?
5   What parallels may be drawn between the processes of development in urban seaside resorts and tourism destinations in the countryside?

Further reading

An outstanding account of the development of tourism from the earliest phases of tourism to the outbreak of the Second World War is provided by:
Towner, J. (1996) *An Historical Geography of Recreation and Tourism in the Western World, 1540–1940*, Chichester: John Wiley.

Changing attitudes to the coastline and the sea are perceptively analysed by:
Corbin, A. (1995) *The Lure of the Sea: The Discovery of the Seaside 1750–1840*, London: Penguin.

A comparable discussion of the emergence of rural tourism is provided by:
Andrews. M. (1989) *The Search for the Picturesque: Landscape Aesthetics and Tourism in Britain, 1760–1800*, Aldershot: Scolar Press.

Comprehensive analyses of processes of resort development are provided in:
Soane, J.V.N. (1993) *Fashionable Resort Regions: Their Evolution and Transformation*, Wallingford: CAB International.
Walton, J.K. (2000) *The British Seaside: Holidays and Resorts in the Twentieth Century*, Manchester: Manchester University Press.

For an interesting collection of essays on the cycle of resort development in Britain, see:
Shaw, G. and Williams, A.M. (eds) (1997) *The Rise and Fall of British Coastal Resorts*, London: Pinter.

For a global view of processes of coastal resort development, see:
Agarwal, S. and Shaw, G. (eds) (2007) *Managing Coastal Tourism Resorts: A Global Perspective*, Clevedon: Channel View Publications.

A concise discussion of the manner in which popular demand for excursions and holidays developed in the industrial communities of nineteenth-century Britain is provided in:
Urry, J. (2002) *The Tourist Gaze: Leisure and Travel in Contemporary Societies*, London: Sage, Chapter 2.

An incisive commentary on the social rituals of the popular seaside is provided in:
Franklin, A. (2004) *Tourism: An Introduction*, London: Sage, Chapter 6.

The Butler model of destination development is discussed in many tourism texts but the most recent and most comprehensive critical review is provided in:
Butler, R.W. (ed.) (2006) *The Tourism Area Life Cycle* (2 volumes), Clevedon: Channel View Publications.

3 From Camber Sands to Waikiki: the expanding horizons of international tourism

The choice in the chapter title of Camber Sands and Waikiki as tourism places that are illustrative of the globalisation of tourism was inspired by a line from a 1980's pop song by the British band Squeeze.[1] In the short span of just three verses, the song presents a set of images that, although fleeting in nature, capture a vivid picture of a popular form of tourism – of lazy sunbathing and the casual perusal of 'airport' novels; the *flanerie* of the seafront and harbour; shopping for cheap souvenirs and gifts for family or friends at home; and for some, perhaps, a holiday romance. Camber Sands lies on the English Channel coast of Sussex, Waikiki on the Hawaiian island of Oahu, yet as the song correctly recognises, the essence of tourism within these very different and distant locales is essentially the same.

In Chapter 2 we have examined a sequence of tourism development that established some very familiar patterns of domestic tourism that were often based initially upon coastal resorts, but which have evolved to the extent that a diverse range of rural, urban and coastal environments comprise tourism spaces in most developed nations today. However, one of the distinguishing features of tourism is its fluidity across space and through time, so it is no surprise to find that patterns that appeared to be well-established and secure in – say – the 1970s are being eroded by significant shifts in the location and character of tourist space. This is evident not just in the emergence of new destinations but also in the restructuring of established ones – a process that reflects, and is a product of, the compression of time and space that is fundamental to the process of globalisation.

This chapter aims to examine the development of international forms of tourism and to place that process into the wider contexts of mobility and globalisation. These themes are examined at both the sub-continental scale (particularly as tourism between the states of Europe) and at the intercontinental or global scale. Although international tourism is not a new phenomenon, the rapidity with which it has grown in the post-1945 era and the scale and extent of contemporary international travel demand the attention of geographers concerned with the study of tourism.

Globalisation and mobility

The theme of globalisation was briefly introduced in Chapter 1 but in prefacing the discussion of the development of international tourism, it is helpful if we consider the concept in a little more detail. In essence, globalisation involves heightened levels of connectivity and mutual engagement between different parts of the world and as a consequence, people, capital, goods, services and information circulate at an increasingly global scale. The emergence of these transnational systems of demographic, economic and cultural exchange not only helps to create and sustain the global networks of production that are now hugely influential in structuring economic space, but also exerts significant effects upon patterns of consumption, producing much greater levels of harmonisation and uniformity across cultural space too.

The formation of global systems has been shaped partly by the acceleration of traditional forms of communication such as rail or aeroplane travel, but especially by the development of new modes (such as satellite communications and the Internet) that have connected societies, institutions and countries in novel ways and compressed the previously limiting effects of time and space. Enhanced levels of international trade and investment have brought most countries into the global trading systems, and, at the same time, new institutional arrangements aimed at facilitating such activity or at addressing what are now defined as 'global' issues (such as climate change) have helped to erode some of the traditional strengths of nation states and lowered barriers to movement of people and goods. As people become more mobile, so do the cultures that they espouse, leading – according to some readings of globalisation – to heightened levels of homogeneity in areas such as culture and identity (Lechner and Boli, 2000).

The enhanced level of mobility of people, goods and ideas that is central to the concept of globalisation is, however, not new, as processes of internationalisation have been evident since the spread of the mercantile, colonial economies of sixteenth-century Europe. The distinction between earlier forms of internationalisation and contemporary globalisation, according to Held (2000) (cited in Shaw and Williams, 2004), is essentially qualitative and is primarily reflected in the increasing intensity of interconnections, allied with a dramatic increase in the geographic 'reach' in the global networks of exchange (Urry's 'scapes' and 'flows'). However, importantly, this process is uneven in its development across space and time, leading to both new and continuing patterns of inequality in the levels of connection to global systems.

Globalisation can thus be variously seen as a demographic, economic, social, cultural and political process (Hall, 2005) producing fundamental structural changes in the way that people and places relate to each other. However, although as Harvey (2000) has argued, globalisation has become a key concept in organising the way that we understand how the world works, it is also a term that is characterised by imprecision (Shirato and Webb, 2003). Outwardly it is a concept that implies a unifying process within contemporary life, but in practice the effects of globalisation at a local level are anything but unified or consistent in their effect. One of the paradoxes of globalisation is that the advance of global systems has been paralleled by rising levels of local resistance to many of their generalising effects. This is evident in several contexts, including the reassertion of local senses of place and identity, as well as the assertion of individuality through patterns of consumption (Hall, 2005).

Furthermore, like 'development', globalisation is both a process and a condition (Hall, 2005), such that international tourism is simultaneously 'global' (in terms of scale, extent and structure) but is also 'globalising' (in that through the process of its development, tourism helps to draw new places into the global networks of exchange). The global *condition* of tourism is evident in, for instance, the significance of multinational firms in key tourism sectors such as transportation and hospitality that have developed and now maintain global strategies that position companies such as American Express and Holiday Inns as world leaders in their sectors. Tourism as a globalising *process* is evident in:

- the pattern of tourism development that is often a primary mechanism for channelling investment capital to new destinations in developing countries and which, through the creation of employment, contributes to international movement of both permanent and, especially, seasonal labour;
- the extended geographical 'reach' of the modern tourist – as seen, for example, in the development and marketing of long-haul destinations in distant places (such as the Far

East) or the promotion of new market areas in proximity to existing tourism regions (such as Eastern Europe);
- the role of tourism as a primary area of cultural exchange – contact between tourists and the people who live in the locations that are toured is often held to contribute to the dissemination of global cultural values and behaviours.

As a concept, therefore, globalisation is multi-dimensioned, sometimes elusive and occasionally contradictory, but is also of undeniable importance in understanding how world systems – such as international travel – are structured and operate.

Subsequent sections of this book will pursue many of the impacts of globalised tourism in greater detail but in this chapter the focus will fall especially upon understanding some of the basic structural and organisational frameworks that have enabled global patterns of tourism to develop, starting with – as an example – an outline consideration of how international tourism emerged in Britain and Europe.

Origins of international tourism

In Chapter 2 we saw how the development of domestic tourism often followed a clearly defined sequence in which several processes were prominent:

- A spatial diffusion through time of tourist places, from an initial position in which tourism was centred in a limited number of small resorts to an eventual pattern of large-scale development of coastlines and rural hinterlands in which many tourist places may be located.
- A change in motives for travel to resorts from (in the British case, at least) a quest for health to the pursuit of pleasure.
- A process of democratisation of tourism whereby what originates as the exclusive practice of a social elite diffuses down the social ladder to become an important area for mass forms of popular participation.

The development of international tourism also reflects these key processes as what were once exclusive and selective forms of travel have become widely accessible, widely practised and popularised.

Many writers place the origins of modern international tourism in the Grand Tour of the seventeenth and eighteenth centuries (e.g., Pimlott, 1947) which established what Towner (1996: 97) describes as an international 'travel culture' in major tourist generating regions in Europe. The primary objective of the Grand Tour was to provide young men of wealth and high social status with the basis of a classical education, by sending them on an extended visit to cultural centres in Europe – in France, Germany, Austria and, especially, Italy. The Renaissance in Europe endowed several nations with a pre-eminence in matters of arts, science and culture, but Italy combined a classical heritage with contemporary ideas and inventions, and its position as an intellectual centre in Europe ensured that for young men of wealth and power, an education could not be considered complete without an extended visit to its main cities. Thus Venice, Padua, Florence and Rome formed a basis to an itinerary that, when extended to include other capitals of culture such as Paris and Vienna, provided the geographical structure for the Grand Tour.

The golden age of the Grand Tour is generally held to be the period between about 1760 and 1790, but references to similar journeys occur much earlier. The Elizabethan courtier Sir Philip Sydney embarked on a tour in 1572, the architect Inigo Jones went to Europe in 1613, the philosopher Thomas Hobbes in 1634 and the poet John Milton in 1638. These tours were probably comparatively short, but by the middle of the eighteenth century a tour might commonly occupy several years. Although the primary objective remained the completion of a formal education, there were evidently important elements of sightseeing too. Those undertaking the tour would have visited sites of antiquity, art collections, great houses, theatres and concert halls. It also became fashionable to combine travel with the purchase and collection of artefacts: paintings, sculpture, books and manuscripts. Here there are tempting parallels between these early patterns of visiting with their 'souvenir' collecting and later styles of modern tourism in which the garnering of memorabilia is a conspicuous part of tourist behaviour.

However, what had started as the preserve of a social elite did not remain so for too long, and by the end of the Napoleonic Wars in Europe in 1815 there was already clear evidence of the emergence of new classes of international traveller, drawn not from the aristocracy but from the bourgeoisie. Because of their more limited budgets, the journey patterns of these new tourists were inevitably shorter and their activities more intensified. Sightseeing became more important than the cultivation of social contacts or the experience of culture. The emergence of new attitudes and ideas at this time also focused the attention of the tourists onto new resources and new tourist places. For example, regions such as the Alps would previously have been characterised as wild and dreadful places, populated by uncivilised peoples and forming major obstacles to travellers *en route* to the important attractions of Italy. However, as noted in Chapter 2, the romantic movement of the early nineteenth century and the popularising of the picturesque transformed public attitudes towards mountain landscapes and quickly promoted new international tourist destinations in Switzerland and the Alpine zones of France, Italy and Austria. The popularity (and accessibility) of these locations was enhanced still further once organised tours to these regions became established by entrepreneurs such Thomas Cook after about 1855.

As large parts of mainland Europe became populated by an ordinary class of tourist, new areas of exclusive tourism inevitably emerged. Amongst these, the most significant was the French Riviera between Nice and Monte Carlo. Lacking the centres of culture that preoccupied the Grand Tourist, the French Mediterranean coast had escaped the attention of the first tourists, but its attractive coastline and equable climate prompted a process of development that, by the end of the nineteenth century, had established the area as the new pleasure reserve of the European aristocracy. People from the colder climes of northern Europe, in particular, used the French Riviera as a winter retreat, and its visitors numbered most of the crowned heads of Europe and the entourage that always followed people of status.

However, the process of social displacement that is such an apparent aspect of tourism development ensured that ordinary tourists eventually followed. The First World War destroyed the old social orders that had sustained areas such as the French Riviera as exclusive places and from the 1920s onwards there is a visible process of social and functional transformation of the French Riviera to a pattern of coastal tourism that was eventually to become widely established along the northern shores of the Mediterranean, drawing both domestic and (especially) international visitors. Initially, the colonisation of the Riviera by influential groups of writers, artists and the new breed of American film stars gave the area an allure that was hard to resist. Then, new forms of beach leisure (such as sunbathing – previously a highly unfashionable practice) helped to promote a summer season in an area that had by custom been deemed climatically too oppressive for summer-time visits, whilst new styles of leisure clothing (especially swimwear) reflected a liberalising of attitudes that would soon be adopted by ordinary people. By 1939, the establishment of paid annual holidays in France had brought an influx of lower-class French holidaymakers to the Mediterranean and the exclusivity of the Riviera had been replaced, in a very short time, by the apparently simple forms of tourism based around sun, sea and sand (Turner and Ash, 1975; Soane, 1993; Inglis, 2000).

Post-1945 development of international tourism

The most pronounced developments in the geography of international tourism have, however, been largely confined to the period since the end of the Second World War. During this time there has been unparalleled growth in the number of foreign tourists, a persistent spread in the spatial extent of activity and the associated emergence of new tourist destinations.

The growth of international tourism

According to the World Tourism Organization (WTO, 1995), in 1950 international tourism (as measured in tourist arrivals at foreign borders) involved just 25 million people worldwide – a figure that was no greater than the number of domestic holidays taken in a single country, Great Britain, at the same time. From this point, international tourism has risen to an estimated 900 million arrivals in 2007 (WTO, 2008). Figure 3.1 charts the upward trend and suggests two basic features in the pattern of growth.

First, the expansion of international tourism has been almost continuous, reflecting not just the growing popularity of foreign travel but, more importantly, the increasing centrality of tourism within the lifestyles of modern travellers. At a global scale, at least, international tourism appears largely immune to the effects of events that might reasonably be expected to exert an impact. Neither the oil crisis of the mid-1970s, the economic recessions of the 1980s, the wars in the Persian Gulf in the early 1990s and again in 2003, nor the succession of natural disasters and terrorist events in the early 2000s, appear to have deterred the international tourist. Travel flows may reveal temporary shifts in direction but the overall impact of these various crises on global levels of tourism has been negligible (WTO, 2005c).

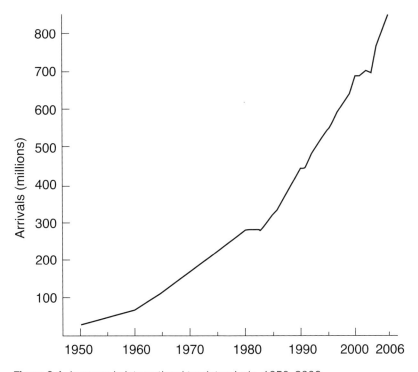

Figure 3.1 Increase in international tourist arrivals, 1950–2006

Annual rates of increase do occasionally show signs of deflection in response to world conditions, especially economic conditions (thus there occurred a temporary stabilisation of demand in the early 1980s before economic recovery encouraged a further round of growth) but, overall, the expansion of international tourism seems irresistible and quite able to withstand pressures of inflation, currency fluctuations, political instability or the incidence of unemployment in most of the countries that generate the principal flows of international tourists.

Second, data show that, unlike many domestic tourism markets which have stabilised or even shown signs of decline (see Figure 2.6, for example), the increase in international travel (when measured in absolute rather than relative terms) is accelerating. Thus in the ten years between 1965 and 1974, the market expanded with an extra 92.8 million arrivals; between 1975 and 1984 it grew by a further 94.7 million; between 1985 and 1994, by an estimated 200.9 million; and between 1995 and 2004 by a further 225 million (WTO, 1995; 2005a). With annual growth rates in world tourism since 1990 running at an average of 4.4 per cent, international tourist arrivals are expected to exceed 1.0 billion by 2010 and perhaps to reach 1.6 billion by 2020 (WTO, 2001).

The spatial spread of international tourism

Aggregate descriptions, however, conceal a great deal of variation within the basic patterns and these repay closer attention. Historically (and indeed at present), international tourism has been dominated by Europe, both as a receiving and as a generating region. This pre-eminence reflects a number of factors including:

● an established tradition in domestic tourism that converts quite readily into international travel;
● a mature and developed pattern of tourism infrastructure, including transportation links, extensive provision of tourist accommodation and organisational frameworks such as travel companies;
● a wealth of tourist attractions including diverse coastal environments, major mountain zones, as well as world-ranked sites of historic or cultural heritage and urban tourism;
● a sizeable industrial population that is both relatively affluent and mobile and thus an active market for international travel;
● a range of climatic zones that favour both summer and winter tourism.

However, although Europeans do possess a higher propensity to travel, the geopolitical structure of the region inflates the level of international travel (Jansen-Verbeke, 1995). In particular, the juxtaposition of relatively small nations creates a large number of international borders that are often routinely crossed by tourists undertaking quite short journeys. In contrast, vacationers in the USA may travel very much further within their home country than the international travellers of Europe, but unless they cross into Mexico or Canada, they fail to register as international travellers.

The extent to which Europe dominates the international tourism market has tended to decline in relative terms in recent years, but European destinations still figure prominently in Table 3.1. This lists the top ten destinations according to visitor arrivals and tourist receipts, alongside the major generators of international travel (as indicated by expenditure levels). In terms of percentage shares of the world market, in 2004 European countries attracted almost 55 per cent of visitor arrivals and 52 per cent of international tourism receipts, whilst 38 per cent of the receipts from tourism at the world level are generated by the ten leading West European countries (WTO, 2005a).

Table 3.1 *International tourism: major receiving and generating countries, 2004*

Country (by rank)	Arrivals (millions)	Country (by rank)	Receipts (US$ billion)	Country (by rank)	Expenditure (US$ billion)
France	75.1	USA	74.5	Germany	71.0
Spain	53.6	Spain	45.2	USA	65.6
USA	46.1	France	40.8	UK	55.9
China	41.8	Italy	35.7	Japan	38.1
Italy	37.1	Germany	27.7	France	28.6
UK	27.8	UK	27.3	Italy	20.5
Mexico	20.6	China	25.7	Netherlands	16.5
Germany	20.1	Turkey	15.9	Canada	16.0
Austria	19.4	Austria	15.4	Russian Fed.	15.7
Canada	19.1	Australia	13.0	China	15.2

Source: WTO (2005a)

Within the European area, however, there are marked spatial variations in the levels of international tourism and, in some situations, a striking imbalance between inbound and outbound tourist flows. Figure 3.2 provides a simple representation of the spatial patterning of international tourism in the European states in 2004, based on a ranking according to the number of arrivals (WTO, 2005b). Four categories are differentiated:

● Primary destination areas that are, in general, large states (or states with large populations) and possess well-developed summer and winter tourist seasons with a diverse range of attractions. There is a strong presence of Mediterranean tourism in this category.
● Second order destinations that also comprise important Mediterranean destinations and a number of central European states from both sides of the former divide between west and eastern Europe.
● Third order destinations that are dominated by the Scandinavian countries which perhaps suffer from high costs of living and a short summer season.
● Fourth order destinations that are characteristically small and often peripheral to the European core.

However, when the data are reworked to show the balance between inbound and outbound tourism, a rather different pattern emerges. Figure 3.3 illustrates the pattern of European states according to whether they are in 'surplus' or 'deficit' on their balance of tourist incomes and expenditures, that is, whether or not they earn more from foreign tourism than their own citizens expend on visits abroad. This reveals an essentially clear divide between, with the exception of Romania, a cluster of northern European states that are in 'deficit' and an extended swathe of central and southern European states that are in 'surplus'.

This pattern reflects what has long been recognised as the predominant tourist flow in Europe – a north–south movement from the high concentrations of urban–industrial populations in the cooler, northern parts of Europe towards the much warmer areas that fringe the Mediterranean (Burton, 1994). This helps to establish a Mediterranean 'core' area centred in France, Spain and Italy which dominates the European holiday tourism market and which draws disproportionately upon Germany, the United Kingdom and the Scandinavian countries as sources for visitors. Superimposed upon primary north–south movements are secondary flows to the mountain regions of Europe (both for winter and for summer holidays) and all-season flows between the major European cities for cultural,

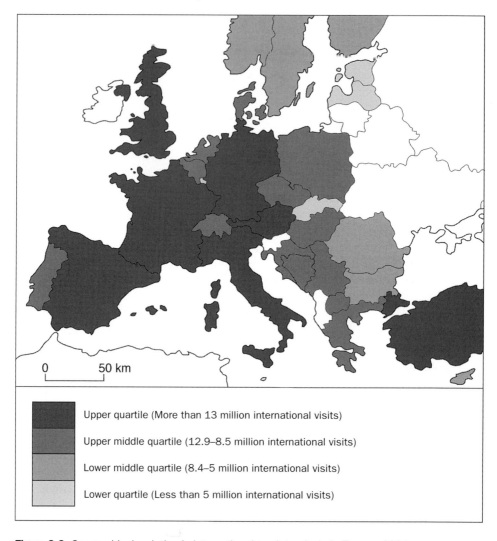

0   50 km

Upper quartile (More than 13 million international visits)

Upper middle quartile (12.9–8.5 million international visits)

Lower middle quartile (8.4–5 million international visits)

Lower quartile (Less than 5 million international visits)

Figure 3.2 Geographical variation in international tourist arrivals in Europe, 2004

historic and business tourism (Williams, 2003). The former trend helps to position countries such as Austria in the top ten world destinations, whilst tourism to Britain is strongly dependent upon the latter.

Part of the explanation for differences in levels of tourism in Europe lies in the way in which the activity has spread. As we have seen, the French Mediterranean coast has a history of tourism which extends back over a hundred years, and the activity appears to have diffused from this region. Thus in the early 1960s, large-scale development spread westwards into Spain and eastwards to the Italian Adriatic coasts. By the early 1970s, the former Yugoslavian coast was an emerging holiday region and tourism to the Greek islands was becoming well established. In the 1980s, package-based coastal tourism reached Turkey. As an example, Figure 3.4 illustrates how patterns of European air charter tourism originating within one state – the Irish Republic – evolved during this critical period of change (Gillmor, 1996). This study shows how additional destinations were developed as new markets for packaged tourism (such as Cyprus and Turkey); how some existing markets

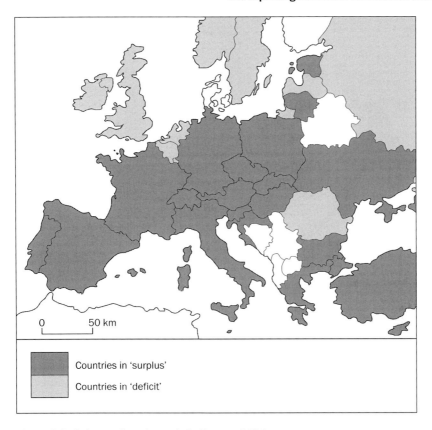

Figure 3.3 Balance of tourist trade in Europe, 2004

were consolidated (e.g., France, Greece and Portugal); whilst countries that formed the initial foci for foreign travel tended to take proportionately smaller shares as the tourist space expands (e.g., Spain).

More recently, tourism has also begun to develop strongly in several of the former Communist states of eastern Europe. The collapse of Communism across eastern Europe in the 1990s led to a relaxation of controls on travel which has enabled higher levels of movement between states such as Poland, Hungary and the Czech Republic, as well as rising levels of visiting from outside the former Soviet bloc. Johnson (1995), for example, shows how proximity to western Europe has assisted the development of tourism to Poland and the Czech Republic and major urban centres such as Prague have become popular, affordable destinations for visitors from Germany, the Netherlands and Britain. On a smaller scale, tourism to the Black Sea coast of Romania has attracted growing numbers of visitors from the Russian Federation (Light and Andone, 1996). However, although states such as Poland now figure in the top twenty world destinations, their position as centres of tourism is not entirely clear. Several studies have shown that the introduction of free market economies across eastern Europe has often triggered inflation, shortages of essential goods or services and unemployment (see, *inter alia*: Bachvarov, 1997; Balaz, 1995; Johnson, 1995). Thus, as Williams (2003) notes, much of the 'tourism' between states in eastern Europe is presently undertaken as day visits for purposes such as buying and selling goods, or seeking work, rather than leisure travel. This does not, though, detract from the longer-term prospects for a region that is rich in historical, cultural and landscape resources.

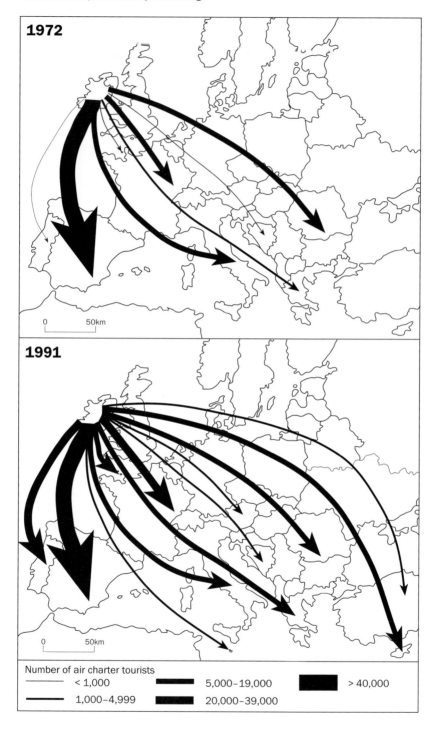

Figure 3.4 Changing patterns of air charter tourism from Ireland to Continental Europe, 1972–1991

New tourist areas

The spatial spread of tourism and the emergence of new tourist areas that may be seen within the European area are also clearly evidenced at the global scale. Although the European share of the world travel market is by far the largest, the trends over the past thirty years show a significant reduction in that share as new destinations begin to attract the attention of tourists. The nature and extent of these spatial shifts are reflected in Table 3.2 and show clearly how the horizons of international tourism have extended since 1960. Key trends to note are:

- the progressive decline in the share of the world market of 'established' tourism regions in Europe and the Americas (although the latter grouping is a confusing amalgam of prosperity and growth in tourism in the USA and Canada and almost total under-development of tourism across most of South America);
- the poorest regions – Africa and South Asia – are characterised by slow expansion on comparatively low absolute figures, although growth of tourism to post-apartheid South Africa has helped to stimulate the African market since the mid-1990s;
- the strong performance of tourism in the East Asia and Pacific region, which reflects the emergence of China as a top ten destination and the rising levels of intra-regional tourism between economical prosperous states such as Thailand, Singapore, Indonesia, Japan, Australia and – most recently – China, as well as continuing growth in long-haul tourism from Europe and North America.

The scale of the expansion of tourism to these distant locations is emphasised when these percentage shares of a rapidly developing global market are translated into actual visitor figures. Expressed in this form, tourist arrivals in East Asia and the Pacific have increased from around 690,000 in 1960 to 145 million visitors in 2004 (WTO, 2005a), providing a compelling illustration of the extent to which modern tourism has been able to take advantage of the wider process of globalisation.

Factors shaping the development of international tourism

How have these substantial transformations in the scale and spatial extent of international tourism come about? The development of international tourism has been shaped by a wide range of factors, but the main elements are here summarised under six key headings.

The role of capital in the production of tourist space

Underpinning much of the physical development of international tourism and the evolving spatial patterns of demand and, especially, supply, is the role of capital in the production of tourist space. As is explained in much greater detail in Chapter 4, tourism is dependent upon the provision of a range of goods and services – in destination areas and along the pathways that link generating and receiving locations – that define the tourism 'product' and enable the activity to take place. Hence decisions taken by suppliers of those goods and services are strongly influential on patterns of development and the resulting geographies of tourism. Moreover, as Shaw and Williams (2004) emphasise, under most situations the dominant mode of tourist production is a capitalist one. This has a number of implications but in understanding how spatial patterns of tourism evolve we need to recognise, first, how the investment decisions and the related quest for profit by independent firms and businesses determines the form and location of tourism development and, second, how the nature and extent of regulation exercised by governments over their territories and the firms that are operating within them may modify or in other ways influence those decisions.

Table 3.2 Change in regional distribution of international tourism arrivals, 1960–2004

Region	1960		1980		1990		2000		2004	
	No. (million)	% share	No. (million)	% share	No. (million)	% share	No. (million)	% share	No. (million)	% share
Africa	1.8	1.1	7.0	2.5	15.2	3.4	28.2	4.1	33.2	4.4
Americas	16.5	24.1	59.2	21.3	92.8	21.0	128.2	18.8	125.8	16.5
Europe	50.0	72.5	183.5	66.0	264.8	60.0	384.1	56.4	416.4	54.6
Middle East	0.6	1.0	5.8	2.1	10.0	2.3	25.2	3.7	35.4	4.6
South Asia	0.2	0.3	2.2	0.8	3.2	0.7	6.1	0.9	7.5	1.0
East Asia & Pacific	0.6	1.0	20.3	7.3	54.5	12.4	108.8	16.0	145.0	19.0
TOTALS	69.7	100.0	278.0	100.0	440.5	100.0	680.6	100.0	763.3	100.0

Source: WTO (2005a)

The tourist industry tends to reveal a polarised structure with a relatively small number of international-scale companies (especially in sectors such as transportation, accommodation and inclusive tours) operating alongside a much larger number of independent small and medium-scale enterprises. However, within the processes of globalisation that were outlined earlier, the trend through time has been for the large, multinational firms to assume a greater significance and the investment budgets and purchasing power of these companies is often critical in shaping change in the patterns of international tourism.

Their influence is felt in at least three ways. First, because tourism firms operate in a highly competitive market, there is a natural tendency to develop destinations and markets that enable the companies in question to maintain (and ideally strengthen) their market position. This may encourage companies to look at destinations that are perceived as being:

- attractive – typically by offering something different;
- capable of delivering high volume business at comparatively low cost;
- relatively unfettered by regulatory constraints.

Consequently, part of the spatial expansion of international tourism can be seen as a response to these types of requirement, so that – for example – the development of Mediterranean package tourism in the 1960s was shaped by the same commercial imperatives that encouraged the spread of long-haul tourism to destinations such as Thailand in the 1990s.

Second, the commercial power of the major multinational firms enables them to adopt, when necessary, a pioneering role in developing new destinations, whether through the formation of strategic alliances with other related firms; through mergers and acquisitions; or simply as the sole investor. Foreign investment in emerging destinations, especially in the developing world, is often critical to the establishment of new spaces of tourism. The development of self-contained luxury resorts in parts of the Caribbean, particularly through investment by American-based multinationals, is one illustration of the process.

Third, the marketing and promotional decisions of major tourism firms – especially tour companies – are highly influential on consumer decisions. Competitive pricing aligned with alluring images of destinations constitute powerful marketing tools that subsequently exert strong influences on the actual patterns of tourist travel.

Development of the travel industry

Linked with the role of capital in producing tourist space is the wider development of the modern travel industry. One of the main prerequisites for the growth of international tourism has been the establishment of a mature travel industry, especially since about 1960. Initially the industry focused upon the promotion and provision of the basic components of accommodation, transport and local entertainment. Subsequently it has also developed new structural forms of international travel based on packaged tours and has acquired a degree of professionalism in its service that, as Williams (2003: 71) notes, 'has brought a level of flexibility, sophistication and simplification to the provision and promotion of international tourism that has largely eliminated many of the risks and difficulties – both real and perceived – that were once attached to foreign travel'.

The development of package (or inclusive) holidays has been particularly influential. The essential feature of the package tour is that it 'commodifies' foreign travel by creating inclusive holidays in which travel, accommodation and the primary services at resorts are all purchased in advance through a single transaction in which the customer buys the holiday as if it were a single product or commodity. Costs are driven down by offering standardised packages and further economies of scale are achieved in the bulk purchase by tour companies of essential components such as airline seats and hotel rooms. Although, as we have seen, packaging of tours was a nineteenth-century innovation, the modern package tour based around air travel is very much a post-1945 innovation – commonly attributed to a Russian emigre named Vladimir Raitz, who was the founder of one of the first specialist package tour companies – Horizon Holidays. But under the Fordist conditions of mass production and consumption that prevailed throughout the 1950s and 1960s, the new provision of standardised, low-cost foreign holidays was an assured way of developing new markets (Williams, 1996).

The growth in the popularity of packaged foreign travel was further assisted by several associated developments, in particular:

- the widespread provision of high street travel agencies at which foreign travel and holidays may be arranged and purchased has helped to make the organisation of foreign travel convenient and accessible;
- the routine provision in destination areas of local tour and holiday guides who liaise between visitor and host and in so doing often remove or minimise problems that foreign tourists might have with language, custom or just simply orientation;
- active promotion of destinations through free brochures and advice services, especially through travel agencies, magazines and newspapers.

These services form part of the rising levels of professionalism in the organisation and delivery of foreign travel to potential tourists. Professionalism is partly about ensuring reliability and efficiency in the delivery of travel to tourists, but, less obviously, it is also about fostering public confidence in the industry. Urry (1994a: 143) explains that this is a matter of significance since the willingness of people to travel requires that they have 'faith in institutions and processes of which they possess only limited knowledge'. By enabling people to travel with confidence, the modern industry contributes to the demystification of international tourism. More practically, perhaps, the professionalism of the industry is also linked to increasing sophistication in marketing strategies for tourism that have exerted positive effects by reducing the relative costs of international travel and widening the range of products and services that are available. Williams (2003) draws attention to several key developments, including:

- increased use of market segmentation, in which travel services are tailored to the specific needs of particular categories of tourist – for example family holidays or educational tours;
- wider use of strategic alliances in which independent companies pool resources or align their operating procedures by agreement, in order to develop new markets or services – the introduction of Europe-wide rail travel tickets is an example;
- development of competitive pricing strategies such as early booking discounts or APEX fares.

(For further discussion of these developments, see Horner and Swarbrook, 1996; Knowles et al., 2001; Shaw and Williams, 2004.)

The impact of technology

Technology exerts a number of direct impacts upon the development of foreign tourism but the most significant technological factors relate to the progressive globalisation of transport and telecommunications. In the area of transport, foreign travel has been shaped especially by the development of commercial air services and also by the acceleration of international rail services and the extension of motorway links within major international destination areas such as Europe. In telecommunications, the primary effects have been felt through the development of global, computer-based information technology systems.

At the global level, air travel is particularly important and the compression of space and time that the aeroplane has produced has had far-reaching consequences for patterns of tourism, ensuring that no part of the globe is now more than 24 hours' flying time from any other part. The advent of jet airliners such as the Boeing 707, and particularly the subsequent generations of wide-bodied jets with their increased passenger capacities and extended ranges (such as the Boeing 747 and now the Airbus A380), reduced both journey times and real costs of air travel and has provided essential under-pinning to the development of large-scale, foreign travel. It seems inconceivable that tourism to distant destinations would have grown to the extent that it has if passengers were still being offered the fares, travel times and comfort of the airways of the 1950s. In 2004, 79 per cent of foreign visits made by UK residents were made using air services, whilst 72 per cent of overseas visitors to the UK relied on air travel. Unsurprisingly, travel from the UK to long-haul destinations such as Australia, China, Japan and New Zealand is entirely dependent upon air travel (ONS, 2006).

However, the influence of air travel on international tourism patterns is far from consistent as in some major destination areas, air travel holds only a secondary share of travel markets. Page (1999) shows that in Europe, for example, although there are strong links between air travel and some sectors of tourism – in particular package tourism from northern European states to low-cost Mediterranean destinations – air travel between European destinations accounts for only 30 per cent of international tourist arrivals. This market share will certainly increase as the rising levels of low-cost, no-frills air services provided by airline companies such as Easy Jet create new opportunities for affordable travel (Child, 2000; Donne, 2000; Mintel, 2003a), but it remains the case that intra-European travel patterns are dominated by the car. However, once Europeans travel outside the continent, over 85 per cent of journeys are made by plane.

Part of the reason for the secondary significance of air travel in the European area is the convenience of other forms of travel for shorter international journeys. Travel by road in Europe has been aided by developments to international motorways and improvements to the Alpine passes into countries such as Italy. Similarly, high-speed rail services in France (TGV), Spain (AVE), Germany (ICE) and between Britain and mainland Europe (Eurostar) have added an extra element of competition, especially in sectors such as short-break/city tourism between centres such as London, Paris, Brussels and Amsterdam.

Whilst aeroplanes and high-speed road and rail links accelerate the pace and extend the distances over which people can travel, innovations in global information systems have revolutionised the communication of information. Information is fundamental to the business of international tourism and it is clear that the impact of new forms of information technology has not only been a primary factor in facilitating growth in international tourism, but has also revolutionised how the industry operates and how providers relate to consumers.

Buhalis (1998) has proposed a three-stage summary of the development of information technology applications in tourism:

● computer reservation systems (1970s);
● global distribution systems (1980s);
● the Internet (1990s onwards).

Computer reservation systems (CRS) were first applied by airlines as a means of centralising control over the sale of airline seats but the benefits of direct access to information on the availability of services and the ability to confirm bookings instantly by use of credit cards was quickly recognised by other tourism sectors such as hotel groups, car rental firms and even some areas of entertainment (Go, 1992; Knowles and Garland, 1994). Global distribution systems (GDS) are a development of CRS but with enhanced capacities to handle simultaneously a much wider range of information from which travel agents can tailor packages to meet the specific needs of tourists. GDS systems typically offer information, booking and ticketing services on airline and rail travel, ferry services, car hire, hotel reservations and other forms of accommodation, as well a wide range of entertainments. Four systems – Galileo, Amadeus, Sabre and Worldspan – dominate the market.

However, the most significant development in information technology has been the advent of the Internet. This innovation offers tourists (whether in domestic or international markets) a number of clear advantages and, in the process, is reworking the relationship between tourists and the industry in some quite fundamental ways. Most importantly, the Internet empowers individuals to seek information and to make bookings online in ways that effectively removes the role of the travel agent as an intermediary. Because the Internet is a multi-media facility it is capable of presenting destination and travel information in attractive and influential ways, which often negates the need for conventional travel brochures, whilst online booking generally provides the service (whether it an airline seat or a hotel room) at preferential rates, thereby driving down the costs that the tourist has to bear.

Economic development and political convergence

The sustained growth of international forms of tourism has also been highly dependent upon rising levels of economic prosperity and geopolitical stability in both generating and receiving areas. Tourism has always been subject to the constraints of cost and until quite recently the expense of foreign travel was a most effective barrier to popular forms of participation. But across large areas of the developed world, general levels of prosperity have risen throughout the post-1945 period and as levels of disposable income have increased, so foreign travel has become more affordable.

The impacts of prosperity are reinforced when placed in the wider context of the economic realignment of nation states. In particular, the creation of trading blocs, such as the European Union in which protective barriers to trade and commerce are generally absent, has enabled the growth of travel companies with international portfolios of products and services (Davidson, 1992). However, because areas such as the EU comprise a free-trade zone with significant levels of deregulation (such as air travel), levels of competition remain high with beneficial impacts upon the pricing of tourism products. The establishment across most of the EU zone of the Euro as a single currency has brought similar benefits on prices by removing the costs of currency exchange transactions and removing the uncertainties that were previously associated with fluctuations in monetary exchange rates. Globally, the development of international credit card services (such as Mastercard, Visa and American Express) brings a similar level of benefit to currency transactions and reduces the need for tourists to carry and exchange large amounts of currency.

Political convergence has exerted a positive influence too. One of the reasons why international tourism in Europe has developed so strongly since 1945 has been the almost total absence of major political and military conflict in the region since the end of the Second World War. The one significant divide that did arise from that war – the division between a largely Communist Eastern Europe and a capitalist West – actually produced a clear demarcation in the geography of tourism, with rapid development in the West and relatively little international travel in the East. As soon as Communist control of East European states began to crumble, tourism both to and from these areas followed. Within the European Union the progressive removal of controls on movements between member states – especially following the Schengen Treaty of 1995 – now permits largely unregulated travel by tourists between member states, whilst outside the EU a dwindling number of states now require a visa as a prerequisite for entry by tourists.

Lifestyle changes

The developments discussed in the previous four sections have all been essential to the expansion of international travel, but perhaps the most critical factor has been those changes in lifestyle that have led to foreign travel becoming an embedded feature of contemporary life in the developed and, increasingly, the developing world. These lifestyle changes are reflected in several areas:

● in the rising incidence of multiple holidays;
● in the fashion for foreign travel and an awareness of its attractions;
● in the competence (or experience) of tourists as foreign travellers.

The rising incidence of patterns of multiple holiday-taking is a product of the general enhancement in levels of affluence, mobility, awareness and the public appetite for tourism and is central to the embedding of tourism in lifestyle. Frequent engagement in travel helps to make it a routine behaviour and reinforces the notion of tourism as constitutive of daily experience rather than a separate entity (see Chapter 1). As an indicator of changes in behaviour, data in Table 2.3 show that by the end of the twentieth century, the proportion of UK residents who took more than one holiday of at least four nights stood at 25 per cent. When people who took no holiday are discounted, the proportion of UK holidaymakers who took more than one holiday rises to over 42 per cent.

International tourism has developed because it has become fashionable. The connections between tourism and fashion, as we have seen, have often been close, but in contemporary societies that have become characteristically cosmopolitan and mobile (Urry, 2000), there seems little doubt that a foreign holiday has become a mark of status. The fashion for international travel reflects a greater level of public awareness that is actively formed and reformed by promotion of travel in the media, through newspapers and magazines, on radio and television, as well as through the travel industry itself. Promotion has made people more aware of distant places and, through the construction and dissemination of exotic images of foreign lands, directly shapes new levels of public awareness of the pleasures and experiences that such places may provide. Hence, part of the problem that many domestic resorts now confront is the perception that foreign places will offer an experience that in many ways will be superior – whether through the enjoyment of a better climate, different landscapes or different places of entertainment, culture, historic or political significance.

International tourism has expanded, too, because tourists, in general, are more competent at the business of international travel. People who travel abroad regularly as part of a mobile lifestyle soon acquire confidence through experience, but even the more occasional traveller can often travel in confidence because of developments within the industry that actively enable the mobile lifestyle. Some have already been mentioned, but additionally we may note the positive impacts of:

● post-1945 improvement in educational levels and better training of personnel within the hospitality industries that mean that language is less of a barrier;
● travel procedures (customs, airport check-ins, etc.) that are – within many destination areas – becoming minimised, standardised and familiar;
● global telecommunications systems that make it simpler to keep in touch with home whilst travelling abroad;
● standardised forms of accommodation and other services – in international hotels, restaurant chains and car hire offices – that reduce the sense of dislocation that foreign travel might otherwise generate.

The confidence that such familiarity creates is one of the factors promoting the increased tendency to personalised forms of independent foreign tourism. The growth of independent travel reflects rising levels of public resistance to packaged styles of tourism and the increasing diversity of tourist demand. Poon (1989) has argued that the wider exercise of consumer choice characterises leisure lifestyles in post-industrial societies and is reflected in a willingness of tourists to contemplate visiting distant and more exotic locations or to pursue specialised forms of active tourism that reflect particular lifestyle choices and preferences. As Williams (2003: 85) observes, the 'flexible, often eclectic nature of contemporary tourism and its wider integration into personal lifestyles, has become one of its defining features'.

Travel security

Most of the factors discussed above exert a broadly positive effect on patterns of inter-national tourism, but to conclude this section some consideration of the essentially negative impact of issues of travel security on international tourism patterns is worthwhile. The security and safety of tourists has become a more prominent concern in the wake of an increase in the general incidence of political instabilities, local war and particularly terrorism, from the 1970s onwards and culminating in the attack on the World Trade Centre on September 11, 2001.

As Sonmez and Graefe (1998) observe, it is intuitively logical for tourists to compare potential destinations according to perceived benefits and costs, but embedded within the notion of cost is the element of risk that tourists take in travelling to a foreign destination where they may come to some harm that they would not encounter in their day-to-day lives. Typical risks will include natural disasters, disease and illnesses, food safety and crime (all of which have been hazards associated with travel for centuries), as well as the impacts of political turmoil and, especially, terrorism (which are comparatively recent, but, in the case of terrorism, is now a very real and influential concern). The impact of terrorism on tourism may be seen in both a direct and an indirect sense. Direct impacts occur where tourists themselves become the subject of terrorist actions. Tourism may attract terrorist attention because of its symbolic position (as an expression of capitalism); because of its economic value to many nation states or regions; or through cultural or ideological opposition to the activity. Indirect impacts occur where tourists become the incidental victims of activity targeted at others.

Risks to personal safety – whether real or perceived – exert significant impacts on international tourism patterns, particularly through the avoidance by tourists of destinations that are perceived as risky and through their substitution with 'safe' alternatives. In the 1990s, for example, terrorism and related political instability was held to account for adverse impacts on the number of foreign visitors to a range of destination areas including China, Egypt, Israel, Northern Ireland, Spain, Turkey and Zimbabwe (Arana and Leon, 2008; Sonmez, 1998). A study by Sonmez and Graefe (1998) of travel patterns by a sample of US international tourists found that:

- 88 per cent avoided politically unstable countries;
- 57 per cent stated that the possibility of terrorism constrained their travel choices;
- 77 per cent would only travel to countries they believed were safe.

Following the attack on the World Trade Centre, the US State Department issued travel advice to US citizens not to travel to twenty-eight countries in the Middle East and Africa, but travel by Americans to other destinations – especially in Europe – was also reduced significantly (Goodrich, 2002). The same author also noted as secondary impacts:

- increased levels of security at major transport hubs, especially airports, causing delays in processing arrivals and departures;
- short- to medium-term reductions in the numbers of people flying leading to loss of revenue and cut-backs in employment;
- loss of tourist trade in the hotel and catering sectors, with a reduction of over 50 per cent in hotel reservations in the USA over the six months following the terrorist attack;
- increased costs of actions such as higher security being passed on to tourists through higher prices.

All of these impacts exerted a marked downward pressure on tourist demand for travel to and from the USA immediately following September 11, but similar responses, albeit often on a lesser scale, typify how international tourism adjusts to issues of travel security.

However, because problems such as political instabilities or the threat of terrorism are seldom a permanent feature of the conditions in particular countries, so there is an 'ebb and flow' in the impact of security issues on tourism. Typically the short-term impacts of, say, a terrorist incident will be significant but unless there is a reoccurrence the perception of risk will diminish and the negative impacts on aspects such as destination image will recede quite quickly. Hence, for example, foreign visiting to China all but vanished after the incidents in Tiananmen Square in 1989, but, as the case study in the following section reveals, the contemporary picture of tourism to China is now one of sustained expansion. In a similar fashion, the resolution of conflicts that may have underlain political turmoil or terrorism will often produce significant growth in tourism by releasing latent demands that were previously deflected elsewhere. For example, a study by O'Neill and Fitz (1996) of tourism to Northern Ireland reported significant increases in the levels of tourist demand following the ceasefire by paramilitary organisations whose actions had made the province an unattractive destination since the onset of sectarian violence in the late 1960s.

Variations in patterns of development

However, it is important to appreciate that the factors that have shaped the growth of international tourism vary in their effect through time and across space, producing quite uneven patterns of growth and development. To illustrate this point – and to conclude this chapter – two case studies of an established major tourism destination (Spain) and an emerging one (China) are presented.

Spain

Spain is an outstanding example of the impact of post-1945 growth in affordable international tourism and, with an estimated 53.6 million tourist arrivals in 2004 (WTO, 2005a), illustrates a mature destination that will contain many locations and attractions that have already reached the final phases of Butler's model. Spain therefore exemplifies many of the problems that resort areas encounter as they reach their capacities and encounter the resulting tendency for tourism places to drift down-market, setting in motion a process of spatial displacement of some groups of tourists to new destinations.

Although from the mid-nineteenth century there was a tradition of small-scale local tourism by wealthy Spaniards both to the Atlantic coast of northern Spain and to Mediterranean coastal resorts such as Malaga, Alicante and Palma de Mallorca, the modern Spanish tourist industry is a visible product of the age of the aeroplane and the international package tour. Spain has benefited from being an early entrant into the field of mass international travel and the period since 1960 has seen rapid and sustained expansion in the numbers of visitors. From a base of less than 1 million visitors in 1950, international tourist arrivals in Spain reached 30 million by 1975 and almost 54 million by 2005 (Albert-Pinole, 1993; WTO, 2006). In 2004 the industry earned US$45.2 billion in foreign currency (WTO, 2005a) and generated 6 per cent of Spanish gross domestic product (GDP) (Garin Munoz, 2007).

The key factors contributing to the rise of mass forms of tourism to Spain have included:

- the attractive climate;
- the extensive coastline, which includes not just the mainland but also the key island groups of the Canaries and the Balearics;
- the accessibility of Spain to major generating countries in northern Europe, especially by air, but – through recent improvements in motorway and rail links – by land-based modes too;
- the competitive pricing of Spanish tourism products, particularly accommodation, which enabled the extensive development of cheap package holidays to Spanish resorts;
- the distinctive Spanish culture.

However, although Spain is a major international destination that ranks second only to France in tourist arrivals (Table 3.1), the recent development of tourism has highlighted several problems. At a macro level some stagnation in the growth of international arrivals has been noted, particularly as new destination areas such as Greece, Turkey and now Florida provide alternative locations for the package tourists that sustained much of the expansion of Spanish tourism between, perhaps, 1970 and 1995. The Spanish tourist industry continues to expand but recent reports from the Spanish national statistical office (Instituto Nacional de Estadistica) suggest that growth in key indicators – such as hotel occupancy – is a reflection of rising levels of travel amongst Spaniards rather than amongst international visitors (INE, 2006a).

Second, Spanish tourism illustrates both sectoral and spatial imbalances. In 2004, just over 66 per cent of foreign visitors to Spain came from just three countries (the United Kingdom, Germany and France) and they revealed a clear preference for a particular style of low-cost holiday centred on sun, sea and sand. Consequently, there are pronounced spatial imbalances in Spanish tourism with significant concentrations of international visitors in the Mediterranean coastal regions and the two off-shore archipelago (the Balearics and the Canaries), but, with the exception of Madrid, significant levels of under-development in the Spanish interior and northern regions. Rural tourism by foreign visitors is particularly poorly developed, especially in comparison with Spain's northern neighbour France (Mintel, 2003b). Data from INE (2006b) on the availability of serviced accommodation shows that 78 per cent of hotel rooms are located in just six of the Spanish regions – Andalucia, Balearics, Canaries, Cataluna, Madrid and Valencia (Figure 3.5). The tendency for other major forms of accommodation provision (apartments, villa developments, second homes and time-share properties) to focus in the same regions simply exacerbates the spatial unevenness in tourism development.

The third problem is that the rapid pace of development and its spatial concentration have commonly promoted a disorderly pattern of growth. There is some evidence to show that this has begun to undermine the attractiveness of the location, leading to movement down-market. This is especially true of coastal resorts that were prominent in the initial phases of expansion between 1960 and 1975, when low-cost package tourism was shaping tourism development. For example, Pollard and Rodriguez (1992) show how a failure to plan the popular resort of Torremolinos has been one of the elements in its gradual loss of image. Torremolinos, prior to about 1960, was a small fishing village and a resort for a select group of local tourists together with a handful of foreign writers and artists. However, the popularisation of the town as a package tour destination led to rapid and uncontrolled

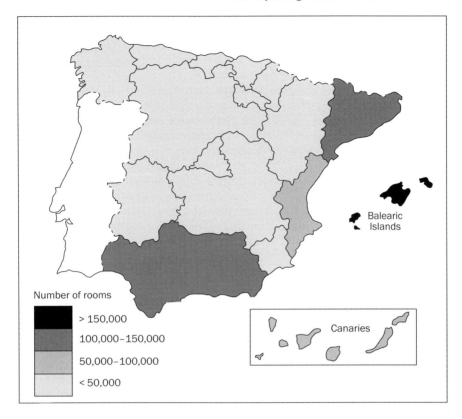

Number of rooms

> 150,000

100,000–150,000

50,000–100,000

< 50,000

Balearic
Islands

Canaries

Figure 3.5 Regional distribution of hotel and hostel rooms in Spain, 2003

developments which created a formless and untidy built-up area, visually polluted by characterless buildings, lacking public open spaces, limited by poor car parking and with an ill-defined and rather inaccessible sea frontage.

Unfortunately, Torremolinos is not an isolated case and a general incidence in popular Spanish resorts of over-development, commercialisation, crowding of bars, beaches and streets, pollution of sea and beaches as key infrastructure such as sewage treatment has failed to keep pace with expansion, and localised incidence of drunkenness and petty crime have all begun to alter popular perceptions of Spain as a destination. As noted in Chapter 2, Agarwal and Brunt (2006) identify how growing numbers of older Spanish resorts have had to confront problems such as a downturn in visitor numbers, over-dependence upon particular market sectors and deterioration in both built and natural environments. A report by EIU (1997) observes that far too many old town centres have been surrounded and drowned by relentlessly ugly new buildings and disfigured by gaudy shop fronts and signboards.

Such problems have become a major source of concern within the Spanish tourism industry and it is evident that future development will, of necessity, need to find ways to reinvest in the traditional resort areas whilst simultaneously exploring ways of diversifying the tourism market and promoting tourism in areas other than the Mediterranean coast (e.g., through city and rural tourism). The gradual development of a more structured and effective system of planning in Spain will assist this process. In the post-Franco era, Spanish planning has moved from a centralised to a decentralised approach (based on the autonomous regions) and with a stronger focus on the linkages between tourism development and the wider context of land planning. However, the impact of these changes varies significantly between regions and in the view of some critics, still lacks the level of coordination that is required for effective regulation of tourism development (Baidal, 2004).

China

If Spain exemplifies many of the problems of a maturing destination for international travel, China provides a contrasting perspective on an emerging tourism region. Not only does the growth of Chinese tourism illustrate the globalisation of international travel, but it also reflects the rapid emergence of international tourism within the WTO East Asia and Pacific region in which China is located and in which it is already a dominant player. As Zhang and Lew (2003) describe, China possesses a wealth of resources around which to develop tourism (which include some unique attractions such as the Great Wall and the Terracotta Army) and the industry has benefited directly from both the sustained economic growth that China has experienced over the last two decades and the active support of government. Ranked fourth in the world in 2004 as a receiving destination (WTO, 2005a), it is forecast to become the first-ranked destination by 2020 and a top four country in generating outbound tourism (WTO, 2001).

It is impossible to comprehend the development of tourism in China without reference to the political context. Although there is a lengthy tradition of travel to China (Xiaolun, 2003), a protracted series of military conflicts in the 1930s and 1940s, followed by the Communist revolution of 1949 (which established the People's Republic of China), ensured that China remained untouched by processes of development in international travel that affected Europe and North America at this time. Under the leadership of Mao Zedong (and particularly during the so-called 'Cultural Revolution' between 1966 and 1976), tourism came to be viewed as the antithesis of the socialist revolution that Mao wished to pursue. Domestic forms of tourism became virtually non-existent (Zhang, 1997) and where international tourism did occur it was considered solely as a political instrument enabling very small numbers of special foreign visitors to witness the achievements of Communist China (Hanqin et al., 1999; Zhang, 2003). China, effectively, became a closed country.

Unusually, the development of modern tourism in China can be traced to a specific date – 1978. This pivotal year marked a significant departure from the tenets of Maoism under the new leadership of Deng Xiaoping (Honggen, 2006). In practice, the legacy of the Cultural Revolution had proved to be endemic backward economic conditions and a chronic shortage of investment capital. Through what has since been termed the 'open door' policy, Deng moved to address the impoverished condition of China through reconnection with the international community. This he aimed to achieve by a progressive shift in policy towards a new blend of socialism with capitalism and a gradual removal of the economic, social and political barriers behind which China had been concealed under Mao. The status of tourism, through its capacity to attract foreign exchange and investment, has thus been transformed from an activity that was widely denigrated as representing the worst excesses of decadent, Western, capitalist society (under Mao) to a central component in economic and social planning in modern China.

Under these changed political conditions, the scale of the industry and the rapidity with which it has developed has been spectacular. According to the China National Tourist Office (CNTO, 2006), some 120 million visitors arrived in China in 2005. These data are somewhat misleading as they contain very large numbers of what the Chinese authorities describe as 'compatriots'. These are Chinese residents from the provinces of Hong Kong, Macao and Taiwan who are taking advantage of the new political relationships between these territories (Yingzhi et al., 2006) to enter China routinely, often on a day basis, to visit relatives, to shop or to work. The number of truly foreign visitors in 2005 was recorded as 20.3 million. However, as Figure 3.6 shows, the levels of visiting by this sector of the market have developed at an accelerating rate since the 'opening of the doors' in 1978.

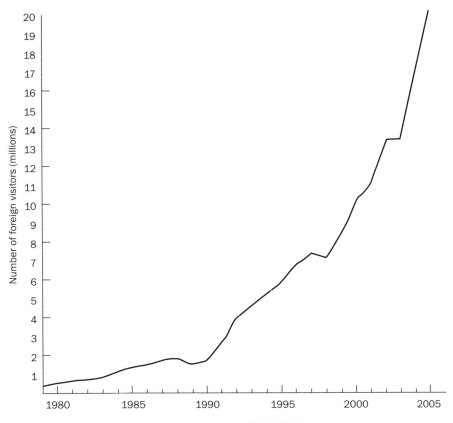

Figure 3.6 Increase in foreign visitors to China, 1979–2005

Although some of the foreign visitor market is accounted for by long-haul tourism from Europe and North America, the majority of foreign visitors presently originate within the WTO East Asia and Pacific region. Figure 3.7 maps the pattern of inbound tourism from major source areas and reveals the particular importance of Asian countries (led by Japan and South Korea) which accounted for 12.5 of the 20.3 million foreign visitors received in 2005. European nations (led by the UK, Germany and France) contributed a further 2.6 million visitors, whilst North America and Russia each added a further 2.2 million. Interestingly, in recent years the growth rates within the Asian and Pacific market have generally exceeded those in North America and Europe, so although absolute numbers are rising, the share of the Chinese market that is accounted for by long-haul visitors from countries such as the USA and the UK is actually falling. For example, in 1990, 17.4 per cent of foreign visitors to China came from the USA, but by 1997, that figure had fallen to 11.7 per cent.

However, the speed with which tourism in China has developed (bearing in mind that the Chinese domestic tourism market – which is already the largest in the world – adds an estimated 870 million travellers to the levels of foreign visiting (Messerli and Bakker, 2004)), has created a number of significant problems. These have included:

● The speed of development has often outstripped the capacity to manage change. This has been evident in key areas such as tourism planning where significant gaps between the expectations of the planners and the actual performance at implementation have been observed (Kun et al., 2006).

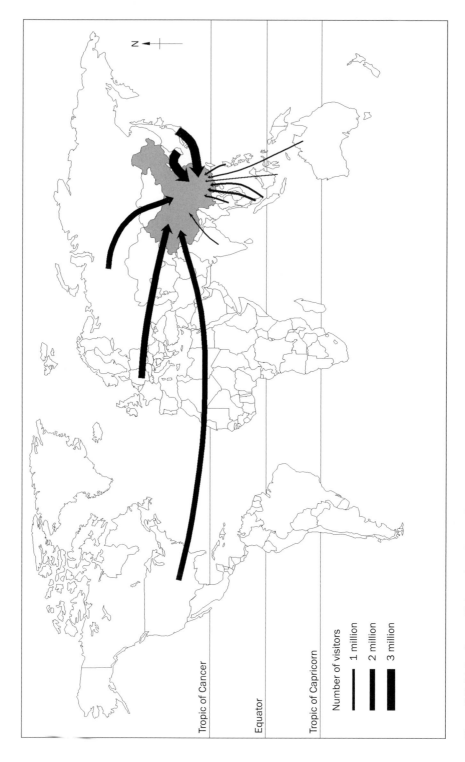

Figure 3.7 Major flows of inbound tourism to China, 2005

Tropic of Cancer

Equator

Tropic of Capricorn

Number of visitors

—— 1 million

—— 2 million

━━ 3 million

- Infrastructural weaknesses in key sectors such as accommodation and transport. In 1978 there were only 203 hotels in the entire country and whilst a rapid programme of hotel construction had raised that number to over 7,000 by 1999 (Zhang, 2003), the supply of accommodation is still not well-matched to tourist needs (Messerli and Bakker, 2004). Similarly, despite major physical expansion, China's internal transport system has continually faced problems of inefficient use of its trains and planes, poor management, lack of maintenance, inadequate facilities and safety problems (Mak, 2003).
- Organisational problems that have centred around shortages of skilled labour at both the managerial and service grades and problems in supporting services such as travel agencies. Travel agencies in China are largely state-owned, but the service is highly fragmented and the operating environment in which they work has been described as 'chaotic' (Qian, 2003).
- Problems of image (Xiaoping, 2003). China has often suffered from an image as a poor, isolated country with a suppressed people governed by a regime with a dubious record in human rights – a view that was reinforced with disastrous short-term consequences on foreign tourism by the Tiananmen Square incidents in 1989.
- Marked spatial imbalances in the distribution of tourism with over-concentrations of activity in urban environments and the eastern provinces and a relative lack of tourism development in rural areas and across large swathes of western China (Jackson, 2006). This is partly a consequence of the natural limitations of time, money and accessibility that confront travellers who propose to explore a country as vast as China (Xiaoping, 2003), but is also a reflection of the way in which the Chinese authorities have been slow to allow new destinations to be opened to foreign tourists (although most areas are now accessible to foreigners).

However, notwithstanding the economic transformation that has affected China over the last twenty years, it remains a developing country and it is unrealistic to expect a seamless integration of a new activity such as tourism into a new economic order. But as might be expected, given the alacrity with which the Chinese tackle major projects, there has been no lack of response to the problems that have been identified. This has included:

- Modernisation of the air transport system through major investment in new and upgraded infrastructure, modernisation of aircraft fleets and development of new routes both within and outside China. In 1978 there were only 70 domestic and international routes that were flown regularly by Chinese airlines, by 1999 over 1,100 routes were in use (Mak, 2003).
- Implementation of programmes of education and training for employees in the hospitality and travel industry.
- Progressive opening of Chinese markets to foreign investment, including joint ventures and, since the admission of China to the World Trade Organization in 2001, wholly foreign-owned companies in sectors such as hotels.
- Active promotion of new tourist sites and areas. This includes major programmes of restoration of historic and cultural sites from China's imperial past (such as Beijing's Forbidden City) that were widely vandalised during the excesses of the Cultural Revolution but which now – paradoxically – are seen as essential to defining China's national identity (Sofield and Fung, 1998).
- Gradual reduction in the levels of regulation exercised centrally and wider devolution of responsibility.

According to recent estimates (Messerli and Bakker, 2004), some 5.64 million people now work directly in China's tourism industry and foreign tourism (including compatriots) earned a total of US$29 billion in 2005 (CNTO, 2006). The potential for tourism to play a transformative role in China is, therefore, not in doubt, provided the Chinese can find sustainable solutions to the challenges that they currently face. China's ability to absorb and respond to global competition will be one of the keys to the future success of its tourism industry, but it is evident that success will also be dependent upon structural and organisational reforms, together with the development of tourism products that match the expectations of modern, international travellers (Yu et al., 2003).

Conclusion to case studies

The experience of international tourism development in Spain and China illustrate most of the key themes that have shaped discussion in this chapter. Each shows just how rapidly international tourism has tended to develop – in the case of China especially so – and each case illustrates the importance of an organised travel industry to that process. Both examples, but especially China, show how economic growth and geopolitical stability are essential prerequisites to tourism development, and also how government has an ongoing role in regulating what is a diffuse industry in which unchecked and spontaneous forms of development occur all too easily and where the positive benefits that tourism can bring are often outweighed by negative impacts. As a socialist state, China exemplifies perhaps an extreme level of governmental control over the development of international tourism, but we have also seen that in a modern, capitalist democracy such as Spain, regulation of development through an ordered planning process is important and when such regulation is applied poorly, as in the first phase of Mediterranean resort development, significant problems may be created.

It is to the broader themes of development and associated impacts that we now turn in Part II of this book.

Summary

The theme of this chapter is the spatial expansion in tourism as evidenced in the development of international travel. Although rooted in a history of travel that extends over several centuries, international tourism is shown to be primarily a product of post-1945 patterns of leisure where growth has been aided by a range of factors. These include:

- the development of a structured travel industry;
- the impact of technological innovation in transportation and communications;
- economic and political stability;
- the fashionability and ease of foreign tourism.

Whilst Europe still dominates international tourism markets, the recent development of new tourist areas in so-called 'long-haul' destinations in East Asia and the Pacific suggest that new tourism geographies are already emerging.

Discussion questions

1   What are the key factors that shape the geographical patterns of tourism in Europe?
2   To what extent are spatial patterns in international tourism a reflection of changing tastes and fashions?
3   How has the growth of international tourism been affected by technological developments in transportation and information transfer?

4   In what ways does the development of tourism in China since 1978 demonstrate the impact of processes of globalisation?
5   What are the essential points of similarity and contrast in the development of tourism in Spain and China since 1980?

Further reading

Despite its age, one of the best accounts of the development of international travel remains:
Turner, L. and Ash, J. (1975) *The Golden Hordes: International Tourism and the Pleasure Periphery*, London: Constable.

An excellent recent discussion of development up to 1940 is provided by:
Towner, J. (1996) *An Historical Geography of Recreation and Tourism in the Western World, 1540–1940*, Chichester: John Wiley.

For a more recent overview of the processes of development in international tourism, see:
Williams, S. (2003) *Tourism and Recreation*, Harlow: Prentice Hall.

Analyses of some of the primary factors affecting structural change in international tourism are given in:
Buhalis, D. (1998) 'Strategic use of information technologies in the tourism industry', *Tourism Management*, Vol. 19 (5): 409–21.
Page, S.J. (1999) *Transport and Tourism*, Harlow: Addison Wesley Longman.
Shaw, G. and Williams, A.M. (2002) *Critical Issues in Tourism: A Geographical Perspective*, Oxford: Blackwell.
Shaw, G. and Williams, A.M. (2004) *Tourism and Tourism Spaces*, London: Sage.

Useful analyses of development of international tourism within major world regions are provided by:
Harrison, D. (ed.) (2001) *Tourism and the Less Developed World: Issues and Case Studies*, Wallingford: CAB International.
Lew, A.A., Yu, L., Ap, J. and Zhang, G (eds) (2003) *Tourism in China*, Haworth: New York.
Williams, A.M. and Shaw, G. (eds) (1998) *Tourism and Economic Development: European Experiences*, Chichester: John Wiley.

Note

1   'Pulling Mussels (from the Shell)', Glenn Tilbrook and Chris Difford, Rondor Music.

Part II
Tourism relations

In Part I of the book we have examined and explained some broad chronologies in the spatial development of tourism, as an essential component in understanding tourism geography. Part II moves the discussion forward to consider how tourists and tourism relate to the places that are toured and, particularly, how those places and their populations relate to tourism. As is true of resort development and of spatial change in the patterns of tourism, this is an area of interest that has also been at the heart of geographical readings of tourism for several decades. This reflects an early recognition of the fact that these relationships are generally contingent upon the local contexts in which tourism occurs and are consequently highly variable across geographic space. So once again, geography is of central importance to any appreciation of the relations between tourism and place.

Much of the formative writing by geographers working in this area has approached the themes that shape this part of the book (physical and economic development; environmental change; the influence of tourism on socio-cultural practices; and the role of tourism planning) from a perspective of tourism *impact* upon destination areas. However, the tendency to frame discussions around the notion of 'impact' (with its associated risk of formulating simplistic, binary readings of tourism as exerting either positive or negative uni-directional effects) has been largely resisted in these chapters. Instead, the adopted approach strives to develop a relational understanding of tourism that acknowledges that what might be labelled as tourism's impacts are seldom capable of simple categorisation, nor are they consistent with regard to cause and effect from place to place.

Once again, each chapter aims to develop explanation rather than more basic geographic description as the primary focus of the discussion. This is approached in two distinct ways. First, wherever appropriate, the opportunity has been taken to ground the examination of tourism relations into some key themes and concepts within the wider arena of contemporary human geography and to suggest how these critical perspectives illuminate the understanding of tourism and its geographies. Globalisation, the changing relationships between production and consumption, sustainability, and concepts of power relations are each deployed to provide some essential contexts through which readers can connect tourism geography to the wider field of geographic enquiry.

Second, these chapters also exhibit a tendency that has long been evident in human geography to develop explanations by borrowing from cognate disciplines. Hence the discussions that follow draw widely and sometimes in an eclectic fashion from disciplines such as anthropology, cultural studies, economics, politics, planning and sociology in their quest to understand their subject. No apology is offered for adopting this approach, not least because this book addresses its task from the position that a full understanding of tourism geography can never be distilled from perspectives that are purely geographical in origin and that by adopting a multi-disciplinary perspective, the subsequent understanding of tourism geography is actually strengthened.

④ Costs and benefits: the physical and economic development of tourism

Amongst the many impacts that tourism may exert upon host areas, the processes of physical and economic development are perhaps the most conspicuous. These effects may be evident in the physical development of tourism infrastructure (accommodation, retailing, entertainment, attractions, transportation services, etc.); in the associated creation of employment within the tourism industry; and, less visibly, through a range of potential impacts upon GDP, balances of trade and the capacities of national or regional economies to attract inward investment. For developing regions in particular, the apparent capacity for tourism to create considerable wealth from resources that are perceived to be naturally and freely available has proven understandably attractive, but the risks associated with over-development and dependence upon an activity that can be characteristically unstable are negative dimensions that should not be overlooked. There are benefits, but there are also costs attached to the physical and economic development of tourism.

For the student of tourism geography, however, 'development' itself can be a problematic concept. This is due primarily to the diversity of ways in which the term has been applied – describing both a process of change and a state (or a stage) of development (Pearce, 1989). Thus, for example, Butler's model of the resort cycle that was discussed in Chapter 2 (Butler, 1980) essentially defines successive stages of development but does not, in itself, articulate the details of process. Further, not only is there a basic distinction to be made between state and process, but the nature of the process has been subject to a variety of interpretations, including, *inter alia*:

- development as a process of economic growth – as defined in increased commodity output, creation of wealth and a raising of levels of employment;
- development as a process of socio-economic transformation – in which economic growth triggers wider processes of change that alter relationships between locations (particularly between developed and underdeveloped places) and between socio-economic groups – thereby creating fundamental shifts in patterns of production and consumption;
- development as a process of spatial reorganisation of people and areas of production. This may be viewed as a visible product of socio-economic transformations and is a common adjunct of tourism development, with its propensity to focus attention upon resources and resource areas that may previously have been idle or little used (see Mabogunje, 1980).

Within geography, development studies have traditionally tended to explore the particular problems of less developed states and their relationships with the developed world (see, e.g., Potter et al., 1999; Hodder, 2000). Part of this tradition has also transferred to the geographic study of tourism (e.g., Britton, 1989; Harrison, 1992, 2001a; Oppermann and

Chon, 1997; Scheyvens, 2002), but it is important to note that tourism development processes are also highly significant within states that would already be described as 'developed'. Thus, whilst some parts of this chapter will examine tourism in the context of less developed nations, it is important that the discussion addresses physical development and economic impacts within the settings of developed nations too.

The chapter explores two distinct, but related, themes:

- the factors that shape and regulate the physical development of tourism and the contrasting spatial forms that may result;
- the basic relationships between tourism and economic development.

However, it is important that these themes are placed into a wider context that recognises both the distinctive dimensions of tourism that influence the ways in which it develops, and the more fundamental shifts in the development process that are associated with the progress of globalisation and the changing relationships around the so-called 'Post-Fordist' pattern of production and consumption.

Tourism development: characteristics and context

There are three characteristics of tourism production that may be considered as central to understanding geographical relationships between tourism and physical or economic development. First – and perhaps most critically – the production and consumption of tourism is place-specific. Unlike most manufactured goods which are generally distributed to consumers from their points of production, in tourism the consumption of the product takes place at the point of production and the tourist has to travel to these locations (e.g., resorts) to consume the product. (In this way, tourists – as consumers – become a part of the production process, since their presence and actions in the destination area inevitably shapes the nature of the product as it is experienced by others.) Moreover, in tourism sectors that are strongly seasonal in character, there are temporal as well as spatial dimensions to the pattern of consumption and these limitations of space and time combine to create what Shaw and Williams (2004) describe as a defining 'spatial and temporal fixity' to many forms of tourism.

Second, in the production of tourism, labour holds a key position. The organisation of labour is central to the competitiveness of any firm, but this is especially true in tourism. As a service industry with several sectors that are especially labour-intensive (e.g., hotels and catering), labour accounts for a relatively high proportion of total costs. Less obviously – but no less important – the process of hiring and discarding labour is a primary mechanism by which tourist businesses manage the polarised demand conditions that they face. This leads directly to the high incidence of seasonal and part-time work that the tourist industries often exhibit and where such labour comprises migrants – as is commonly the case in tourism – the industry will be directly influential in shaping geographical patterns of labour mobility.

Third, although – as in all sectors of the contemporary global economy – the development of transnational and global corporations is exerting a progressively more influential effect upon the economic organisation of tourism, the industry typically remains fragmented and is dominated by small- and medium-scale enterprises. Tourism is not an undifferentiated product and the large number of niche markets and associated services encourages entry by small firms. This means that the successful development of tourism is often conditional upon the capacity of small and medium-sized firms to align or coordinate their activities to deliver the bundle of goods, services and experiences that the tourist requires (Shaw and Williams, 2004). However, the highly competitive nature of the trading environment in tourism means that the failure rate of small-scale tourist enterprise is often high, which may create local-level instabilities.

As noted in Chapter 3, the firms that provide the goods and services that go to make up the tourism 'product' operate within an essentially capitalist framework in which tourism commodities (and it is also a characteristic of tourism that it is becoming progressively commodified) are produced and sold through a competitive market. However, because individual firms in a capitalist system have no capacity to influence the wider framework of the economy, there is an evident need for regulatory frameworks (such as monetary systems or legal controls over working terms and conditions) to provide overall control and direction. This is a central tenet of regulation theory which argues that there is normally a dominant set of principles that shapes the regulation of capitalist systems and which seek to provide, through what are termed 'regimes of accumulation', the systematic organisation of production, distribution, social exchange and consumption (see, *inter alia*, Ateljevic, 2000; Dunford, 1990; Milne and Ateljevic, 2001; Shaw and Williams, 2004; Tickell and Peck, 1994).

For much of the twentieth century the dominant regime of accumulation under which tourism developed was a Fordist/Keynsian model of production and consumption. This delivered some characteristic forms of development based around mass consumption of standardised packages that were determined more by the producer than the consumer and which offered a narrow range of products and services (Ioannides and Debbage, 1997). However, during the last quarter of the century, the Fordist/Keynsian model has been gradually yielding its position as the dominant regime, to the more flexible and dynamic pattern of production and consumption that is commonly labelled as post-Fordist (Milne and Ateljevic, 2001). In the field of tourism this shift has been associated with much greater differentiation of products and segmentation of the market; the development of new destination areas; and some movement away from standardised packages as a common product (Ioannides and Debbage, 1997). Shaw and Williams (2004), though, sound an important note of caution. The shift to post-Fordist patterns of production and consumption does not represent a linear transition from one regime of accumulation to another since, in practice, most of the production systems in which tourism is being formed still reveal the co-existence of both Fordist and post-Fordist patterns. In many settings, tourism remains a high-volume product, the important transformation is the enhanced levels of flexibility that shape its production and which are seen as essential to maintaining profitability in a market where tastes and preferences now appear to be far more fluid than previously (Poon, 1989).

At the core of regulation theory lies the presumption that nation states (usually in the form of their national governments) play a fundamental role in shaping regulatory frameworks and how they impact upon tourism development. Governments may exert a number of effects, including:

● mediation of relations between the state and the global economy;
● exercise of controls over the movement of labour and capital;
● creation of legal frameworks that regulate production;
● application of regional development policy;
● management of state security (Shaw and Williams, 2004).

However, the progressive integration of tourism into processes of globalisation through, for example, the rising incidence of multinational ownership of tourism businesses or the internationalisation of capital, weakens this relationship. This is partly because globalisation challenges the territoriality around which nation states are built (Shaw and Williams, 2004), but also because the development of global frameworks for exchange and control and their associated institutions (such as the International Monetary Fund and the World Bank), alters the context of regulation. Hence, for example, within the EU the ability of member states to regulate development independently has been diminished by several related processes. In particular, the adoption across most of the Europe Union of the Euro as a common currency, the establishment of an elected European parliament, and the greater centrality of EU Directives in shaping a range of policy areas have all been instrumental in modifying the regulatory capacity of individual member states.

But although the regulatory frameworks are becoming increasingly global in scale, the actual development of tourism remains focused at a regional and local level. There is, therefore, what Milne and Ateljevic (2001) term a 'nexus' at which the global forces that influence tourism development intersect with the localised agendas of the regions and communities that actually deliver the product. This relationship is inevitably both complex and contingent upon the particular conditions under which tourism is being developed. Some of this complexity will become evident if we move to consider how patterns of physical development are shaped.

Patterns of physical development of tourism

Geographically informed discussions of tourism development (together with a range of spatial models) have been established within the tourism literature for some time (see, e.g., Britton, 1989; Miossec, 1977; Pearce, 1987, 1989). These studies generally reveal that the development of tourism in any given location depends upon the existence of a set of prerequisites for growth and that the resulting spatial forms of development and their geographical characteristics will reflect the interplay between several factors that may be conceived as shaping the directions that development may take.

The essential prerequisites are:

● the presence of resources and attractions – which will include the natural attributes of climate, landform, scenery and wildlife; the socio-cultural heritage of the destination area (such as places of interest, historic sites, local cuisine or arts and crafts); as well as attractions such as entertainments, theme parks or leisure complexes that may form part of a built environment;
● infrastructure, primarily in the form of accommodation, transportation services and public utilities such as water supplies, sanitation and electricity;
● sources of capital investment, labour and appropriate structures for marketing and promoting the destination.

The primary factors (or groups of factors) that are seen as shaping the physical development of tourism are identified in Figure 4.1 which attempts an outline summary of what is actually a most complex pattern of interrelations. Five primary factors are proposed:

● physical constraints;
● the nature of tourist resources and attractions;
● the state of the tourism market;
● planning and investment conditions;
● levels of integration.

It is argued here that the differing ways in which these factors exert their influence – both in isolation and in combination – will commonly result in one of four general forms of tourism development: enclaves, resorts, zones and regions. In spatial terms, these different forms are associated with varying levels of concentration or dispersal and may also be

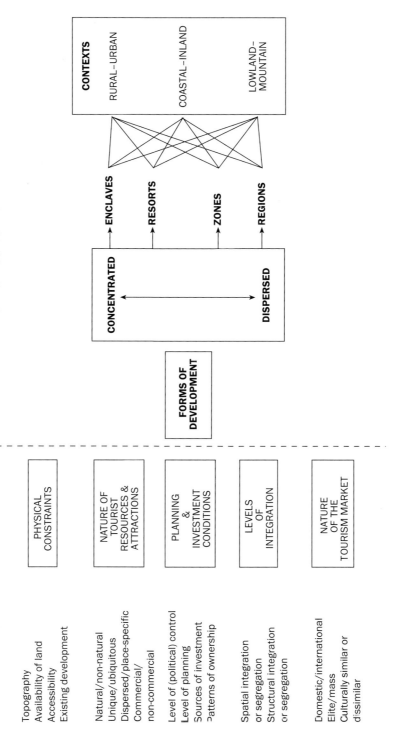

FACTORS OF INFLUENCE

PHYSICAL CONSTRAINTS

Topography
Availability of land
Accessibility
Existing development

NATURE OF TOURIST RESOURCES & ATTRACTIONS

Natural/non-natural
Unique/ubiquitous
Dispersed/place-specific
Commercial/
non-commercial

PLANNING & INVESTMENT CONDITIONS

Level of (political) control
Level of planning
Sources of investment
Patterns of ownership

LEVELS OF INTEGRATION

Spatial integration
or segregation
Structural integration
or segregation

NATURE OF THE TOURISM MARKET

Domestic/international
Elite/mass
Culturally similar or
dissimilar

DEVELOPMENT OUTCOMES

FORMS OF DEVELOPMENT

CONCENTRATED ⟷ DISPERSED

→ ENCLAVES
→ RESORTS
→ ZONES
→ REGIONS

CONTEXTS

RURAL–URBAN

COASTAL–INLAND

LOWLAND–MOUNTAIN

Figure 4.1 Factors affecting patterns of tourism development

located into one of several geographic 'contexts' that are here expressed as simple continua: urban/rural; coastal/inland; and lowland/mountain. In relation to the earlier discussion of the concept of 'development', it is evident that these factors reveal the incidence of development as both state and process. Hence, the influence of physical constraints, the nature of resources and attractions, and the state of the tourism market tend to reflect *states* of development; whilst the influence of planning, investment and integration are much more reflective of *process*.

Physical constraints will often have a direct bearing upon forms of tourism development and consequent geographical patterns. Topography, for example, can influence the availability of suitable sites for construction, levels of access and the ease with which key utilities (water, power, sewage disposal, etc.) may be installed or extended from existing settlements and their infrastructure. 'Difficult' environments might include rugged coastlines (such as the Amalfi coast in Italy) or mountain zones (such as the Alpine zones in Switzerland – see Plate 4.1), both of which tend to fragment and disperse development in a way that is generally untrue of (say) a flat, open coastline which enjoys ease of access (such as the Languedoc-Roussillon region in France).

Second, development patterns will reflect the state and disposition of the resources and attractions around which tourism is based and affect, especially, the extent to which tourism

Plate 4.1
Tourism development in a difficult environment: the mountain resort of Zermatt, Switzerland

becomes dispersed or concentrated. In particular, unique or place-specific attractions, whether natural or non-natural, tend to focus development around the site(s) in question, whereas more ubiquitous or spatially extensive resources (e.g., an accessible coastline or good-quality rural landscapes) may have a dispersing effect. Thus, rural tourism – in which sightseeing is an important pastime – is often characterised by a diffuse pattern of development across a multiplicity of relatively small-scale sites, with activity frequently being absorbed within existing facilities through farm tourism or second homes (where these are conversions of existing properties).

Third, it is suggested that patterns of development will be influenced by the state of development within the tourism market. This will vary according to whether development is targeted at a domestic or an international clientele, but more significant distinctions will normally exist between mass and so-called 'alternative' forms of tourism, because of the contrasting volumes of activity that these sectors deliver.

Although, historically, many forms of tourism development were spontaneous and only loosely controlled, the value of tourism as a tool for regional and national development has tended to mean that the modern industry is far more closely regulated. Local planning and investment conditions will therefore provide a fourth primary influence upon forms of development, and, as Figure 4.1 suggests, important factors include political attitudes towards tourism and the levels of political control (including the extent to which effective land planning procedures are in place); the extent to which investment is local or external to the region; and the levels of corporate interest in tourism and the associated patterns of ownership. In a rapidly globalising area such as tourism, the incidence of external investment and foreign ownership of facilities can be especially influential on resulting patterns of development. Local and regional communities that are anxious to attract inward investment will often accept development conditions that are imposed by outside investors as a price to be paid to ensure that investment is secured.

Planning and investment conditions are closely allied with the final key factor, the level and nature of integration. Discussions of 'integration' of tourism development tend to use the term in two senses. At one level, concerns have focused upon the extent to which tourism development is integrated in a spatial sense with existing, non-tourist forms of development – in other words, is tourism inter-mixed with other functions and land uses, or is it spatially segregated? Alternatively, integration may refer to whether or not a development is integrated in a structural sense. A structurally integrated development will bring together all the key elements – accommodation, transportation, retailing, entertainment and utilities – within a single, comprehensive development. This form contrasts with what are sometimes termed 'catalytic' patterns of development (Pearce, 1989) in which a small number of lead projects that are often externally financed and controlled, stimulate subsequent rounds of indigenous development as local entrepreneurs are drawn into an expanding tourism industry.

Contrasting forms of tourism development

We can exemplify how these different elements interact to produce varying forms of tourism development by examining the three most common development 'outcomes': tourist enclaves, resorts and zones.

Tourist enclaves

Enclaves represent the most highly concentrated form of tourism development and reflect most clearly the influence of:

- the constraints posed by limitations in infrastructure within a locality;
- investment patterns in which there are relatively few entrepreneurs developing provision for tourists and where funding is likely to be external in origin;
- a market which is focused upon a particular segment – usually elite groups – and where the tourist activity is often concentrated upon a particular resource – commonly, although not exclusively, in beach resorts.

Enclave developments, in their purest form, are entirely enclosed and self-contained areas, not just as physical entities, but as social and economic entities too (Pearce, 1989). They will display several features:

- physical separation (and isolation) from existing communities and developments which are generally not intended to benefit directly from the development;
- a minimising of economic and other structural linkages between the enclave and the resident community;
- a dependence upon foreign tourists which is reflected in pricing structures that reinforce the exclusivity of the enclave;
- pronounced lifestyle contrasts between the enclave and its surroundings (Jenkins, 1982).

Enclave developments are often a reflection of immaturity (or a pioneering stage) within a local tourism industry that has yet to evolve to the point where it can support a wider base of provision. In this sense, Regency Brighton, for example (see Case Study 2.1), represented a leisure enclave – a socially exclusive space that through real and symbolic boundaries was accessible only to a favoured few. However, in modern tourism, enclaves are most commonly found in developing nations, although this is not exclusively the case. The recent development in temperate parts of Europe (e.g., Britain, Belgium and the Netherlands) of high-quality indoor holiday villages with integral and comprehensive facilities set in artificially regulated 'exotic' environments marks a reworking of the enclave idea that is very much the product of a developed rather than a developing economy (Gordon, 1998). Similarly, modern theme parks such as Alton Towers (UK) or the various parks belonging to the Disney Corporation might be construed as a form of enclave development.

For tourism in emerging nations, enclave developments offer several distinct advantages. First, the concentration of investment into small numbers of contained projects represents a pragmatic response to the problems of how to begin to provide the high-quality facilities that modern travellers expect and how to form and reinforce a distinct and marketable product. Second, the tendency for enclaves to be partially or often entirely financed and owned by offshore companies is seen as a means of attracting inward investment to the developing economy and creating service employment for local people. Third, and less obvious, is the fact that enclaves may be favoured by local governments that are anxious to contain or limit potentially adverse social, cultural or political effects emanating from contact between visitors and host populations.

However, set against these potential benefits are several serious weaknesses, including increased economic dependence on foreign corporate institutions and investors; high levels of 'leakage' from the economy – especially in the form of profit paid to foreign owners or investors; limited levels of dependence upon local supplies of goods and services; and, sometimes, a seasonality in the employment of labour. These problems are explored more fully in the second half of this chapter, whilst Case Study 4.1 provides an illustration of several of the themes discussed above through a case study of the Okavango Delta in Botswana.

CASE STUDY 4.1

Enclave development in the Okavango Delta, Botswana

The Okavango Delta in north-western Botswana (Figure 4.2) comprises a wetland zone that extends over some 550 square miles and provides a wildlife habitat of international importance. It is also home to more than 120,000 people but a combination of a lack of employment, low incomes, limited food sources, periodic crop failures and high incidence of disease contributes to a significant level of local poverty in what is already one of the

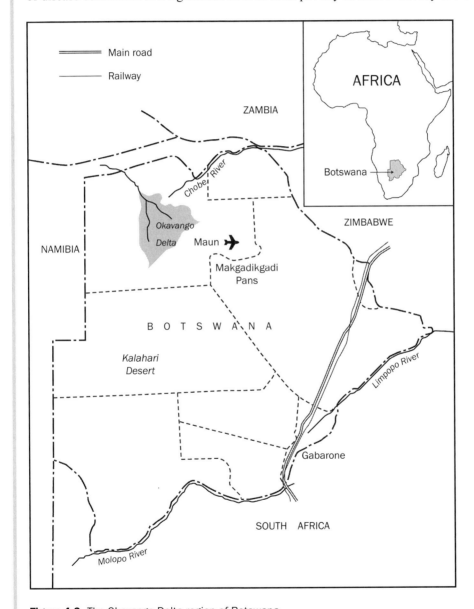

Figure 4.2 The Okavango Delta region of Botswana

continued

poorest countries in sub-Saharan Africa. When the country gained independence in 1966, tourism was almost completely absent from the Botswana economy but it has developed to the point at which it is now estimated to contribute around 5 per cent of the GDP and is the second largest sector after diamond mining. Some 50,000 tourists visit the Okavango Delta each year, primarily to enjoy the wildlife. Almost 90 per cent are foreign visitors (most of whom originate from Europe, the USA, Australia, New Zealand and South Africa) and the visitors stay in accommodation that ranges from luxury lodges to camps.

As is characteristic of many developing countries, tourism was adopted enthusiastically by the Botswanan government as a sector that was expected to bring positive development benefits and new economic opportunities. However, the enclavic nature of development in the Okavango has, in practice, limited the true impact in a number of ways. These include:

- Limited interactions between visitors and local people. Most tourists arrive by air at Maun and are transferred directly by light aircraft or road to the main lodges in the Okavango. There is, consequently, little expenditure by tourists in Maun or the villages of the Okavango and sectors such as cultural tourism are poorly developed.
- Heavy dependence upon foreign services and weak linkages to other sectors of the Botswana economy. Almost 80 per cent of the tourist infrastructure is wholly or partially owned by foreign companies and most of the food supplies to the main lodges are sourced from South Africa.
- Poor retention of revenue from tourism within Botswana. With such high levels of foreign ownership it is inevitable that much of the revenue from tourism is repatriated outside the country – through payments for imports and foreign services; salaries paid to foreign staff; or dividends to overseas investors. The travel companies that handle bookings and other organisational arrangements are also foreign, as are the principal airlines that transport visitors to and from Botswana. Consequently, over 70 per cent of the revenue from tourism in the Okavanga is dispersed elsewhere.
- Limitations on local employment. The development of tourism has directly created more than 1,600 jobs in the region, but the jobs occupied by local people are almost entirely in low-skill, low-pay occupations such as maids, kitchen staff, drivers and porters. As is characteristic of enclave developments, higher-level, managerial posts are generally filled by foreign workers who bring the requisite level of expertise and experience.
- Difficulties in recovering adequate levels of local taxation from companies registered outside of Botswana. In practice, only 11 per cent of the tourist companies operating in the Okavanga Delta area pay local tax.

Positive impacts have therefore been hard to discern and it is perhaps indicative of the difficulty of generating wider benefits through enclave developments that as tourism has increased in the Okavanga Delta, so has the incidence of poverty. The problem has not been helped by government policy in Botswana which (with every good intention) has pursued a programme of low volume-high value tourism as a means of conserving the special character of the Okavanga, but which has limited the scope for cheaper forms of tourism to develop that would probably enable a wider integration of local communities into the industry.

Source: Mbaiwa (2005)

Resorts

The most familiar form of tourism development is the resort and these may occur in a number of contexts. The seaside resort is the most commonplace, but resorts may also develop around inland health spas (e.g., Harrogate, UK), in mountain regions (e.g., La Grande Plagne, France) and even in deserts (e.g., Palm Springs and Las Vegas, USA). Resort developments are perhaps most strongly influenced by the nature of the resources that form the basis of their attraction and therefore a concentrated form of development tends to occur, centred on key resources. At a detailed level, however, resorts will illustrate the effect of accessibility and availability of land, levels of planning and control, sources of investment and varying levels of integration. Prideaux (2000, 2004) has also emphasised the need to understand how resorts develop as economic entities that are regulated by market forces of demand, supply and the associated price of services that the resort offers. In Prideaux's resort development spectrum model, resorts are envisaged as potentially progressing through successive stages of development as a response to evolving levels of demand and associated patterns of supply. However, that progression may, at any stage be influenced by changes in price that are associated with imbalances between demand and supply; by the development of competing destinations; or through investment and/or planning decisions taken within the resident community and by its representatives. Hence, some resorts develop to the point at which they acquire an international stature, whilst others remain as purely local attractions.

As it is the most established form of development, we will concentrate upon the seaside resort. The historic evolution of seaside resorts has already been traced in Chapter 2, and the processes described there have produced a form of resort development that may be considered 'traditional' and which is widely encountered in countries such as Britain. Such resorts reveal the attraction of the sea as a resource, whilst their complex land patterns point to processes of incremental growth, much of it spontaneous and unplanned, and often in the form of small-scale, local investment.

In general, the importance of the sea within these resorts has been responsible for a common pattern of linear development along the sea frontage itself (Plate 4.2), with pronounced 'gradients' of decline in land values and associated changes in land uses with increasing distance from the front. A secondary gradient of change is also evident along the front as one moves from prime locations at the core of the resort towards the periphery. The natural tendency for certain resort functions – accommodation, tourist retailing and entertainment – to group together for commercial reasons then produces quite well-defined spatial zones within the traditional resort, albeit integrated and interspersed with other non-resort activities; for example, local industry or residents' housing. Zoning may also lead to the formation of a distinctive 'recreational business district' (RBD) that is partially or wholly separate from the normal 'central business district' (CBD) that we may expect to find in any urban place. Furthermore, within tourist zones, competition for prime sites will tend to separate larger enterprises from smaller ones, whilst the particular needs of some sectors (e.g., the dependence of bed and breakfast houses on attracting passing trade) will produce particular locational tendencies within sectors. In the case of bed and breakfast houses, the attraction of positions on main routeways is an observable pattern. For different reasons, low-cost functions that require large areas of land (e.g., holiday villages and caravan parks) gravitate to the edge of the resort where cheaper land is most likely to be available. Figure 4.3 provides a diagrammatic summary of these ideas in the form of a simple descriptive model.

However, as tourism has developed, other forms of beach resort have emerged that do not match the 'traditional' model since development conditions may be different and the extended chronology of change that has shaped the traditional resort is absent. Figure 4.4 summarises a model of beach resort development proposed by Smith (1991) and based upon empirical observation of modern resort formation in the Asia-Pacific region.

Plate 4.2 A traditional pattern of linear development of hotels and attractions along a sea front: Eastbourne, UK

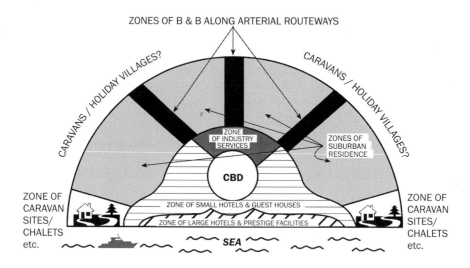

Figure 4.3 Model of a conventional seaside resort

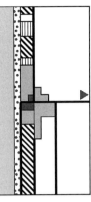
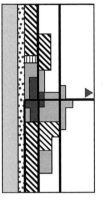
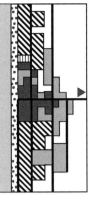
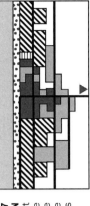
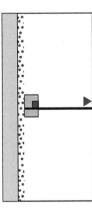
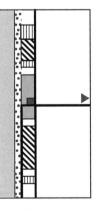

Stage 1
PRE-TOURISM DATUM
No tourism
Settlement in some cases

Stage 2
SECOND HOMES
First tourism development
Low-budget tourism
Second homes along beach
Roads defined
Strip development

Stage 3
FIRST HOTEL
Visitor access improved
First hotel opens
Ad hoc development
High-budget visitors
Jobs in tourism

Stage 4
RESORT ESTABLISHED
More hotels
Strip development intensified
Some houses displaced
Residential expansion
Hotel jobs dominate

Stage 5
BUSINESS AREA ESTABLISHED
More accommodation
Visitor type broadens
Non-hotel business growth
Tourism dominates
Large immigrant workforce
Cultural disruption
Beach congestion and pollution
Ambience deteriorates

Stage 6
INLAND HOTELS
Hotels away from beach
Rapid residential growth
Business district consolidates
Flood & erosion damage potential
Tourism culture dominates
Traditional patterns obliterated
Entrepreneurs drive development
Government master plan

Stage 7
TRANSFORMATION
Urbanised resort
Rehabilitation of natural ambience
Accommodation structural change
Visitors and expenditures change
Resort government fails

Stage 8
CITY RESORT
Fully urbanised
Alternative circulation
Distinct recreational and commercial business districts
Lateral resort spread
Serious pollution
Political power to higher government

Sea   Beach   Businesses   Hotel   Second homes   Residential   Road   ▶ Prime access

Figure 4.4 Smith's model of beach resort formation

This more complex description repays closer attention, but distinctive features that are worth emphasising include:

- the role of second-home development in the early phases of resort development;
- the tendency initially towards linear development along the sea frontage which is reinforced by the first phases of hotel development;
- the processes of displacement of residential properties from the frontage as the tourism industry becomes established;
- the emergence of secondary developments of hotels at inland locations, once the front has become fully developed;
- the eventual separation of a CBD from an RBD in the mature stages of the resort's formation.

Although developed around observation of resort formation in Malaysia, Thailand and Australia, this model can be applied in other areas of recent resort development. Case Study 4.2 presents an outline example of one of the beach resorts studied in the formation of this model, the Thai resort of Pattaya.

CASE STUDY 4.2

Development of a modern beach resort: Pattaya, Thailand

In the 1940s, Pattaya (which lies 140km south-east of Bangkok) was a relatively inaccessible fishing community which contained a handful of second homes established by wealthy Thais. These second homes formed the basis for the development of the resort that was included by Smith (1991) in his study of modern beach resorts and its associated model (see Figure 4.4). Improved road access to Bangkok in the early 1960s coincided with the development of US military bases in the region and created new demands that led to the construction of the first hotels on the beach frontage from 1964. These provided the catalyst to a subsequent expansion of hotel-based tourism catering for both the domestic market and, with time, a progressively more important international market. As Smith's model hypothesises, these developments of hotels and guest houses formed distinctive zones in close proximity to the sea, whilst the demands created by tourism stimulated both the growth of business and increases in the resident community that triggered new construction of residential areas inland.

Smith's (1991) survey of Pattaya revealed a resort that had perhaps reached Stage 6 of his model of beach resort formation (see Figure 4.4) with established zones of coastal hotels, zones of tourism business and more recent zones of hotels at inland locations also in evidence. However, in the twenty years since Smith's original fieldwork was undertaken, Thailand has developed strongly as an international destination with international arrivals having risen from just under 2 million in 1980 to more than 11.6 million in 2004 (Economic Intelligence Unit, 1995; WTO, 2005a). Pattaya has also emerged – along with Phuket and Ko Samu – as a major destination for visitors to Thailand, attracted both by the exotic and often luxurious facilities that are on offer in parts of the resort and by a less commendable reputation for sex tourism. Consequently, the resort has more than tripled in size and, in the process, clearly progressed to the final stage of Smith's model. Figure 4.5 (which was compiled remotely through a combination of online tourist maps and satellite images) shows a generalised land use map of the area in 2006 which conforms to several of the criteria set down by Smith for a 'city resort'. Satellite images reveal a highly urbanised environment with many new road networks and associated infrastructure, whilst the resort shows both

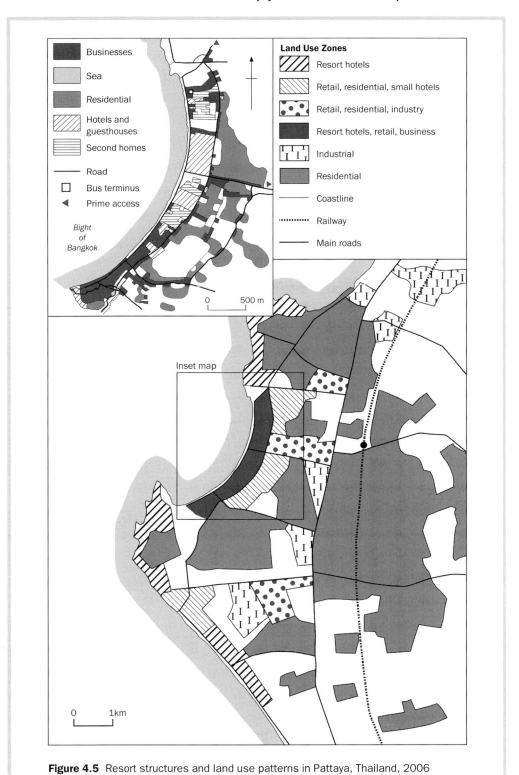

Figure 4.5 Resort structures and land use patterns in Pattaya, Thailand, 2006

continued

lateral expansion into the neighbouring coastal areas of Wong Amat and Jomtien, and a greatly increased zone of inland residential neighbourhoods. The inset in Figure 4.5 shows the resort as originally surveyed in 1991 and reveals, as the model suggests, that many of the functions that marked the formative phases of the resort development have been displaced by new construction of zones of high-quality hotel and condominium developments, the further expansion in tourism-related businesses, and a widening range of attractions that are associated with mature resorts.

Source: Smith (1991)

Tourism zones

In mature tourism destinations, the scale and extent of development will often proceed to the point at which extended zones of tourism emerge. These typically will be formed by combinations of resorts, enclaves and other types of development that form areas of in-filling around primary centres (e.g., villa complexes, holiday villages, caravan sites, attractions and golf courses) to create a landscape that is infused with tourism. In contrast to the other forms discussed above, however, the emphasis in zonal development is upon dispersal rather than concentration, although there may still be concentrations of activity within the zone, usually around urban resorts wherever these are present.

The precise form that such zones may take is variable, reflecting key factors of topography, access, availability of land for development, and planning and investment conditions. However, one of the most characteristic patterns of zonal development is a linear growth along accessible and attractive coastlines. In some instances the topography encourages such growth by creating only narrow coastal strips that are suitable for development, but the attraction of the seashore also tends to encourage linear forms, irrespective of physical constraints. This tendency may then be further reinforced by, for example, construction of coastal roads that link the different elements together. In conditions where local planning control is poorly applied, the negative impacts will often be pronounced. Case study 4.3 provides an example of linear-zonal development from the coast of north-east Wales.

CASE STUDY 4.3

Linear zonal development: the coast of north-east Wales

The primary tourism zone on the coast of north-east Wales extends for some 45km from Point of Air in the east to the Great Orme and Conwy Estuary in the west. Except for an area of more extended lowland around the estuary of the River Clwyd, development has generally been confined to a narrow coastal strip of often no more than 1–2km in width. The southern fringes of this coastal zone are mostly defined by areas of higher ground that quickly attain heights in excess of 200m and which tend to slope steeply towards the sea, providing a significant barrier to urban encroachment.

Tourism development in this region dates from the middle of the nineteenth century. Although steamer services from Liverpool had helped to establish a modest excursion trade to destinations such as Prestatyn during the 1830s, it was the arrival of the Chester and Holyhead Railway in 1848 that enabled the development of a string of new Victorian resorts at Llandudno, Colwyn Bay and Rhyl (see Figure 4.6). Prior to their emergence as

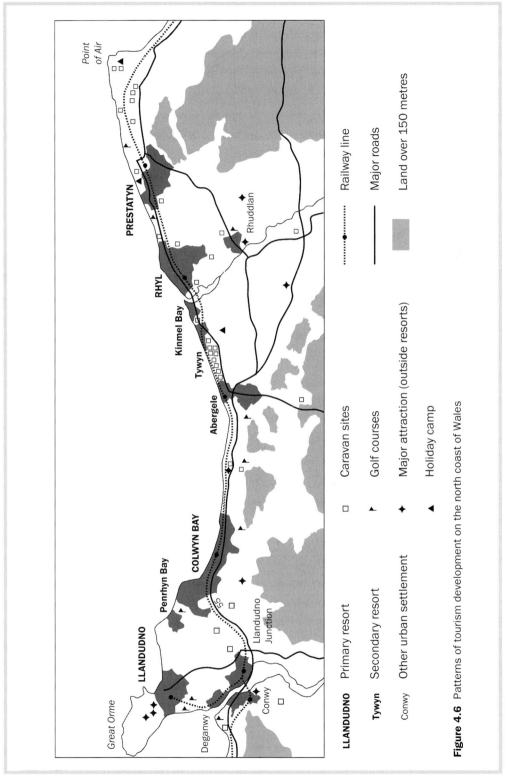

Figure 4.6 Patterns of tourism development on the north coast of Wales

continued

resorts, each of these places comprised little more than a handful of dwellings but under the growing levels of demand from the industrial regions in north-west England, each had become established as a primary coastal resort by the start of the twentieth century with – by local standards – a sizeable permanent population of around 9,000 people each. Alongside the development of these major resorts, a secondary string of smaller tourism places also emerged – often through a 'spill-over' effect – including Kinmel Bay, Tywyn, Abergele, Rhos-on-Sea and Penrhyn Bay, helping to create a semi-continuous zone of urban coastal tourism along the route of the main railway line and the main (A55) coast road.

Subsequent development has served to emphasise the linear qualities of this tourism zone. After 1945, the more-easily developed coastal areas between Abergele and Prestatyn proved to be an attractive location for the development of both permanent and touring caravan sites and, on a more limited basis, holiday camps. Tywyn, in particular, has seen significant development of caravan sites, but as Figure 4.6 illustrates, caravanning is ubiquitous across the region. The impact of these developments – when combined with the parallel development of other tourism amenities (such as golf courses) and the more recent integration of a widening range of heritage and commercial attractions into the framework of tourism in the region – has been to trigger a significant level of in-filling around the older resorts. The composite effect of these different phases of development has been to produce a distinctive zone of tourism that dominates the landscape of this coastal region of the UK.

Tourism and economic development

The physical development of tourism is, of course, linked with a range of environmental and social impacts (see Chapters 5 and 6), but the closest ties are arguably economic in character. Tourism may:

● aid economic development through the generation of foreign exchange earnings;
● exert beneficial effects upon balance of payments accounts;
● create substantial volumes of employment;
● assist in the redistribution of wealth from richer to poorer regions;
● promote and finance infrastructural improvements;
● diversify economies and create new patterns of economic linkage.

Less positively, however, tourism's economic effect may also:

● increase dependence upon foreign investors and companies;
● introduce instabilities and weakness in labour markets;
● divert investment from other development areas.

The relative balance of impact that tourism may exert is strongly dependent upon context and a general note of caution needs to be sounded on tendencies to relate economic effects as being broadly positive or negative in outcome. Identical processes may produce contrasting outcomes in different geographical settings. It is also true that economic impacts are complex and notoriously difficult to isolate and measure. However, there are good grounds to anticipate significant spatial variations in effect according to:

- the geographic scale of development, that is, international, national, regional or local;
- the initial volume of tourist expenditures, which will be primarily shaped by the number of visitors and their market segments – for example, effects differ between low-cost and luxury travel, or mass tourists and independent travellers;
- the size and maturity of the economy, which will affect particularly the ability to supply tourist requirements from within the economy rather than relying upon imports or foreign sources of investment;
- the levels of 'leakage' from the economy. Leakage represents the proportion of revenue which is lost through, for example, the need to import goods and services to sustain the tourism enterprise, or through the payment of profits and dividends to offshore owners or investors. In general, the larger and more developed an economy, the lower the levels of leakage and vice versa.

In addition, there are some general instabilities that affect the performance of tourist economies. Most areas of tourism are subject to the effects of seasonality in demand and whilst some destination areas have developed summer and winter markets (e.g., Alpine Europe) in many tourism regions, climatic constraints produce a pronounced seasonal effect. Figure 4.7 illustrates the seasonality of tourism for a number of destinations and, from an economic perspective, points to the problems of having facilities under-utilised or even closed (and therefore entirely unproductive) for parts of the season. Cutting across such seasonal patterning are more unpredictable fluctuations in demand within the industry which may be seen as a response to a number of potentially disruptive influences, including:

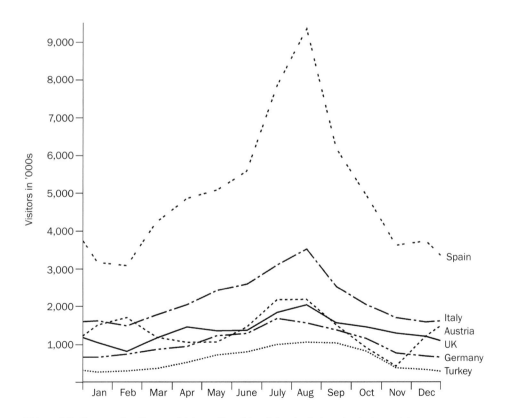

Figure 4.7 Seasonal patterns of international tourist arrivals in selected countries

- economic recession in generating countries;
- changes in the price of holidays consequent upon fluctuations in international monetary exchange rates or price wars within the travel industry;
- changes in costs of transportation, reflecting particularly changes in oil prices and associated costs of aviation fuel;
- short- or medium-term economic and political instability in destination areas;
- warfare and civil unrest;
- negative images stemming from a range of potential problems at destinations, including levels of crime, incidence of illness and epidemics, or even simple decline in fashionability.

With these points in mind, the following section considers some of the primary ways in which tourism affects economic development focusing especially on impacts on balances of trade and investment; economic growth, employment and regeneration.

Tourism impacts on balances of trade and inward investment

The first effect that we may note is the capacity for international forms of tourism to earn foreign currency and to influence a country's balance of payments account (this being the net difference between the value of exports and cost of imports). With a world tourism 'trade' that was valued at US$620 billion in 2004 (WTO, 2005b), the potential for tourism to influence the accumulation of wealth in particular regions is clearly considerable.

Tourism is a sector with high levels of what economic geographers refer to as 'invisible' trade elements, meaning that such trade is not necessarily in tangible (and hence easily measurable) flows of goods. However, we may gain an idea of the spatial patterning in national income and the gains and losses of foreign currency through tourism by comparing what a nation earns through foreign visitors' expenditure with what its own nationals expend when they themselves become tourists to another country. (This is sometimes referred to as the 'travel account'.) Table 4.1 sets out the balance of tourist trade in the top twenty international destinations (as measured by gross tourism receipts) and, by comparing receipts with expenditures, reveals whether countries are in a basic surplus or deficit on their travel account.

Two points are worth noting. First, nations that are conspicuous generators of tourists, especially Germany, Japan, Netherlands and the United Kingdom, tend to be in 'deficit' on this particular form of measurement, Germany, Japan and the UK spectacularly so. Second, there is a marked geographic pattern amongst the European nations, with a clear flow of foreign currency from the northern urban-industrial economies to the Alpine or southern European economies, which, in the latter case, are generally more ruralised and less prosperous. (This illustrates a supplementary advantage that is often claimed to be associated with tourism development, namely that it can be a medium for redistribution of economic wealth from richer to poorer areas.)

However, Table 4.1 tells only a partial story since the raw data take no account of the 'invisible' earnings associated with tourism or of its secondary effects. Invisible earnings are generated through activities such as ownership and control of tour companies, airlines, transport operators, international hotel chains and, less obviously, profits from insurance and banking services that support the international tourism industry. Hence, countries such as Germany and the United Kingdom are able to transform the contribution of tourism to their balance of payments – often from apparent deficit into surplus – through high levels of activity in the invisible sectors of the tourism economy. Similarly, discussions of the contribution of tourism to balance of payment accounts generally fail to take into account secondary effects. These relate to the complex patterns of redistribution of initial

Table 4.1 *International balance of tourism trade, 2004*

Country	Receipts (US$ million)	Gross surplus (US$ million)	Gross deficit (US$ million)
USA	74,481	17,037	
Spain	45,248	36,187	
France	40,842	17,411	
Italy	35,656	15,026	
Germany	27,657		37,032
UK	27,299		20,630
China	25,739	10,552	
Turkey	15,888	13,775	
Austria	15,351	3,590	
Australia	12,952	5,661	
Greece	12,872	10,456	
Canada	12,843		579
Japan	11,202		17,617
Mexico	10,753	4,500	
Switzerland	10,413	2,942	
Netherlands	10,260		4,339
Thailand	10,034	6,539	
Belgium	9,185		2,991
Malaysia	8,198	5,352	
Portugal	7,788	5,378	

Source: WTO (2005a)

expenditures by tourists that occur, for example, through the purchasing of goods and services for consumption by tourists or through the payment of wages and salaries to employees in the tourist industry. In situations where goods, services or labour are purchased from outside the national economy, such transactions have their own impacts on balance of payment accounts (Wall and Mathieson, 2006).

In many contexts, tourism is also valued through its apparent capacity to attract inward investment to finance capital projects. Although the industry is still dominated by small-scale local firms, the trend is towards greater levels of globalisation in the organisation of world tourism and the emergence of large-scale international and multinational operators, each capable of moving significant volumes of investment to new tourism destinations. These firms are distinctive not just because of the manner in which they have extended their horizontal linkages (where firms merge with, take over, or form alliances with other firms operating in the same sector), but especially through the development of vertical linkages in which, for example, an airline purchases or develops its own travel company and takes on ownership of hotels. (The Grand Metropolitan group, for example, has interests in international hotels, holiday camps, travel agencies, package tours and restaurants.)

For developing nations, in particular, the role of foreign investment in initiating a tourism industry through, for example, hotel and resort construction can be an essential first step out of which an indigenous industry may eventually develop. Without foreign investment, start-up capital may not be found locally and although profits from foreign-owned firms will tend to leak out of the local economy, local taxation on visitors and their services may provide initial funding to assist in the formation of new indigenous firms and the development of key infrastructure (roads, water and power supplies) around which further expansion of tourism may then be based.

Set against these presumed advantages, however, are the risks that are associated with increased levels of economic dependence upon foreign companies and investors. Ideally, foreign investment provides a catalyst to growth that will foster the subsequent formation of local enterprise, yet in many emerging nations foreign ownership continues to dominate tourism industries, prompting extensive leakage of revenue and minimising the local economic gain. Problems of dependency are generally greatest in small nations and, especially, small island states (Harrison, 2001b) and there is a wide range of studies of destination areas such as the Caribbean and the Pacific that demonstrate the point (Archer, 1989, 1995; Britton, 1982; Freitag, 1994; Lockhart, 1993; Weaver, 1998; Wilkinson, 1987). For example, amongst some of the Pacific micro-states that are popular with Australian and New Zealand tourists, local ownership of tourism enterprise is minimal. A case study of the Cook Islands conducted in the late 1980s found that tourism businesses belonging to local people received only 17 per cent of tourist expenditure, whilst in Vanuatu over 90 per cent of expenditure was gathered by foreign-owned companies (Milne, 1992). Under these conditions, the beneficial impact of tourism is significantly less than might be imagined from a cursory examination of the flows of tourists.

Tourism and economic growth

Closely associated with the attraction of inward investment is the role that tourism may play in encouraging new economic linkages and increasing the gross domestic product (GDP) of an economy. Tourism's contribution to GDP will vary substantially according to the level of diversity and the extent of economic linkages within an economy. In a developed nation, tourism's contribution to GDP is usually quite small. In contrast, in emerging nations which lack economic diversity or which, perhaps through remoteness, have limited trading patterns, contribution of tourism to GDP can be substantial. Data provided by Wall and Mathieson (2006) show, for example, that in the highly developed economies of Canada and the USA the estimated contribution of tourism to GDP is in the order of 2.4 and 2.2 percent respectively, whereas in the much smaller economies of the Seychelles and the Maldives, the tourism contribution in 2004 was 28.6 per cent for the Seychelles and 41.8 per cent for the Maldives.

The mechanisms by which tourism development may foster new-firm formation and development of new linkages are complex but in simplified form may be envisaged as shown in Figure 4.8. This is a direct adaptation of a model originally developed by Lundgren (1973) to show key stages in the development of entrepreneurial activity in an emerging tourism economy, but is utilised here to show how patterns of economic linkage may evolve in a new destination area. The model reflects a developing-world scenario in which, at an initial stage, local provision is limited and the industry is highly dependent upon overseas suppliers. After a time, numbers of tourism businesses increase and become more spatially spread, profits (or expectations of profit) filter more widely into the local economy and existing or newly formed local firms start to take up some of the supply market. Levels of foreign dependence therefore diminish as these local linkages emerge. Eventually, a mature stage is reached in which a broadly based local tourism economy has been formed with developed patterns of local supply and much reduced dependence on foreign imports to meet tourism demands.

One of the primary ways in which the potential contribution of tourism development to the wider formation of economic growth, inter-firm linkages and the generation of income is assessed is through what is termed the 'multiplier effect'. Multiplier analysis was first applied to tourism by Archer (1973, 1977, 1982) but has since been widely developed as a means to measure the economic impact of tourism within national, regional and even local tourism economies (see, e.g., Archer, 1995; Huse et al., 1998; Khan et al., 1990).

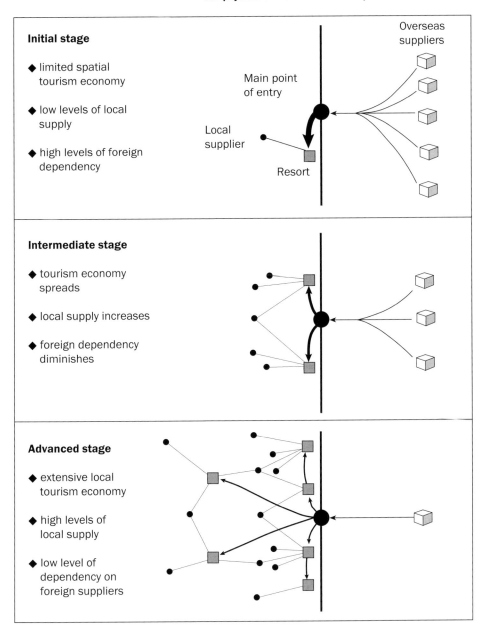

Figure 4.8 Tourism development and the formation of economic linkages

Multipliers attempt to measure the impact of tourist expenditure as it re-circulates within a local economy. Tourist spending is initially introduced as direct payment for goods and services such as accommodation, food, local transport and souvenir purchases. In turn, the providers of these services re-spend a portion of their tourism receipts, for example in making their own purchases, in payment of wages to employees or in taxes to local government. These transactions form further flows of money and extend the indirect linkages of tourism well beyond the immediate core of tourism businesses. This cyclical process is reflected in the recognition of three levels of effect:

- a direct effect, which is the initial injection of revenue to the local economy by the tourist – for example, through payment of an hotel bill;
- an indirect effect, which is represented by a second round of spending by the recipients of initial expenditures in purchasing the goods and services demanded by the tourist – for example, purchase by the hotelier of local supplies of food and drink for the hotel restaurant;
- an induced effect, which is further spending by the beneficiaries of the direct and indirect effects on goods and services for their own consumption – for example, the purchase of clothing by the hotel waiter (see Wall and Mathieson (2006) for a fuller discussion of these mechanisms).

By convention, multipliers are expressed as a ratio in which the expected increase in income associated with a unit of currency is stated. Thus a multiplier of 1.35 would indicate that for every $1 spent, a further $0.35 dollars is generated by indirect and induced effects. However, the scale of the multiplier effect will vary – dependent upon the level of development within the economy, the type of tourism and the extent to which the local economy can supply the tourism industry from its own resources and, thereby, the extent to which the leakage effects may be minimised. In developed destinations multiplier values are generally high whereas in emerging destinations – and especially within the developing world – much lower multipliers are generally encountered. Wall and Mathieson (2006) provide contrasting examples which show tourism multipliers of 1.92 and 1.73 for the USA and the UK respectively, whereas in Fiji and the British Virgin Islands the equivalent values are only 0.72 and 0.58.

However, although there are identifiable pathways through which tourism development can trigger a diffusion of benefits into the wider economy of a destination, the presumed capacities of tourism to generate regional development, redistribute wealth and benefit local economies have also been questioned. Tourism has been widely associated with localised inflation of the price of land, labour costs and prices of goods in the shops, whilst studies of tourism destinations as diverse as the UK and Malaysia suggest that rather than diffusing economic gain into less wealthy, peripheral regions, development tends to refocus growth into areas of existing development. In Britain, traditional domestic tourism regions such as Devon and Cornwall have seen a loss of business with domestic holidaymakers flocking abroad. Ideally, such losses would be counterbalanced by new flows of foreign tourists into the same regions, yet the primary focus for foreign visitors to Britain is London, and only tiny numbers of overseas visitors venture as far as Cornwall. So at a regional level, the gains in foreign tourism in one locality are not compensating for losses in the domestic market elsewhere, thereby sustaining rather than eroding regional disparities (Williams and Shaw, 1995). Similarly, Oppermann's (1992) study of tourism in Malaysia found that although state planning initiatives had prompted some redistribution of tourism, growth remained strongly focused in just three of the country's fourteen regions (Kuala Lumpur, Penang and Pahang) which together accounted for 67.5 per cent of international tourist visits.

Tourism and employment

Many of the wider economic benefits of tourism stem from the fact that the industry can be a significant source of employment. In contemporary consumption-based economies, sectors such as tourism that form central components in processes of consumption create work in both direct employment within tourism businesses (e.g., hotels) and indirect employment in enterprises that benefit from tourism (e.g., general retailing). Moreover, as Shaw and Williams (2004) acknowledge, the process of engaging and shedding labour is a primary mechanism by which tourism businesses adapt to the fluctuating nature of demand for tourism products and is thus a central component in the internal regulation of the industry. This is because labour represents a relatively high proportion of production costs in most sectors of tourism.

In comparison with many modern industries, tourism retains a relatively high demand for labour, based particularly around service work in the accommodation sector, food and beverages, and local transportation. Data on the Canadian tourism industry, for example, suggests that direct employment accounts for over 540,000 jobs – which is indicative of the capacity of the industry to create employment – and that 62 per cent of these jobs were in the three sectors identified previously (Wall and Mathieson, 2006). Much smaller numbers work in the travel industry – within agencies, as couriers and guides, or in tourist information services. In the case of Canada, this sector accounts for just 2 per cent of the tourist jobs. However, across all sectors the dominant component in the workforce is often comprised of temporary or part-time workers who reinforce the roles of smaller groups of permanent employees and a comparatively small number who exercise managerial roles within the industry.

This arrangement of tourism employment has been conceptualised by Shaw and Williams (2002), drawing on earlier work on core and peripheral labour markets proposed by Atkinson (1984). In this model, tourism labour markets are centred around a relatively small core group of permanent, skilled managers and workers that form a primary labour source that is capable of a range of tasks (i.e., it is functionally flexible). Alongside the core are much larger secondary and tertiary groups that are more likely to be composed of relatively low-skilled personnel with more limited capabilities (i.e., functionally inflexible), but probably working part time and therefore in sectors that are numerically flexible in their size and composition. Flexibility in the secondary labour market typically extends to importation of labour, and hence employment migration is often a distinctive geographic dimension in tourism economies. As tourism – and the labour markets on which it draws – have become more globalised in their composition and operation, so the geographical range over which migrant tourist labour may travel to secure work has become similarly extended.

These structural characteristics are important since it means that major elements in tourism labour forces may be formed relatively quickly, with only modest levels of training and, equally, may be rapidly adjusted (through hiring or shedding of labour) to reflect fluctuations in the market. From the perspective of developers and employers these represent considerable advantages but the fact that many sectors of tourism offer comparatively easy entry to employment may also be perceived as advantageous for workers.

The issue of the strengths and weaknesses of tourism as an area of employment has been debated in the tourism literature for many years (see, e.g., Mathieson and Wall, 1982; Choy, 1995; Thomas and Townsend, 2001; Riley et al., 2002; Wall and Mathieson, 2006). On the negative side, tourism work has been widely represented as:

- low-paid;
- menial and unskilled;
- part-time and seasonal;
- strongly gendered through an over-dependence upon female labour.

The seasonality of tourism employment is an inevitable product of the natural rhythms of tourism activity whilst the high incidence of part-time work is, as has been noted above, a primary mechanism in enabling the supply of labour to be matched to fluctuating levels of tourism demand. Low wages reflect two key factors. First, because labour comprises a relatively high proportion of total production costs in tourism, there is a consequent tendency to produce considerable downward pressure on wages (Shaw and Williams, 2004). Second, there is often a direct link between wage levels and skill, so that basic tasks such as room cleaning will tend to attract low wages. Whilst such descriptions over-simplify a complex labour market and disregard the presence of a core of employees who fit none of these categories, many tourism jobs do suffer from some (or all) of these characteristics. Studies of tourism employment in Africa, for example, show a recurring pattern with local labour placed into the low-pay, low-skill jobs, whilst positions with responsibility, higher earnings and prospects for advancement tend to go to foreign workers who possess appropriate skills and training (Dieke, 1994, 2002; Poirier, 1995). But these problems are not confined to developing world scenarios, as low pay, temporary employment and poor working conditions have been widely associated with tourism employment in Europe and North America too (Baum, 1996; Thomas and Townsend, 2001).

However, although issues of low pay, low skill and the high incidence of part-time work are widely encountered in many sectors of tourism employment, there are still evident benefits. This is especially true when development is placed in a local context and viewed from the perspective of the employee. Thus, for example, Choy's (1995) study of tourism employment in Hawaii found very high levels of dependence upon tourism in an area in which alternative work was often hard to locate and, consequently, a high level of job satisfaction amongst people working in the industry. Similarly, a comparative study of Florida and Fiji conducted by Pizam et al. (1994) revealed positive perceptions amongst local communities of the employment that tourism created, particularly where alternative work in sectors such as agriculture actually offered poorer rewards and prospects. Riley et al. (2002) also point out that the perceived deficiencies of low pay may be off-set in the minds of employees by the non-material rewards that might be associated with working in attractive locations or smart hotels and restaurants. Of course, not all locations are attractive and not all hotels are smart, but in a globalising tourism industry with high levels of labour mobility, it is important to appreciate that the perceptions of the quality of tourism employment need to reflect the conditions in which workers originate as well as those in which they work.

In a similar vein there has often been a tendency to problematise the incidence of part-time work and the reliance on female labour, yet this overlooks the fact that for a growing number of people, part-time work is often a preferred mode of employment within lifestyles that are shaped more by practices of leisure and consumption than by the world of work (Haworth, 1986; Reid and Mannell, 1994; Franklin, 2004). The impacts of the gendered nature of tourism employment has also been reappraised in light of a growing body of evidence which reveals that although employment of women in many low-skill, low-pay sectors of tourism is prevalent, tourism has also created new employment opportunities and new levels of independence from traditional family roles for women through tourism work. This has been noted especially in Mediterranean countries such as Greece and Cyprus, as well as the Caribbean (Tsartas, 1992; Leontidou, 1994; Sinclair and Stabler, 1997).

Tourism as an agent of regeneration

Finally we should note that alongside the macro-level effects on national economies and the potential for tourism to create significant opportunities for work, tourism may play a key role in processes of economic regeneration or in the provision of new support for marginal economies through diversification. These effects have been evident for some time within rural economies. In Britain, in regions such as Wales or Devon and Cornwall, less profitable hill farm economies, or dairy farms where profits have been limited by EU production quotas, have been widely sustained by development of farm holidays and tourism activities: fishing, riding, shooting, self-catering facilities, bed and breakfast businesses, caravanning and camping (Busby and Rendle, 2000; McNally, 2001).

More recently, however, tourism-based regeneration and diversification have been recognised in new forms of urban tourism. The need for urban regeneration has been widely experienced across the many cities in the developed world that have fallen victim to the loss of manufacturing industries and more general processes of deindustrialisation. Regeneration projects that include a strong tourism dimension are seen as bringing a number of tangible and intangible benefits, including:

- job creation;
- new-firm formation and increased investment opportunities;
- enhancement of image and associated opportunities for place promotion;
- creation of new economic spaces within regeneration zones;
- environmental enhancement.

Although regeneration schemes have sometimes been criticised for problems such as a failure to provide adequate returns on investment or a lack of recognition of the interests of local people, the popularity of tourism-led regeneration with urban developers remains undiminished.

One of the most interesting consequences of tourism-led regeneration has been the way in which the active promotion of urban business tourism (conferences and conventions, etc.), sport and event-related tourism and the development of new attractions centred around leisure shopping or industrial heritage, has permitted places with no tradition of tourism to develop a new industry that has revitalised flagging local or regional economies (Goss, 1993; Hiller, 2000; Jackson, 1991; Page, 1990; Robinson, 1999). The capacity of tourism to contribute to regeneration was first demonstrated in the USA in some high-profile redevelopments such as the Inner Harbour at Baltimore (Blank, 1996) but has since been widely applied in deindustrialising regions in other parts of the world (Williams, 2003). In the UK, for example, tourism and visiting to Manchester (which, as a city, has been revitalised by several major regencrative projects) brings an estimated £400 million annually from more than 3.2 million staying visitors (Law, 2000). This theme is pursued in greater detail in Chapter 9.

Summary

Processes of physical and economic development are the most visible ways in which tourism affects host areas. This chapter attempts to define the primary factors that will shape patterns of tourism development and show how they may combine to produce spatially contrasting forms. However, such developments not only alter the physical environments of destinations but also exert a range of economic effects too. These will vary from place to place, depending upon the form of tourism development that is occurring and the nature of the national or local economy in question, but could include a range of impacts upon balance of payments accounts, national and regional economic growth, and the creation of employment. Unfortunately, the instabilities of tourism that make it vulnerable to a range of influences (e.g., exchange rate or oil price fluctuations, political crises, changes in fashion) mean the industry is not always able to provide a firm basis for economic development. For Third World countries, tourism may increase levels of foreign dependence and in many contexts the quality of employment that the industry creates is low, although it is important to appreciate that the presumed strengths and weaknesses of tourism's economic effect are often highly contingent upon local conditions and, especially, the perspective from which the effects are judged.

Discussion questions

1   What are the principal elements that are needed to secure the physical development of a tourism destination?
2   How do variations in local conditions produce contrasting spatial patterns of tourism development?
3   Using examples of both established and emerging resorts, examine the validity of the models of resort structures provided in Figures 4.3 and 4.4.
4   What are the main strengths and weaknesses of tourism as a means of economic development?
5   Evaluate the potential of tourism as a source of local employment.

Further reading

Although not a recent publication, excellent coverage of the processes of physical development of resorts and tourism zones is contained in:
Pearce, D.G. (1989) *Tourism Development*, Harlow: Longman.

Excellent overviews of the range of economic effects associated with tourism are provided by:
Shaw, G. and Williams, A.M. (2004) *Tourism and Tourism Spaces*, London: Sage, especially Chapters 2 to 4.
Wall, G. and Mathieson, A. (2006) *Tourism: Economic, Physical and Social Impacts*, Harlow: Prentice Hall.

General discussions of tourism economics and development in a range of settings in both the developed and the developing world, together with good case studies, are provided in:
Go, F.M. and Jenkins, C.L. (eds) (1998) *Tourism and Economic Development in Asia and Australasia*, London: Pinter.
Harrison, D. (ed.) (2001) *Tourism and the Less Developed World: Issues and Case Studies*, Wallingford: CAB International.
Williams, A.M. and Shaw, G. (1998) *Tourism and Economic Development: European Experiences*, Chichester: John Wiley.

Issues around tourism employment are explored in detail in:
Riley, M., Ladkin, A. and Szivas, E. (2002) *Tourism Employment: Analysis and Planning*, Clevedon: Channel View.
Shaw, G. and Williams, A.M. (2002) *Critical Issues in Tourism: A Geographical Perspective*, Oxford: Blackwell, Chapter 7.

Useful case studies of more specific aspects of tourism and economic development in contrasting destinations include:
Church, A. and Frost, M. (2004) 'Tourism, the global city and the labour market in London', *Tourism Geographies*, Vol. 6 (2): 208–28.
Diagne, A.K. (2004) 'Tourism development and its impacts in the Senegalese Petite Cote: a geographical case study in centre-periphery relations', *Tourism Geographies*, Vol. 6 (4): 472–92.
Mbaiwa, J.E. (2005) 'Enclave tourism and its socio-economic impacts in the Okavango Delta, Botswana', *Tourism Management*, Vol. 26 (2): 157–72.
Nepal, S.K. (2005) 'Tourism and remote mountain settlements: spatial and temporal development of tourist infrastructure in the Mt. Everest region, Nepal', *Tourism Geographies*, Vol. 7 (2): 205–27.
Simpson, P. and Wall, G. (1999) 'Consequences of resort development: a comparative study', *Tourism Management*, Vol. 20 (3): 283–96.

5 Tourism, sustainability and environmental change

'The environment, be it predominantly natural or largely human-made, is one of the most basic resources for tourism and a core element of tourism products' (Wall and Mathieson, 2006: 154). It is crucial to determining the attractiveness of most destination areas and forms an essential 'backdrop' for the majority of tourist activities (Farrell and Runyan, 1991). From the earliest times, the enjoyment of 'environments' – whether defined in physical or in socio-cultural terms – has had a major impact in shaping a succession of tourism geographies and as public tastes for different kinds of leisure environment have developed through time – for example, through the formation of resorts or the changing preferences for scenic landscapes in the nineteenth century, or the quest for amenable climates or the attraction of historic heritage in the twentieth century – so new spatial patterns of interaction between people and environments have been formed.

However, tourism–environment relationships are not just fundamental, they are also highly complex, although the level of complexity has probably evolved through time as levels of activity and the spatial extent of tourism has increased. Page and Dowling (2002) suggest that the relationship between tourism and the environment might initially be characterised as one of '*coexistence*'. This implies that whilst tourist activities were not necessarily fully compatible with their environments, neither did they initiate damaging impacts and might actually have delivered some benefits. By the 1970s, though, the expansion of mass forms of international tourism had raised growing levels of awareness of the role of tourism in promoting environmental change and its considerable capacity to destroy the resources upon which it depends. Under these conditions the relationship between tourism and environment has often evolved from one of coexistence to one of '*conflict*' (Page and Dowling, 2002). The notion of tourism and environment in conflict spawned a burgeoning literature on tourism impacts that emerged during the 1980s and the 1990s and which is conveniently and well-summarised in Mathieson and Wall's (1982) benchmark text and in subsequent work around the same theme (e.g., Hunter and Green, 1995).

However, as Butler (1991) reminds us, because tourism is not a homogeneous activity, different types of visitor create contrasting demands and impacts upon resources and areas. Moreover, because the places that tourists visit are themselves highly variable in their capacity to withstand use, the character of the relationship between tourism and environment is seldom consistent from place to place (Wall and Mathieson, 2006). This has encouraged recognition of a third form of relationship between tourism and environment – one of '*symbiosis*' – in which mutual benefits for both tourism and the environment are derived from the relationship (Romeril, 1985). In England and Wales, for example, the designation of national parks came about partly because these high-quality environments were seen as potentially valuable areas for tourism and the argument for their conservation was strengthened accordingly (MacEwan and MacEwan, 1982). Similarly, there is no doubt that the cause of wildlife preservation in East Africa has been assisted by the growing popularity of safari holidays to the same region and the realisation of the benefits that tourism can bring to local communities (Sindiga, 1999).

Although the characterisation of the tourism–environment relationship in terms of co-existence, conflict and symbiosis is useful in drawing some important distinctions, Dowling (1992) proposes – quite correctly – that in reality all three conditions are likely to occur in parallel, the balance of emphasis and effect being largely dependent upon how the relationship is managed. This concept is important to the overall approach within this chapter which seeks to understand the relationship between tourism and the physical environment not in terms of a simple, bi-polar framework of impacts as being necessarily either positive or negative, or of management approaches as being either sustainable or unsustainable, but rather as a relationship that is infinitely variable and essentially dependent on the local conditions under which development is taking place. It is also essential – given current levels of interest and concern – to ground the discussion of tourism and environmental change in the overarching concept of sustainable development.

The concept of sustainable development

The modern concept of sustainable development originates with the report of World Commission on Environment and Development (1987) (also known as the Brundtland Commission) and which offered the now-familiar definition of sustainability as 'development that meets the needs of the present without compromising the ability of future generations to meet their own needs'. According to Wall and Mathieson (2006: 289) the key elements in the Brundtland approach to sustainable development are that it should:

- maintain ecological integrity and diversity;
- meet basic human needs;
- keep options open for future generations;
- reduce injustice;
- increase self-determination.

Sustainable development principles also support the empowerment of people to be involved in decisions that influence the quality of their lives and enable cultures to be sustained. By these means, developments that are truly sustainable will meet the essential criteria of being economically viable, environmentally sensitive and culturally appropriate.

From certain perspectives, the concept of sustainable development appears to offer little more than a new reading of some well-established practices, especially in so far as some of the principles that it espouses simply articulate a form of prudent resource management that has been widely and effectively practised in areas such as agriculture for many centuries. Butler (1991), for example, observes that royal hunting forests in twelfth-century England were managed in ways that we would now define as 'sustainable' whilst there is an interesting and compelling argument to be made that the growth of urban seaside resorts as centres of mass tourism after 1850 also represented a highly sustainable form of development that was not only able to absorb a rapidly expanding market, but also to both maintain and contain its activity over many decades.

However, the proponents of sustainable development will surely point to the modern concept as presenting a much more holistic vision of how development should be organised (embracing, in an integrated fashion, the political, social, cultural, economic and ecological contexts), and being informed both by a stronger ethical dimension and by possessing a clearer emphasis upon the adoption of long-term views of developments and their potential impacts (Sharpley, 2000). The concept implicitly recognises that there are basic human needs (e.g., food, clothing, shelter) that processes of development must match and that these needs are to be set alongside aspirations (e.g., to higher living standards, security and access to discretionary elements such as tourism) that it would be desirable to match. But since there are environmental limitations that will ultimately regulate the levels to which development can actually proceed (and if principles of sustainability are also to embrace implicit notions of equity in access to resources and the benefits that they bring), then the achievement of sustainable development requires a realignment in attitudes and beliefs that mark this approach out as being fundamentally different.

Although there is inherent logic to the sustainable development approach, the concept has nevertheless been subject to some quite significant criticisms. The outwardly simple definition of sustainability provided by the Brundtland Commission conceals much controversy and debate over who defines what is, or is not, sustainable and what sustainable development might therefore mean in practice. Whilst it is generally recognised that the term has become an essential item in the vocabulary of modern political discourse, it has also come to be used in 'meaningless and anodyne ways' (Mowforth and Munt, 2003: 80). For some critics the lack of conceptual clarity is compounded by a basic ambiguity in the term itself: the concept of 'sustainability' implying a steady-state, whereas 'development' implies growth and change (Page and Dowling, 2002). Wall and Mathieson (2006) suggest that the reconciliation of this apparent tension can only be achieved by placing an emphasis on one or other of the component words to help clarify the approach, and perhaps for this reason, the concept of sustainability has acquired a diversity of interpretations. These range from, at one extreme, a 'zero-growth' view that argues that all forms of development are essentially unsustainable and should therefore be resisted, to very different perspectives that argue for growth-oriented resource management based around the presumed capacities of technology to solve environmental problems and secure a sustainable future. Such flexibility in interpretation whilst, at one level, constituting a weakness, may also be seen as a strength if it allows differing perspectives to co-exist under the broad umbrella of 'sustainability'. Imprecision can be easily translated into flexibility. Both Hunter (1997) and Sharpley (2000) have therefore suggested that the idea of sustainable development can be conceived as what they label 'an adaptive paradigm' that establishes a set of meta-principles within which contrasting development approaches may legitimately co-exist.

From the preceding discussion, the relevance of sustainable forms of development to tourism should be obvious, given that it is an industry with a high level of dependence upon 'environments' as a basic source of attraction but also one that has, as we will see, a considerable capacity to stimulate a significant degree of environmental change. Tourism therefore needs to be sustainable even though – as with the wider concept of sustainable development – there are difficulties of definition and, especially, significant challenges in turning a theory of sustainable tourism into practice.

A number of difficulties around the concept of sustainable tourism have been noted. First, Wall and Mathieson (2006) emphasise that since sustainable development is an holistic concept, any approach that deals with a single sector (such as sustainable tourism) raises the risk that one system is sustained at the expense of another. Hunter (1995) offers a similar criticism of many early sustainable tourism initiatives that, he asserts, failed to place tourism development into the wider contexts of development and environmental change.

In developing this point, second, Sharpley (2000) notes that holistic approaches are difficult to implement in sectors (such as tourism) that are characteristically fragmented and therefore dependent upon large numbers of small, independent enterprises (not to mention their customers) adopting sustainable principles and practices in a coordinated fashion. Perhaps for this reason, many sustainable tourism projects have been implemented only at a local level, characterised by Wheeller as 'micro solutions struggling with a macro problem' (cited in Clarke, 1997). Indeed, it is perhaps a more damning criticism of some sustainable tourism approaches that they implicitly seem to reject the notion that mass forms of tourism can be sustainable, even to the extent that mass and sustainable forms of tourism have been represented in some readings not only as polar opposites, but also as being characterised as – respectively – 'bad' and 'good' forms of tourism.

Third, and perhaps because of this type of dichotomised reading of sustainable tourism, the concept has become widely confused with a plethora of alternative forms of tourism and their associated labels. As we will see later in this chapter, a diversity of alternatives – many of which are focused around the enjoyment of nature – have emerged over the last two decades or so and with which sustainable tourism has been widely confused. 'Responsible tourism', 'soft tourism', 'green tourism', 'ecotourism', 'nature tourism', 'ethical tourism' and, of course, 'sustainable tourism' are all epithets that have been applied to new styles of travel. But whilst many of these forms of tourism may indeed embrace most of the preferred attributes of sustainable tourism, sustainability is not confined to alternative travel, nor is it necessarily a characteristic, as the growing body of research literature on the unsustainable nature of alternative tourism makes clear. (These issues are considered more fully in the final section of this chapter.)

Finally, it may be noted that one of the primary practical barriers to the development of sustainable forms of tourism is embedded in the nature of tourism consumption itself. As we have seen in Chapter 1, tourism is widely perceived by tourists as a means of escape from routines and typically as a hedonistic experience in which behavioural norms are frequently suspended in favour of excessive patterns of expenditure and consumption. In this context, the prudence and social responsibility that is implicit in most understandings of sustainability sits uneasily. This has prompted a number of writers, for example McKercher (1993a), to argue that there is little evidence of a widespread propensity amongst tourists to adopt sustainable tourism lifestyles, even though the encouragement of changed patterns of behaviour on the part of tourists is often an integral objective of sustainable tourism policies.

However, although doubts about the true sustainability of tourism have been widely aired in the academic literature (see, *inter alia*: McKercher, 1993b; Weaver, 2000; Hardy and Beeton, 2001), the concept remains very much at the forefront of current thinking around the theme of tourism and environmental change and although there are divergent perspectives, there is still a consensus that proposes that a sustainable approach – one which manages growth within acknowledged resource conservation limits – offers the best prospects for continued tourism development. Sustainable tourism needs therefore to develop in ways that:

- ensures that renewable resources are not consumed at a rate that is faster than rates of natural replacement;
- maintains biological diversity;
- recognises and values the aesthetic appeal of environments;
- follows ethical principles that respect local cultures, livelihoods and customs;
- involves and consults local people in development processes;
- promotes equity in the distribution of both the economic costs and the benefits of the activity amongst tourism developers and hosts (Murphy, 1994).

Tourism and environmental change

The challenge to tourism that is presented by the sustainable development agenda will become clearer if we move to consider in more detail how tourism relates to environmental change. It has already been intimated that the basic complexities of tourism–environment relationships are compounded by the diverse nature of those impacts and the inconsistencies through time and space in their causes and effects. But it is also important to note that the effects of tourism upon the physical environment are often partial, and one of the practical difficulties in studying those impacts is to disentangle tourist influences from other agencies of change that may be working on the same environment. So, for example, the beach and inshore water pollution that developed as a serious environmental problem along parts of the Italian Adriatic coast in the late 1980s was partly attributable to the presence of tourists but was also a consequence of the discharge of considerable volumes of urban, agricultural and industrial waste into the primary rivers that drain to this sea (Becheri, 1991).

The diversity of environmental impacts of tourism and the seriousness of the problem vary geographically for a number of reasons. First, we need to take account of the nature of tourism and its associated scales of effect. Impact studies often make the erroneous assumption that tourism is a homogeneous activity exerting consistent effects, but, as we have seen in Chapter 1, there are many different forms of tourism and types of tourist. The mass tourists who flock in their millions to the Spanish Mediterranean will probably create a much broader and potentially more serious range of impacts than will small groups of explorers trekking in Nepal, although paradoxically, where mass forms of tourism are well planned and properly resourced, the environmental consequences may actually be less than those created by small numbers of people visiting locations that are quite unprepared for the tourist. For example, depletion of local supplies of fuel wood and major problems of littering have been widely reported along the main tourist trails through the Himalayan zone in Nepal (Hunter and Green, 1995).

Second, it is important to take account of the temporal dimensions. In many parts of the world, tourism is a seasonal activity that exerts pressures on the environment for part of the year but allows fallow periods in which recovery is possible. So, there may be short-term/temporary impacts upon the environment that may be largely coincident with the tourist season (such as air pollution from visitor traffic) or, more serious, long-term/permanent effects where environmental capacities have been breached and irreversible changes set in motion (e.g., reductions in the level of biodiversity through visitor trampling of vegetation).

Third, diversity of impacts stems from the nature of the destination. Some environments (e.g., urban resorts) can sustain very high levels of visiting because their built infrastructure makes them relatively resilient or because they possess organisational structures (such as planning frameworks) that allow for effective provision for visitors. In contrast, other places are much less robust, and it is perhaps unfortunate that a great deal of tourist activity is drawn (by tastes, preferences and habits) to fragile places. Coasts and mountain environments are popular tourist destinations that are often ecologically vulnerable, and even non-natural resources can suffer. Historic sites, in particular, may be adversely affected by tourist presence and in recent years attractions such as Stonehenge in England, the Parthenon in Greece and the tomb of Tutankhamen in Egypt have all been subjected to partial or total closure to visitors because of negative environmental effects.

In exploring the environmental impacts of tourism, it is helpful to adopt a holistic approach to the subject. Environments, whether defined as physical, economic or social entities, are usually complex systems in which there are interrelationships that extend the final effects of change well beyond the initial cause. Impact often has a cumulative dimension in which secondary processes reinforce and develop the consequences of change, so treating individual problems in isolation ignores the likelihood that there is a composite impact that may be greater than the sum of the individual parts. As an illustration of this idea, Figure 5.1 shows how the initial effects of trampling of vegetation by tourists become compounded through related processes of environmental change that may, in extreme circumstances, culminate in the collapse of local ecosystems.

A second advantage of a holistic approach is that it encourages us to work towards a balanced view of tourism–environment relationships. The temptation is to focus upon the many obvious examples of negative and detrimental impacts that tourism may exert, but,

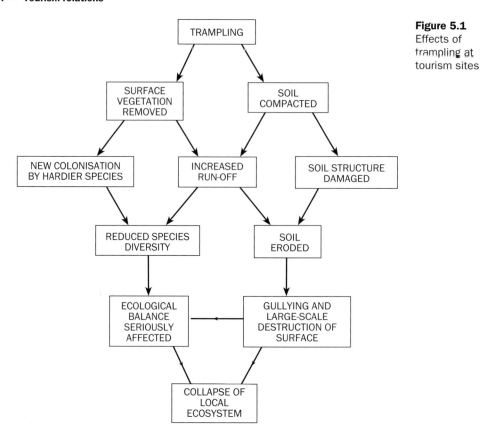

Figure 5.1
Effects of trampling at tourism sites

as the concept of a symbiotic relationship makes clear, there are positive effects too. These might be represented in the fostering of positive attitudes towards environmental protection/ enhancement or be reflected more practically in actual investment in environmental improvement that restores localities for resident populations as well as providing support for tourism.

The third advantage of a holistic approach is that it recognises the breadth (some might say the imprecision) of the term 'environment' and the fact that different types of impact are likely to be present. As is perhaps implicit in the preceding discussion, the term can embrace a diversity of contexts – physical ecosystems; built environments; or economic, social, cultural and political environments – and tourism has the potential to influence all of these, in varying degrees. The economic and socio-cultural dimensions of sustainability are discussed elsewhere in the book, so the discussion that now follows focuses upon the influences that tourism may have upon physical environments, ecosystems and the built environment, together with a consideration of ways in which symbiotic relationships between tourism and the environment may be sustained through managed approaches.

Table 5.1 summarises a representative cross-section of pathways through which tourism may promote environmental change in the physical environment (both natural and non-natural). This classification is informed by several key works on this subject (Mathieson and Wall, 1982; Hunter and Green, 1995; Wall and Mathieson, 2006) and, it may be noted, retains – purely as a convenient frame for the structuring of ideas – the characterisation of change as being essentially positive or negative. It is perhaps appropriate, therefore, to remind readers of earlier cautionary comments about the risks of simple categorisations and the need to recognise that impacts are seldom consistent in their nature across all contexts

Table 5.1 A 'balance sheet' of environmental impacts in tourism

Area of effect	Negative impacts	Positive impacts
Biodiversity	Disruption of breeding/feeding patterns Killing of animals as leisure (hunting) or to supply souvenir trade Loss of habitats and change in species composition Destruction of vegetation	Encouragement to conserve animals as attractions Establishment of protected areas to meet tourist demand
Erosion and physical damage	Soil erosion Damage to sites through trampling Overloading of key infrastructure (e.g., water supply networks)	Tourism revenue to finance ground repair and site restoration Improvement to infrastructure prompted by tourist demand
Pollution	Water pollution through sewage or fuel spillage and rubbish from pleasure boats Air pollution (e.g., vehicle emissions) Noise pollution (e.g., from vehicles or tourist attractions such as bars Littering	Cleaning programmes to protect the attraction of locations to tourists
Resource base	Depletion of ground and surface water Diversion of water supply to meet tourist needs (e.g., swimming pools) Depletion of local fuel sources Depletion of local building material sources	Development of new/improved sources of supply
Visual/structural change	Land transfers to tourism (e.g., from agriculture) Detrimental visual impact on natural and non-natural landscapes through tourism development Introduction of new architectural styles Changes in urban functions Physical expansion of urban areas	New uses for marginal or unproductive land Landscape improvement (e.g., to clear urban dereliction) Regeneration or modernisation of built environment and reuse of disused buildings

Sources: Mathieson and Wall (1982); Hunter and Green (1995); Wall and Mathieson (2006)

of development. Five key headings under which tourism effects may be grouped are proposed.

Biodiversity

Under the heading of biodiversity are located a number of effects that broadly impact upon the flora and fauna of a host region. The potential areas of positive influence on environmental change through tourism mainly relate to the ways in which tourism provides both the impetus and the financial means to further the conservation of natural areas and the species they contain through the designation of protected zones and the implementation of new programmes of land management (Wall and Mathieson, 2006). The scope for tourism to provide economic support for conservation has been illustrated in areas as diverse as, for example, Australia, Brazil, Greece and Kenya (Craik, 1994; de Oliveira, 2005; Okello, 2005; Svoronou and Holden, 2005).

However, arguably the more commonplace patterns are those associated with varying forms of damage to biodiversity. Most widely, processes of tourism development (construction of hotels and apartments, new roads, new attractions, etc.) can result in a direct loss of habitats. In the Alps, extensive clearance of forests to develop ski-fields and the loss of Alpine meadows with particularly rich stocks of wild flowers to new hotel and chalet construction has significantly altered ecological balances and, in the case of deforestation, greatly increased risks associated with landslides and snow avalanches (Gratton and van der Straaten, 1994).

At a more localised scale, other impacts become apparent. Destruction of vegetation at popular visitor locations through trampling or the passage of wheeled vehicles is a common problem. Typically, trampling causes more fragile species to disappear and to be replaced either by bare ground or, where regeneration of vegetation is possible, by more resilient species. The overall effect of such change is normally to reduce species diversity and the incidence of rare plants which, in turn, may impact upon the local composition of insect populations, insectivorous birds and possibly small mammals for which plant and insect populations are key elements in a food chain.

Larger animals may be affected in different ways by tourism, even within environments that are protected. Reynolds and Braithwaite (2001) have developed a detailed summary of how tourist engagement with wildlife can initiate important behavioural changes in animals and alter the structure of animal communities. Some of the effects relate to modification of habitats through actions such as land development, reduction in plant diversity or pollution, but perhaps of greater importance is the potential to modify behaviours and introduce new levels of risk to animal communities. Behavioural modifications that Reynolds and Braithwaite associate with increased levels of tourist engagement include:

- disruption of feeding and breeding patterns;
- alterations to dietary patterns where animals are fed by tourists;
- increased instance of animal migration;
- increased levels of aberrant behaviour;
- modification of activity patterns, such as a raised incidence of nocturnalism;

whilst noted risks include:

- reduced levels of health and conditioning;
- reduced levels of reproduction;
- increased levels of predation, especially of young animals where parents are suffering frequent disturbance through the presence of tourists.

Erosion and physical damage

The impacts of tourism upon the diversity of flora and fauna link with the second area of concern, erosion and physical damage, and this illustrates how environmental problems tend to be interlinked. Erosion is typically the result of trampling by visitors' feet, and, whilst footpaths and natural locations are the most likely places for such problems to occur, extreme weight of numbers can lead to damage to the built environment The Parthenon in Athens, for example, not only is under attack from airborne pollutants but also is being eroded by the shoes of millions of visitors. However, in such situations, tourism can have positive impacts, for although the activity may be a major cause of problems, revenue generated by visitors may also be a key source of funding for wider programmes of environmental restoration.

A more common problem is soil erosion, and Figure 5.1 shows how the systematic manner in which the environment operates actually transmits the initial impact of trampling to produce a series of secondary effects which may eventually exert profound changes upon local ecosystems, leading to fundamental change. Localised examples of such damage can be spectacular. In north Wales, popular tourist trails to the summit of Snowdon now commonly reveal eroded ground that may extend to 9m in width, whilst localised incidence of soil erosion and gullying has lowered some path levels by nearly 2m in a little over twenty years.

Case Study 5.1 illustrates a range of observed impacts upon both vegetation profiles and ground conditions in tourist camping areas of Warren National Park, Western Australia.

CASE STUDY 5.1

Biophysical impacts of camping in Warren National Park, Western Australia

Warren National Park covers an area of some 3,000ha in the south-west of the state of Western Australia. The park straddles a section of the Warren River and contains significant tracts of established eucalyptus forest that draw in excess of 120,000 visitors per annum from both domestic Australian and international tourism markets. The focus of visitor activities is upon informal outdoor recreations in which the enjoyment of the natural environment is a dominant motive and alongside popular recreations such as walking and picnicking there is provision for overnight camping in three formal and nine informal camp sites. The formal sites generally offer easy access via gravelled tracks and contain cleared ground on which to pitch tents together with basic infrastructure such as tables, benches and toilet facilities. In contrast, the informal sites are only accessible by use of off-road vehicles and tents are pitched on ground that has already been trampled or cleared by previous users. There is no other infrastructure.

A key part of this study was to examine the nature and extent of biophysical impacts on the environment of camping activity, considering both types of sites and with comparisons with a third (control) location on land that was unused. The research concluded that even at the low levels of visitor use that were encountered in some of the informal sites, the activity of campers introduced significant changes to the environment in the vicinity of all sites. The principal findings were that:

- all sites showed significant increases in the level of compaction of the soil (which increased on average by 304 per cent at formal sites and 172 per cent at informal sites);

continued

- soil compaction was further adjudged to be responsible for higher levels of surface run-off of rainfall creating areas of gullying and exposure of tree roots, especially along riverbanks;
- there was significant loss of native vegetation cover (61 per cent at formal sites and 51 per cent at informal sites) and a commensurate increase in either bare ground or ground occupied by weeds;
- due to campers collecting firewood (even when wood was provided at managed sites), there was a general absence of woody debris on the ground (which affects habitats for some species of insect) and a higher level of damage to trees;
- campsites adjacent to the Warren River revealed degradation of riverbanks (including root exposure and bank collapse) as well as changes in the species composition of riverbank plants arising from trampling.

Although these findings relate to a specific location in Australia, the authors draw a number of direct parallels between their findings and similar studies conducted in the USA and Canada that suggest issues of compaction, increased run-off and deleterious effects upon vegetation through the use of natural areas by tourists are essentially ubiquitous.

Source: Smith and Newsome (2002)

Pollution

The environmental impacts of which the tourist is probably most aware are those associated with pollution, particularly the pollution of water. With so much tourism centred in or around water resources, pollution of water is a major concern. Poor-quality water may devalue the aesthetic appeal of a location and be a source of water-borne diseases such as gastro-enteritis, hepatitis, dysentery and typhoid. Visible water pollutants (sewage, organic and inorganic rubbish, fuel oil from boats, etc.) will also be routinely deposited by wave action onto beaches and shorelines, leading to direct contamination, noxious smells and visually unpleasant scenes.

Pollution of water also has a number of direct effects upon plant and animal communities. Reduced levels of dissolved oxygen and increased sedimentation of polluted water diminish species diversity, encouraging rampant growth of some plants (e.g., various forms of seaweed) whilst discouraging less robust species. In some cases, such changes have eventually impacted upon tourists. In parts of the Mediterranean, and particularly the Adriatic Sea, the disposal of poorly treated sewage (supplemented by seepage of agricultural fertilisers into watercourses that feed into the sea) has created localised eutrophication of the water. (Eutrophication is a process of nutrient enrichment.) This has led directly to formation of unsightly and malodorous algal blooms that coat inshore waters during the summer months, reducing the attractiveness of the environment and depressing demand for holidays in the vicinity (Becheri, 1991).

Water pollution has been especially commonplace in areas of mass tourism where the industry has developed at a pace that is faster than local infrastructures have been able to match (e.g., the Spanish Mediterranean coast), but even in long-established tourism locations, where local water treatment and cleansing services ought to be adjusted to local needs, water pollution is still a problem. In 2004, for example, the European Environment Agency reported that although 96 per cent of beaches in Europe complied with EU minimum mandatory standards governing faecal contamination of bathing waters, only 87 per cent met the more demanding guide standard. Inevitably, there is some significant variation in the level of compliance with the required standards from country to country. Thus, for example, whilst both the Netherlands and Greece achieved 100 per cent compliance with mandatory standards for coastal bathing beaches, France achieved only 87 per cent compliance. In terms of compliance with guide standards, the performance was much more variable. Greece attained 98 per cent compliance but the UK managed only 75 per cent whilst Belgium achieved just 18 per cent (EEA, 2006). Variation is also evident within countries. In the UK in 2005, for example, the north-west region – which covers some of the most popular British holiday beaches at resorts such as Blackpool and Southport – included just one beach that merited the coveted 'Blue Flag' for beach cleanliness, compared with 43 in Wales and 30 in the south-west region.

Alongside water pollution, tourism is also associated with air pollution and, less obviously, noise pollution. Pollution resulting from noise is usually highly localised, centring upon entertainment districts in popular resorts, airports and routeways that carry heavy volumes of tourist traffic. However, the dependence of tourism upon travel means that chemical pollution of the atmosphere by vehicle exhaust fumes is more widespread and, given the natural workings of the atmosphere, more likely to travel beyond the region in which the problem is generated. Nitrogen oxides, lead and hydrocarbons in vehicle emissions not only threaten human health but also attack local vegetation and have been held to account for increased incidence of acid rain in some localities. The St Gotthard Pass, which lies on one of the main routes between Switzerland and Italy, is one location where atmospheric pollution from tourist traffic has been held responsible for extensive damage to vegetation, including rare Alpine plants (Smith and Jenner, 1989).

Since perhaps the mid-1990s more significant areas of concern have arisen around global warming and the contributions that emissions of 'greenhouse' gases – principally carbon dioxide (CO_2) – are making to this emerging problem. Whilst greenhouse gases are produced by a wide range of processes and activities, the rising levels of air travel (to which tourism is a major contributor) has been identified as an emerging and serious problem. A recent study of the energy use associated with air travel to New Zealand (Becken, 2002), for example, cited forecasts from the Intergovernmental Panel on Climate Change that suggested that by 2050 air travel might account for as much as 7 per cent of global CO2 emissions. However, because aircraft generally emit pollutants into the upper troposphere, these emissions remain active in the atmosphere for longer periods, creating a cumulative effect that may be as much as four times greater than equivalent emissions at ground level.

However, these impacts are not uni-directional and there is a growing body of evidence that the global warming to which tourism is – in part – a contributor is damaging the capacity of some tourism resource areas to meet demand. For example, higher winter temperatures are now being held to account for reduced levels of snow fall in mountain areas in Europe and North America which, in turn, is raising questions over the long-term viability of some winter ski resorts (Konig and Abegg, 1997; Hamilton et al., 2003). Similarly, higher summer temperatures in regions such as the Mediterranean have been associated with increased incidence of droughts, heat-waves and – as a related problem – forest fires (Perry, 2006).

Resource base

A fourth area of concern centres on tourism impacts upon the resource base. Whilst tourism may be an agency for local improvement in supplies of key resources or the promotion of resource conservation measures from which everyone benefits, it will exert negative effects associated with depletion or diversion of key resources. The attraction of hot, dry climates for many forms of tourism creates particular demands for local water supplies, which may become depleted through excessive tourist consumption or be diverted to meet tourist needs for swimming pools or well-watered golf courses. In parts of the Mediterranean, tourist consumption of water is as much as six times the levels demanded by local people. Tourism may also be responsible for depletion of local supplies of fuel or perhaps building materials and the removal of sand from beaches to make concrete is not uncommon. Case Study 5.2 illustrates some of the issues surrounding the use of water resources, through a study of the problems created through increasing demands for water from tourism on the Spanish holiday island of Mallorca.

CASE STUDY 5.2

Water supply and tourism development on Mallorca

Mallorca, which lies in the Mediterranean and comprises the principal island of the Spanish Balearics, illustrates well the dilemma that faces a growing number of tourist destinations in warm or hot climates. On the one hand, their climatic regimes are a primary attraction to mass tourists who – in the case of Mallorca – originate largely in northern European countries such as the UK and Germany, yet, on the other hand, the general absence of rainfall during the summer months renders the task of meeting tourist demand for water increasingly problematic. With current visitor levels running consistently above 10 million per annum, the tourists create a level of demand for water that far exceeds that of the 700,000 people who comprise the resident population.

As the authors of this study note, 'the environmental conditions on Mallorca make the island predisposed towards water shortages' (Essex et al., 2004: 10). Geologically, large areas of the island comprise highly permeable limestone, so such rainfall as does occur (which is mostly between October and December) produces very little surface water. Consequently, aquifers supply most of the island's requirements. Furthermore, there are significant spatial variations, with much higher levels of rainfall (*circa* 1,000–1,200mm per annum) falling over the more sparsely populated northern mountain zone of the Serra de Tramuntana, whilst urban resort areas on the southern coast receive as little as 350mm per annum. Spatial problems of supply are compounded by the seasonality of demand, as peak levels of tourist (and resident) demands generally coincide with the irrigational requirements of the island's farmers.

The problems of balancing demands against supply have, however, been exacerbated by recent trends in tourism development which have had the effect of increasing the already high demand for water from tourism. In particular, a conscious effort to enhance the image of Mallorca and attract a more up-market clientele has seen a major expansion in provision for golf and a more concerted effort to smarten resort areas (and especially new developments) with green open spaces and gardens. Tourist demand for local fruit and vegetables has increased too, placing added pressures on Mallorca's irrigated agricultural systems.

The management of water has therefore come to occupy a central position in the sustainability of the island's tourist industry. Excessive usage of water has, it is estimated, been responsible for a fall of over 120 metres in the level of the water table in the main aquifers since 1973, leading to increased incidence of saline intrusion and salt water contamination of water supplies in some coastal areas. Local water shortages have become a more common problem with forecasts suggesting that by 2016, supply will be meeting only 87 per cent of anticipated demand. Accordingly, the regional government of the Balearics has brought forward a strategy for water management aimed at addressing the expected deficits. Central to this strategy are several key actions including:

- the introduction of measures to make more effective use of existing supplies through the reuse of waste water for agricultural and golf course irrigation;
- more effective repair of distribution systems to limit losses through leakage;
- encouragement of conservation through new pricing policies and universal use of water metering;
- development of new sources of supply through additional bore holes and an extension of existing (though expensive) systems of desalination of contaminated groundwater;
- artificial recharge of aquifers that have been over-exploited by transfer of water from new sources or perhaps through piping of water from areas that are presently in surplus.

However, despite the confidence of the regional government that issues of water management can be resolved, the authors of the study are rather less sanguine, noting that 'a technical approach to water management in order to sustain the tourist industry may simply be postponing the inevitable outcome' and that 'the reality may be that a new balance will have to be achieved between environmental capacities, in this case available water supplies, and the provision for tourism' (Essex et al., 2004: 24).

Source: Essex et al. (2004)

Visual or structural change

The final area of environmental impact concerns visual and structural changes, and it is here that there is perhaps the clearest balance between negative and positive impacts of tourism. The physical development of tourism will inevitably produce a series of environmental changes. The natural and non-natural environment may be exposed to forms of 'visual' pollution prompted by new forms of architecture or styles of development. Land may be transferred from one sector (e.g., farming) to meet demands for hotel construction, new transport facilities, car parks or other elements of infrastructure. The built environment of tourism will also expand physically, whether in the form of accretions of growth on existing urban resorts, new centres of attraction or second homes in the countryside.

However, set against such potentially adverse changes, there are significant areas of benefit. First, tourist-sponsored improvements to infrastructure, whether in the form of enhanced communications, public utilities or private services, will have some beneficial effects for local residents too. Second, tourism may provide a new use for formerly unproductive and marginal land. The rural environments in the west of Ireland, for example, have been partly sustained and regenerated by the development of rural tourism under the EU LEADER programme which was introduced to assist in the replacement of declining agricultural economies with alternative forms of development. Almost half of the funded projects conducted in Ireland under the first phase of the LEADER programme were associated with rural tourism (Hall and Page, 2005). Third, tourism to cities has helped to promote urban improvement strategies aimed at clearing dereliction. Examples include the programme of national and international garden festivals held in several British cities during the 1980s which took derelict industrial sites and created new tourist attractions out of the wasteland (Holden, 1989). In Britain, continental Europe, the USA and Canada, the regeneration through reuse of redundant areas – dockland and water frontages being favoured targets – has been a recurring theme in contemporary urban development (see, e.g., Couch and Farr (2000); Law, (2000); and for a fuller discussion of this theme, Chapter 9 of this book).

Managing tourism and environmental change

The evident capacity of tourism to create significant levels of environmental change has led to the formulation of a range of management responses to the perceived opportunities and difficulties created by tourism development. These approaches have been conveniently summarised by Mowforth and Munt (2003) under eight broad headings which they describe as 'tools of sustainability' (Table 5.2). Some of these approaches are concerned with establishing regulatory frameworks (such as area protection, regulation of industry and codes of conduct); some are concerned with ways of managing visitors; whilst the remainder are concerned with ways of understanding and assessing impacts (such as the use of environmental impact assessment, carrying capacity, public consultation processes and local participation, or the development of indicators of sustainability). Constraints of space preclude a full consideration of all the tools of sustainable management, but the following sections review a cross-section of established and emerging approaches.

Table 5.2 *The 'tools' of sustainability*

Technique	Typical responses
Area Protection	Designation of national parks, wildlife or biological reserves
Industry Regulation	Government legislation Professional association regulations Voluntary regulation
Codes of Conduct	Tourist codes Industry codes Best practice
Visitor Management	Zoning Honeypots Visitor dispersion Pricing and entry restrictions
Environmental Impact Assessment	Cost-benefit analyses Mathematical modelling Environmental auditing
Carrying Capacity	Physical carrying capacity Ecological carrying capacity Social carrying capacity Limits of Acceptable Change
Consultation	Public meetings Attitude surveys Delphi technique
Sustainability Indicators	Efficiency gains in resource use Reduced levels of pollution Better waste management Increased local production

Source: Mowforth and Munt (2003)

Visitor management

There is now a lengthy history of visitor management in tourism and an established repertoire of management techniques that can help to deliver sustainable forms of tourism through the regulation of the visitors. These include:

- spatial zoning;
- spatial concentration or dispersal of tourists;
- restrictive entry or pricing.

Spatial zoning is an established land management strategy that aims to integrate tourism into environments by defining areas of land that have differing suitabilities or capacities for tourism and then attempting to align levels of tourist access with those assessments of capacity by use of controls. Hence zoning of land may be used to exclude tourists from primary conservation areas; to focus environmentally abrasive activities into locations that have been specially prepared for such events; or to focus general visitors into a limited number of managed locations where their needs may be met and their impacts contained and managed. As an example, Figure 5.2 shows a zoning policy developed within the Peak District National Park in northern England.

Zoning policies are often complemented by strategies for concentrating tourists into preferred sites (sometimes referred to by recreational planners as 'honeypots') or, where sites are under pressure, deflecting visitors to alternative destinations. Honeypots are commonly provided as interceptors – planned locations that attract the tourist by virtue of their promotion and on-site provision (e.g., information, refreshment, car parking, etc.) and which then effectively prevent the further penetration of tourists into more fragile environments that may lie beyond. (Commercial tourist attractions, tourist information and visitor centres, country parks and heritage sites are all examples of locations that can act as honeypots and assist in the wider environmental management of tourism.) In contrast, where conditions require a redistribution of tourist activity, devices such as planned scenic drives or tourist routes may have the desired effect of taking people away from environmental pressure points.

In some locations, regulation of environmental impacts of tourism is now being achieved via pricing policies and/or exclusions and controls. The nature and scope of such practice varies considerably from place to place. In the USA, for example, entry into many of the national parks is subject to payment of an entry toll, whereas in the UK, entry is free. However, although access to British national parks is unrestricted, policies of exclusion and control are commonplace in local level management of these areas, as the pressures of tourism grow. In the Dartmoor National Park in south-west England, for example, planning restrictions are now widely deployed to protect the open moorland environment that lies at its core, with explicit presumptions against development that threatens the special character of the moorland landscape and with physical development mostly limited to existing settlements (DNPA, 2002). Visitors are encouraged (through patterns of access and planned provision of facilities) towards a relatively small number of higher-capacity sites, whilst movement of vehicles is subject to a park-wide traffic policy which both restricts and segregates vehicles to prescribed routes according to size and weight (Figure 5.3).

The concept of carrying capacity

The concept of carrying capacity is a well-established approach in attempting to understand the ability of tourist places to withstand use and is inherent in the notion of sustainability. In simple terms the concept proposes that for any environment, whether natural or

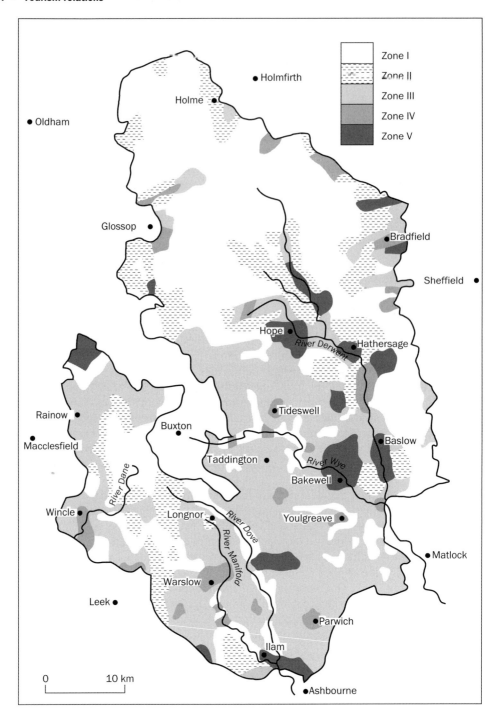

Figure 5.2 Spatial zoning strategies in the Peak District National Park, UK

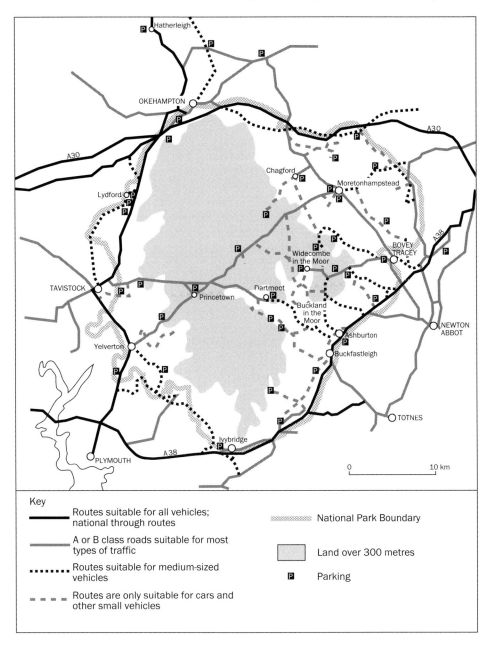

Figure 5.3 Traffic management strategies in the Dartmoor National Park, UK

non-natural, there is a capacity (or level of use) which when exceeded is likely to trigger environmental change and promote varying levels of damage and/or be associated with reduced levels of visitor satisfaction (McIntyre, 1993; Wahab and Pigram, 1997). Carrying capacity has been visualised in several distinct ways. For example:

● as physical carrying capacity – which is normally viewed as a measure of absolute space, such as the number of spaces within a car park;

- as ecological capacity – which is the level of use that an environment can sustain before (long-term) damage to the environment is experienced;
- as perceptual (or social) capacity – which is the level of crowding that a tourist will tolerate before he or she decides a location is too full and relocates elsewhere.

Although carrying capacity is outwardly an attractive concept that gains a measure of credibility through its grounding in observations that intuitively make sense, the notion that issues of capacity may be reduced to simple numbers has attracted a growing body of criticism that has brought its practical value as a tool for measurement of impact into question (Lindberg et al., 1997; Lindberg and McCool, 1998; Buckley, 1999; McCool and Lime, 2001). For example, it has been noted that ecological capacities are difficult to anticipate and may often emerge only after damage has occurred, whilst perceptual carrying capacities – as personalised responses – are prone to variation both between and within individuals or tourist groups, depending very largely upon circumstance and motives. It is clear that the capacity of any location will vary according to factors such as tourist behaviours and the types of use, ground conditions, seasonal variations, as well as prevailing management practices and objectives (McCool and Lime, 2001). This implies that any location will have multiple capacities – not just in terms of physical, ecological or social capacities – but in terms of variation *within* these categories too.

More importantly, perhaps, the concept has been criticised as overly reductionist in approach (Wagar, 1974), in distilling complex issues into the quest to identify critical thresholds of use in terms of numbers, and omitting essential qualitative considerations. Indeed, McCool and Lime (2001: 373) argue that by adopting the traditional approach of carrying capacity we become guilty of asking the wrong question. Thus rather than asking how many people an area can sustain, we should be addressing the social and biophysical conditions that are desired or are appropriate at the destination – in other words, how much change is acceptable given the goals and objectives for an area?

The limits of acceptable change

As a result, perhaps, of the recognised limitations of the concept of carrying capacity, alternative approaches to impact assessment have become more popular. The limits of acceptable change (LAC) technique was developed in the USA as a means of resolving development-related conflicts in conservation areas. The central features of the method (which is summarised as a set of key stages in Table 5.3) are:

- the establishment of an agreed set of criteria surrounding a proposed development;
- the representation of all interested parties within decision-making;
- the prescription of desired conditions and levels of change after development;

Table 5.3 *Key stages in the Limits of Acceptable Change approach*

- Background review and evaluation of conditions and issues in development area
- Identification of likely changes and suitable indicators of change
- Survey of indicators of change to establish base conditions
- Specification of quality standards to be associated with development
- Prescription of desired conditions within zone(s) of development
- Agreement of management action to maintain quality
- Implementation, monitoring and review

Source: Adapted from Sidaway (1995)

● the establishment of ongoing monitoring of change and implementation of agreed strategies to keep impacts of change within the established limits.

The LAC approach therefore embodies several key aspects of sustainable forms of tourism development. It recognises that change is an inevitable consequence of development but asserts that by the application of rational planning, overt recognition of environmental quality considerations and broad public consultation, sustainable forms of development may be realised. However, the approach does suffer practical weaknesses too. There are technical difficulties in agreeing and assessing qualitative aspects of tourism development and the process is dependent upon the existence of a structured planning system and sufficient resources in expertise and capital to operationalise the monitoring and review stages. Hence the contexts in which LAC would be most beneficial – for example, in shaping tourism development in Third World nations – often prove the least suited to the technique in practical terms (Sidaway, 1995).

Environmental impact assessment

The same constraint is also true of the use of environmental impact assessment (EIA). EIA is becoming a widely used method for evaluating possible environmental consequences of all forms of development and is potentially a valuable tool for translating sustainable principles into working practice. In particular, EIA provides a framework for informing decision-making processes that surround development, and a widening number of industries are routinely required (or advised) to undertake EIAs and produce written environmental impact statements (EISs). Table 5.4 summarises the four key principles to which most EIAs will conform.

The methodologies of EIA are diverse and may embrace the use of key impact checklists, cartographic analysis of spatial impacts, and simulation models or predictive techniques (Hunter and Green, 1995). Their strengths are that when properly integrated into the planning phases of a project, they should help developers anticipate environmental effects, enable more effective compliance with environmental standards and reduce need for subsequent (and expensive) revision to projects. The overall goal of a sustainable form of development is also a more achievable object when environmental impacts have been evaluated in advance. However, EIA has also attracted criticisms, which have included tendencies:

● to focus on physical and biological impacts rather than the wider range of environmental changes;
● to application on a project-specific basis and/or at the local geographic level, thereby overlooking wider linkages and effects;
● to require developed legislative and institutional frameworks in which to operate;

Table 5.4 *Key principles of Environmental Impact Assessment*

- Assessments should identify the nature of the proposed and induced activities that are likely to be generated by the project
- Assessments should identify the elements of the environment that will be significantly affected
- Assessments will evaluate the nature and extent of initial impacts and those that are likely to be generated via secondary effects
- Assessments will propose management strategies to control impacts and ensure maximum benefits from the project

Source: Adapted from Hunter and Green (1995)

- to require a range of scientific and other data as a means of assessing likely impacts;
- to advocate technocratic solutions to environmental problems, which some advocates of sustainable development view as inappropriate (Craik, 1991; Hunter and Green, 1995; Mowforth and Munt, 2003).

Thus, as with LAC, EIA has practical limitations that will inhibit its application in many tourism development contexts, especially those that might benefit most from the technique.

Consultation and participation

EIA has also been criticised as a professionalised approach that excludes or disempowers local people in ways that conflict with some of the principles of inclusion that sustainable development is intended to reflect (Mowforth and Munt, 2003). Ashley (2000) has argued that community-based approaches to resource management are essential if both the conservation needs of the tourism industry and the local development agendas are to be met, whilst Scheyvens (2002: 55) states that 'too many efforts at implementing environmentally sensitive tourism have focused on conservation of resources and failed to embrace the development imperative, thus neglecting the livelihood needs of local communities'. Such realisations are reflected in an increasing number of projects that have placed greater emphasis upon management that involves consultation and community participation in developing sustainable solutions to problems of tourism development. Case Study 5.3 illustrates one application of this principle in recent attempts to establish a community-based approach to the management of tourism and conservation in KwaZulu-Natal, South Africa, although as this case study also demonstrates, the success or failure of community initiatives are highly dependent upon local conditions. In particular, it shows how participatory approaches are dependent upon the assumption that community members have shared interests and aspirations and where such convergence is absent (and where more fundamental concerns shaped around poverty, personal safety and political stabilities are present), sustainable development through consultation and participation can prove an elusive goal.

CASE STUDY 5.3

Tourism and community-based conservation in KwaZulu-Natal

The province of KwaZulu-Natal lies on the Indian Ocean coast of South Africa and contains some of the country's most important wildlife parks, marine reserves and wetlands. Almost 30 per cent of the coastline and just under 10 per cent of the land area falls within protected reserves, of which the Natal Drakensberg Park, the Hluhluwe-Umfolozi Park and the Greater St Lucia Watersports Park are the largest (Figure 5.4). Recent data suggest that over 6 million domestic tourists and over 400,000 international tourists visit the province each year to enjoy the all-year sunshine, the spectacular landscapes and, especially, the wildlife (Brennan and Allen, 2001).

The conservation of these environments is, however, compromised by the grinding poverty of many of the rural communities that live in proximity to the parks and for whom the animal and plants of the reserves represent enticing – and much-needed – resources. Poverty and conservation are seldom compatible and in the case of KwaZulu-Natal these difficulties are compounded by the legacy of the South African system of apartheid and its

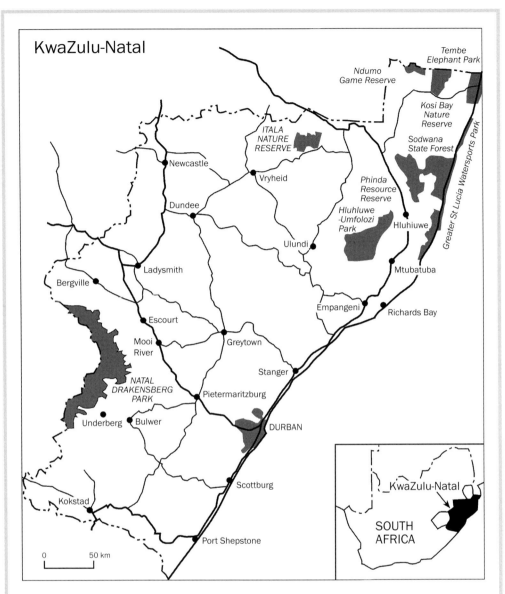

Figure 5.4 Major conservation zones in KwaZulu-Natal, South Africa

many injustices. In northern areas of the province, in particular, this legacy includes restricted access to land, lack of work, a general absence of welfare provision, ethnic rivalries, gender inequalities, corruption and violence.

Against this unpromising background, the Natal Parks Board and its successor (since 1998) the KwaZulu-Natal Conservation Board have sought to develop conservation-based tourism and in an important change in policy that reflects new agendas in post-apartheid South Africa, with a remit to consult and involve local communities in developing sustainable management of these reserves. Key initiatives have included the establishment of an ambitious Community Conservation Project aimed at supporting a range of local

continued

improvement schemes – such as health and sanitation projects – and a system of tourist levies that are intended to support a fund on which communities may draw to meet the costs of local projects. Environmental education programmes (especially through neighbourhood forums) have been established in over 80 locations across the province. In this way the conservation agencies hope to persuade more local communities of the benefits of tourism and the value of conserving the protected areas as a means of attracting visitors.

Opinions on the impact of these initiatives are, however, divided. Scheyvens (2002) provides a broadly positive reading noting, for example, how controlled harvesting of resources such as reeds and wood fuel have been introduced in several parks and how communities in close proximity to entry points to parks have been able to develop small-scale tourist enterprises around the sale of craft goods. She also maps over 160 community projects and notes – as a positive development – moves to involve tribal representatives on park management boards. In contrast, Brennan and Allen (2001: 218) offer a much less optimistic view, concluding that 'in KwaZulu-Natal, parks and other protected areas are painful reminders of apartheid's injustices, and of the continuing privilege of whites who enjoy looking at wildlife while Africans suffer from land starvation. Community-based ecotourism has achieved little in securing the protected areas for the future, and divided rural communities have few grounds for optimism over plans of the conservation sector.'

Sources: Brennan and Allen (2001); Scheyvens (2002)

Sustainability and alternative tourism

To conclude this chapter, we return to the subject of alternative tourism and, in particular, the question of how far do the so-called 'alternative' forms of tourism provide templates for sustainable tourism in general? There is perhaps a natural temptation to view the mass forms of packaged tourism as the least sustainable and the style of tourism that is most likely to bring widespread environmental change. In contrast, alternative forms of tourism (which are often characterised by their smaller scale, the involvement of local people, a preference for remoter areas and a predilection to place enjoyment of nature, landscape and cultures at the centre of the tourism experience) outwardly appear more in tune with principles of sustainability. Further, the alluring names that are commonly given to alternative forms of tourism – 'green tourism', 'eco-tourism', 'soft tourism' and 'responsible tourism' – tend to reinforce a popular belief that sustainability can only be equated with alternative tourism. Such views do, however, need to be accorded considerable caution, for although the underlying philosophies of alternative tourism may strongly reflect the concept of sustainability, the experience of alternative tourism in a growing number of places suggests that such forms may be highly potent as agents of change and generators of impact. In fact, alternative tourism can be just as problematic, in development terms, as mass forms of tourism.

The risks attached to a simple equation of alternative tourism with sustainability may be illustrated by a closer look at what is arguably the most prominent of the so-called alternatives – ecotourism. Fennell (1999: 43) has defined ecotourism as 'natural resource-based tourism that focuses primarily on experiencing and learning about nature and which is ethically managed to be low impact, non-consumptive and locally-oriented'. Ecotourism thus strives to minimise impacts upon environments and local people, increase awareness and understanding of natural areas and their cultural systems, and contribute to the conservation and management of those areas and systems (Wallace and Pierce, 1996).

According to Page and Dowling (2002), ecotourism has become one of the fastest growing sectors of the global tourism industry with some WTO estimates suggesting that as much as 20 per cent of the world market is based around ecotourism. Outwardly this estimate appears generous but is probably a consequence of the fact that there are differing forms of ecotourism which reflect the core values mentioned above to varying degrees. For example, Page and Dowling (2002) draw an important distinction between 'hard' and 'soft' forms of ecotourism:

- Hard ecotourism emphasises intense, personal and prolonged encounters with nature, that generally occur in undisturbed settings and where local communities offer none of the conventional trappings of modern tourism. Participants are usually highly committed environmentalists seeking what Cohen (1979) would categorise as 'existential' experience.
- Soft ecotourism centres on short-term exposure to nature where the experience is often one component in a range of tourist activities and experiences, where visits tend to be mediated by guides and facilitated by a higher level of facility provision. Participants in soft forms of ecotourism are less likely to be committed environmentalists.

This subdivision is helpful in developing a critique of ecotourism (and other alternative forms) as an intrinsically sustainable, low impact form of tourism. The concept of 'soft' ecotourism, in particular, captures a style of travel that a number of writers have characterised as mass tourism in a different guise (Wheeller, 1994) whilst many of the assumptions about the ethical superiority of ecotourism are often misplaced when it is clearly being deployed as a marketing label to represent traditional tourism products in an outwardly new way. Butler (1994) has also developed a persuasive critique that many types of alternative tourism – such as ecotourism – simply represent the pioneering stages in new practices of mass recreation – in Butler's words, 'the thin end of the wedge'. In this way, alternative tourism becomes a mechanism for constructing new geographies of mass travel and its associated impacts, centred on the exotic and the distant, that whilst initially experimental and low impact in character, develop into larger scale, more organised forms of visiting. Moreover, wherever small-scale forms of ecotourism develop into a larger enterprise, there is a heightened risk that the economic benefits of tourism that may initially accrue locally, will quickly be lost as large-scale travel companies from outside the local area take up the new business opportunities (Page and Dowling, 2002).

But even in the more specialist areas of 'hard' ecotourism, it is difficult to escape the conclusion that tourism promotes change and that much of this change is not necessarily beneficial. Wall (1997) sees many forms of ecotourism as instigating change in areas that previously were seldom visited and which thus place new demands on host communities and their environments. Similarly, Butler (1991) argues that it is difficult for even the most low-key forms of ecotourism to avoid creating impact and whilst some writers have noted that ecotourism may promote local identities and pride (Khan, 1997), those forms of ecotourism that involve deep immersion by participants in the places they are visiting means that alternative tourism often penetrates far deeper into the personal lives of residents than the more aloof forms of mass tourism, with similarly enhanced capacities to generate a range of environmental, economic, social and cultural impacts.

These criticisms suggest two important conclusions. First, although alternative forms such as ecotourism outwardly appear as environment-friendly and sustainable, the reality is often rather different. Second, alternative tourism – whilst perhaps embracing many principles of sustainability – does not in itself provide a model for sustainable forms of mass tourism and nor is it a replacement for mass tourism. It lacks the physical capacity, logistics and organisation to meet the growing levels of demand, it lacks the economic scale that has become so important to many national, regional and local economies, and the style of alternative tourism fails to match the tastes and preferences of many millions of holidaymakers and travellers worldwide.

So, whilst there are aspects of alternative tourism that certainly provide lessons in how to forge sustainable relationships between tourism and the environment, alternative tourism is not a natural (sustainable) replacement for the supposedly problematic mass forms of travel. Solutions to the problem of sustainability therefore need to be forged within the context of mass tourism, and that suggests that if the symbiotic relationship between tourism and the environment is to be maintained, careful management and planning of tourism development – whether guided by sustainable principles or not – must be a central component in the future growth of tourism.

Summary

Many forms of tourism are dependent upon the environment to provide both a context and a focus for tourist activity, yet those same activities have a marked capacity to devalue and, occasionally, destroy the environmental resources upon which tourism is based. Environmental effects of tourism are broadly experienced in impacts upon ecosystems, landscapes and the built environment, although specific impacts vary spatially – reflecting differences in the nature of the places that tourists visit, the levels and intensity of development, and the skills and expertise of resource managers. As the environmental problems associated with tourism have become more apparent, greater attention has been focused upon ways of producing sustainable patterns of development and alternative forms of tourism that produce fewer detrimental effects upon the tourist environment. However, truly sustainable tourism has often proven to be elusive, whilst there are evident risks that alternative tourism, in time, develops into mass forms of travel, with all the attendant problems that such practices tend to produce.

Discussion questions

1   What are the main factors that will lead to spatial variation in the environmental impacts of tourism?
2   Critically assess the value of the concept of sustainable development as an approach to understanding the environmental impacts of tourism.
3   With reference to a tourism destination area of your choosing, identify the range of environmental effects that tourism development has created in the area.
4   In what ways, and with what consequences, may global warming affect international tourism?
5   How far can conceptual tools such as carrying capacity, limits of acceptable change and environmental impact assessment actually help us to create sustainable forms of tourism?
6   Is ecotourism intrinsically sustainable?

Further reading

An excellent understanding of the environmental impacts of tourism is provided by:
Wall, G. and Mathieson, A. (2006) *Tourism: Change, Impacts and Opportunities*, Harlow: Prentice Hall.

The following provides a convenient critique of links between tourism and sustainability:

Hunter, C. and Green, H. (1995) *Tourism and the Environment: A Sustainable Relationship?*, London: Routledge.

Although not confined to discussions of the environment, a very insightful critique of tourism and sustainability is provided by:

Mowforth, M. and Munt, I. (2003) *Tourism and Sustainability: Development and Tourism in the Third World*, London: Routledge.

Other useful essays on tourism and sustainability include:

Hunter, C. (1995) 'On the need to reconceptualise sustainable tourism development', *Journal of Sustainable Tourism*, Vol. 3 (3): 155–65.

Hunter, C. (1997) 'Sustainable tourism as an adaptive paradigm', *Annals of Tourism Research*, Vol. 24 (4): 850–67.

Sharpley, R. (2000) 'Tourism and sustainable development: exploring the theoretical divide', *Journal of Sustainable Tourism*, Vol. 8 (1): 1–19.

There are several very readable texts devoted to the theme of alternative tourism and, in particular, ecotourism:

Boo, E. (1990) *Ecotourism: the Potentials and Pitfalls*, Washington DC: World Wildlife Fund.

Fennell, D. (1999) *Ecotourism: an Introduction*, London: Routledge.

Page, S.J. and Dowling, R.K. (2002) *Ecotourism*, Harlow: Prentice Hall.

Smith, V.L. and Eadington, W.R. (eds) (1994) *Tourism Alternatives: Potentials and Problems in the Development of Tourism*, London: John Wiley.

6 Socio-cultural relations in tourism

The relations between tourism, society and culture are characteristically complex. Societies and cultures are simultaneously objects of the tourist gaze – 'products' to be consumed by global travellers – as well as arenas of interaction in which social and cultural attributes are modified by the practices of tourist consumption and the varying forms of contact between visitors and host communities that modern tourism enables. However, these relations are not only complex, but they have also become contested and there is a growing body of empirical evidence acquired through case studies that reveals significant variation and inconsistency in the effect of tourist relations with the societies and cultures that are toured. More importantly, perhaps, this is a field in which critical thinking has developed in some significant ways. Traditional understandings tended to locate tourist relations with society and culture in Smith's (1977) rather comfortable conception of 'hosts and guests' and with an assumed directionality to the relationship in which tourism primarily created *impacts* on host societies and cultures. More recent thinking around the concept of power relations has, however, challenged traditional representations of tourism destinations and their communities as comprising predominantly passive recipients of tourism impacts (Mowforth and Munt, 2003). Instead, alternative interpretations have been substituted that characterise social and cultural contact as a negotiated relationship (Crouch et al., 2001) in which influences are far from unidirectional and affect all parties in differing ways. Urry's recent work around mobilities (Urry, 2000; Sheller and Urry, 2004) adds further intriguing dimensions to this subject by highlighting how – in increasingly mobile societies – the issue of who is local and who is a visitor is often less than apparent.

The social dimensions of tourism and the attraction of different cultures as a motivation for travel are now well established. As Hollinshead (1993) notes, the quest to encounter difference, to encounter 'others' and to experience novel environments has become a prime motivator for global travel. This tendency has become more pronounced as tourism has migrated from the often undifferentiated forms of relatively localised mass travel that characterised industrial mobilities of the early twentieth century, to increasingly globalised and differentiated forms of tourism that reflect mobilities of the post-industrial era (Franklin, 2004). For international tourists – at least those who originate in the developed world – the appeal of foreign cultures, with their distinctive traditions, dress, languages, handicrafts, food, music, art and architecture, has become a dominant pretext for tourist engagement with their destinations. Culture, and the societies that create culture, have become central objects of the tourist gaze.

In this chapter we explore several facets of the relations between tourism, society and culture, looking first at some of the key conceptual ideas that have shaped approaches to this area of study and then moving on to consider the nature of the tourist encounter with other cultures and societies and some of the key areas of effect that have been identified as arising out of such encounters. However, it is important to preface these discussions but emphasising that our understanding of the relationships between tourism, societies and cultures is still incomplete and uncertain. It is hampered both by limitations in what is increasingly a contested conceptual basis and by inconclusive or conflicting empirical studies. Several factors may be held to account for this situation:

- Uncertainty arises especially from the complexity of processes of socio-cultural change and the near impossibility of filtering the specific effects of tourism from the general influence of other powerful agents of change, such as globalised television and the media. Culture is not a fixed reality (Hunter, 2001) but evolves in response to a range of influences, *one* of which may be tourism.
- Socio-cultural relations have received less attention than has been paid to economic and environmental consequences of tourism development, partly because most social and cultural beliefs or practices are much less amenable to direct observation and to the conventional forms of measurement through survey-based enquiry of the kind that is so popular in the analysis of tourism.
- It is sometimes assumed that communities will adapt to the socio-cultural changes that tourism may bring to their lives as a price worth paying to realise the economic benefits that the industry can create. Indeed, a number of authors have made an implicit observation that social acceptance of tourism is often tied to its economic impact and to the positionality of the subject in relation to tourist activity (e.g., Wall and Mathieson, 2006). Hence people who work directly in the industry are less likely to acknowledge its negative impacts than those who do not.

Tourism, society and culture: theoretical perspectives

As has been suggested in the opening remarks, critical understanding of tourism relations with society and culture has moved through some important phases of conceptual development. This has influenced understanding of how tourists relate both to the societies and cultures of others and to the mechanisms by which tourist encounters may trigger socio-cultural change. In this section we review several key concepts that attempt to capture the essence of these relations: authenticity, commodification, the demonstration effect and acculturation, as well as noting how recent interest in power relations provides important new perspectives on some established ideas. Although none of these critical concepts are derived from geography as a discipline, their inclusion in this discussion reflects a belief that any attempt to understand socio-cultural relations in tourism from a geographical perspective must – of necessity – engage with these ideas. Authenticity, for example, is often grounded in place and in sectors such as heritage tourism, the geographical patterns of tourists movement and their patterns of concentration will often directly relate to the spatial patterns of authentic destinations that people wish to visit. Similarly, the processes of demonstration and acculturation arise most powerfully through the juxtaposition of people of differing backgrounds in geographical space, in which the dynamics of power relations are then played out.

Authenticity

In seeking to conceptualise the tourist's fascination with other societies and cultures, one of the earliest and most influential ideas has been MacCannell's concept of staged authenticity (MacCannell, 1973, 1989). The basic premise of MacCannell's position is that tourist behaviours are widely shaped by an implicit search for authentic experiences as an antidote to the inauthentic and superficial qualities of modern life. This quest is best realised – so the argument runs – through engagement with the real lives of others, whether reflected in the societies and cultures of other times (through heritage tourism) or of other places (through travel to the distant or the exotic). However, MacCannell draws on earlier work by Goffman (1959) relating to the structuring of social establishments to develop an important idea. Goffman argues that social spaces (of the kind that tourists will occupy) are habitually divided into 'front' areas in which the formalities of the encounter between the host and the guest are played out as a performance, whilst the real lives of local people are lived in 'back' areas that are generally inaccessible to the guest. Thus MacCannell posits that whilst tourists might seek the authentic experiences of 'back' areas through their travels, they are usually confronted with (and are satisfied by) staged versions of authenticity that are presented only in 'front' areas. Inevitably these performances attract a measure of inauthenticity through the process of staging.

Whilst the performative nature of tourism has become quite widely acknowledged (see, for example, Edensor, 2001), the notion that tourists are essentially motivated by a quest for authentic experience has come under a sustained challenge. Both Urry (1990, 1991) and Wang (1999) argue that the diversity of modern tourism clearly exceeds the explanatory capacities of MacCannell's original concept, whilst the actual experience of authenticity is often eroded both by the commodification of the cultures that are being represented and by the simple presence of tourists as an audience, which inevitably compromises authenticity by altering the social dynamics that surround the performance (Olsen, 2001). Postmodern critical perspectives have also undermined the relevance of authenticity as conceived by MacCannell, noting that for many forms of postmodern tourism, authenticity is not an attribute that tourists actively seek. To the contrary, writers such as Ritzer and Liska (1997) argue that many tourists actually seek out the *in*authentic representation as being an infinitely more comfortable and reliable experience than the real conditions that the inauthentic purports to represent.

However, those same postmodern perspectives have also opened some interesting areas of debate around how authenticity is established that suggests that the concept – when re-thought and re-applied – retains a relevance in explaining tourist behaviours. Bruner (1994) notes that in many contexts authenticity is not an inherent property of a particular object or situation but rather it represents a projection onto the toured objects and sites of the tourist's own beliefs (Wang, 1999). Authenticity thus becomes a cultural value that is constantly created, reinvented and negotiated through social processes (Olsen, 2001), so that many forms of tourism can aquire a level of authenticity simply because they are seen by the participants to be what real tourists do. Wang (1999) talks of an 'existential' authenticity through which tourists reconnect with their real selves through engagement with diverse tourism practice. In this way forms of tourism such as beach holidays, sea cruises or touring theme parks, all of which appear to have little to do with authentic experience as conceived by MacCannell, may nevertheless be validated in the minds of the tourist as authentic expressions of those particular styles of tourist behaviour. Wang (1999) explains, for example, how tourist activity such as visiting friends and relatives can be seen as a ritual that celebrates 'authentic' family ties and even the patently inauthentic (in MacCannell's terms) can acquire a form of authenticity. There is no doubt, for example, that for many of the visitors to the Disney theme park at Anaheim, Los Angeles (which is arguably one of the least authentic sites in world tourism), the park acquires a form of authenticity – as well as a degree of status – by virtue of being the original park from which others have subsequently developed. It is an 'authentic' expression of the concept of the theme park and enjoyed by millions of visitors precisely on that basis.

It is therefore the case that in many situations there is no single, authentic form of experience but a plurality of expressions of the authentic. Geography is important here because the manner in which places vary across the globe means that in many sectors of tourism there exist multiple authenticities that are place-dependent. These may be variations on a common theme, but distinctions of place will often serve to differentiate the authenticity of experience across geographical space. Hence, for example, the quintessential tourist experience of the beach holiday is qualitatively different between, say, a northern European country such as Britain, a Mediterranean country such as Greece, or an Indian Ocean state such as Mauritius.

So despite the many critiques of the concept of authenticity and the ambiguities and pluralities of meaning that the term now appears to have acquired, it remains a prominent component in the discourse of tourism relations with society and culture and, as we shall see in Chapters 8 and 9, a notion of some significance in key sectors such as heritage tourism and the wider processes of place promotion.

Commodification

Staged forms of authenticity form part of what Shaw and Williams (2004) describe as 'engineered' tourist experiences – essential components in the circuits of production and consumption that now shape global tourism. Within this framework, tourist consumption often forms a basis for social differentiation and identity formation through the acquisition of what Bourdieu (1984) defines as 'cultural capital' (Ateljevic and Doorne, 2003). In other words, it is argued that people seek to define themselves and their perceived place within social frameworks through the patterns of consumption that they exhibit, a process in which cultural goods and services possess a particular primacy. This is clearly evident in tourism.

However, in order for consumers to acquire such cultural capital through tourism, the tourist experience (as conceived in Figure 1.5, Chapter 1) needs – at least in part – to become a commodity that in addition to any *utility* value, also possesses an *exchange* value, that is, it can be traded through the conventions of the international marketplace (Llewellyn Watson and Kopachevsky, 1994). Although it has not always been recognised in these terms, commodification has been a feature of tourism for centuries. For example, if we consider the process of commodification as transforming elements of tourist experience into something that can be purchased as a good – in the same way that one might purchase, say, an item of clothing from a store – then common tourist practices such as organised excursions from industrial communities to Victorian seaside resorts represent a form of commodification. But in modern tourism, commodification has become a much more pervasive and subtle influence on tourist behaviours, so that it is not only the tangible products of tourism (such as transport, accommodation or souvenirs) that become commodities, but more importantly also the intangibles, such as experience.

In cultural terms, Meethan (2001) argues that commodification occurs in two connected ways: first as an initial representation of the destination in the images that are promoted through travel brochures and the media; and, second, through the ways in which local culture is represented in the tourist experience of the destination. The initial representation is especially significant because if, as is often the case, the representation of the destination to potential travellers wanders freely into areas of myth and fantasy, then images are created that are actually alien to the identities and practices of the host communities (Shaw and Williams, 2004). However, in order that tourists should leave their destinations as satisfied clients, market forces will often determine that the constructed images of local customs and practice must be delivered as performed, commodified experiences, even though they may be unrepresentative. Commodification therefore sets up powerful tensions with MacCannell's notion of authenticity for, as Wall and Mathieson (2006) observe, in the process of commodification the artefacts and practices lose many of their original meanings and hence their ability to represent authentically the cultures from which they are derived. (This theme is expanded in a subsequent section.)

The demonstration effect and acculturation

The concept of commodification has become an important position from which tourism studies seeks to explain the mechanisms through which tourism relates to cultures and societies, but it is not the only perspective from which this question has been addressed. A number of theories and concepts have previously been advanced to attempt to explain how contact between tourists and the communities that are toured advances socio-cultural change and two ideas – the demonstration effect and processes of acculturation – have proven particularly popular.

The demonstration effect is dependent upon the existence of visible differences between visitors and hosts as this theory suggests that changes in the hosts' attitudes, values or behaviour patterns may be brought about through processes of imitation based upon local contact with, or observation of, the tourist. It is argued that by observing the behaviours and superior material possessions of tourists, local people may be encouraged to imitate actions and aspire to ownership of particular sets of goods – such as clothing – that they see in the possession of the visitors and to which they are attracted as markers of status. Fisher (2004) locates the origins of the concept of a demonstration effect in tourism in the work of de Kadt (1979) who observed how local patterns of consumption will often adapt to reflect those of the tourist, but he also notes how the links between consumption patterns, lifestyle and social status have much earlier antecedents in what has been described as the urban cultural renaissance of seventeenth century Europe (Borsay, 1989).

In some cases, the demonstration effect can have positive outcomes, especially where it encourages people to adapt towards more amenable or productive patterns of behaviour and where it encourages a community to work towards things that they may lack. But more typically, the demonstration effect has been characterised as a disruptive influence, displaying a pattern of lifestyle and associated material ownership that is likely to remain inaccessible to local people for the foreseeable future, especially when contact is between First World tourists and Third World communities. This may promote resentment and frustration or, in cases where visitor codes and lifestyles are partially adopted by locals, lead to conflicts with prevailing patterns, customs and beliefs. Young people are particularly susceptible to the demonstration effect and, hence, tourism has occasionally been blamed for creating new social divisions between community elders and the young in host societies, or the encouragement of age-selective migration, with younger, better-educated people moving away in search of the improved lifestyles that the demonstration effect outwardly displays. The migrant, of course, may well benefit from such a move but the social effects upon the community that is losing its younger members will be broadly detrimental.

However, recent critical perspectives have begun to call into question many of the assumptions on which the demonstration effect implicitly resides. Once again the difficulties of isolating the influence of tourism on local society from the wider effects of modernisation raises questions over the actual significance of the demonstration effect. Fisher (2004) notes several areas of influence that are likely to be far more significant agents of change than tourism. Global television, for example, may broadcast directly into MacCannell's 'back' areas and these are the spaces where local people may also read newspapers and magazines that carry international news and the advertisements of global corporations, and where they perhaps read letters or receive telephone calls from friends and relatives who live and work abroad.

But beyond the recurring doubt about the specific role of tourism, there are more focused questions around who is demonstrating what to whom? The concept assumes a largely unidirectional influence and a neo-colonial relationship between 'strong' tourist and 'weak' local cultures. Yet as recent writing by Franklin (2004) makes clear, a widening range of 'foreign' practices (for example in dress styles or tastes for food) are routinely imported by people into their day-to-day lives – as cultural capital and markers of social status or taste – as a consequence of their experience of other places as tourists. These practices suggest, therefore, that the demonstration effect should be more properly considered as a process of cultural exchange rather than a form of cultural colonialism.

The demonstration effect, with its emphasis upon detached forms of influence, is particularly attractive in explaining tourism impacts where contacts between host and visitor are typically superficial and transitory. However, where links between hosts and visitors are more fully developed, acculturation theory offers an alternative perspective. Acculturation theory states that when two cultures come into contact for any length of time, an exchange of ideas and products will take place that will, through time, produce varying levels of convergence between the cultures, that is, they become more similar (Murphy, 1985). The process of exchange will not, however, be a balanced one since a stronger culture will tend to dominate a weaker one and exert a more powerful effect over the form of any new socio-cultural patterns that may emerge. (Interestingly, 'strength' of culture is not necessarily a reflection of cultural distinctiveness or integrity. The USA, for instance, has one of the most pervasive and powerful cultural influences that is spread by tourism, yet American cultural strength is more a reflection of population size, economic power and a growing domination of global media than a particularly well-defined cultural identity, US society being strongly multicultural.)

As with the demonstration effect, processes of acculturation are most easily envisaged in relationships between the developed and the developing world, but such patterns may also be found within developed states. Many European nations contain marginal or peripheral regions that are attractive to tourists and which also contain distinctive cultures. Within the UK, for example, this is the case for large parts of Wales, where a strengthening of local resistance to changes arising from acculturation that have in part been associated with tourism (such as the erosion of the Welsh language) has been noted. Processes of acculturation should therefore be expected to operate in a range of spatial contexts.

Power relations

Implicit in all these critical positions are concepts of power relations. The role of power in shaping tourism relations has been recognised for some time although the interpretation of how those relations map onto the actual world of tourist experience has developed in some interesting ways in recent years. By tradition, tourism relations with the communities that are toured have been widely represented as shaped by inequalities. Mowforth and Munt (2003) capture this perspective in noting how tourism is often envisaged as a contact zone in which disparate cultures meet in asymmetric relations and with sharply differentiated experiences. Moreover, this asymmetry has been shown to be operating at both a global and a local scale of effect. At a global scale, tourism has been characterised as creating neo-colonial patterns of dependency between dominant areas of tourism generation in the First World and subordinate areas of tourism development (in either peripheral zones of First World states or, more typically, emerging destinations in the Third World). Tourism therefore tends to occupy spaces that have been opened through the exercise of power, especially the commercial power of global capital and its market places (Mowforth and Munt, 2003). Meanwhile, at the local scale, the nature of the encounter between the tourist and the people who are visited has been viewed as an unequal meeting in which attributes such as the material wealth of the tourist tend to place them in a position of superiority relative to local populations, who are expected to fulfil subservient roles in relation to the visitors' needs.

Such readings are, of course, not necessarily inaccurate. It has been widely recognised, for example, that processes of commodification implicitly exercise power by imposing foreign expectations onto receiving areas and defining how cultures and cultural artefacts should be represented to visitors, through activity such as place promotion or the staged events that visitors experience at the destination. The power of local communities to resist external commodification of their locales and its culture is moderated by the perceived need to conform to external expectations in order to realise the economic benefits that tourism may bring. Similarly, the increasing tendency for First World ideologies – such as sustainability – to be imposed on others as a condition of participation in processes of tourism development has also been noted (Mowforth and Munt, 2003).

The concession of local power to those who commodify tourism's products forms a part of a much wider issue of how cultural identities are formed. Shaw and Williams (2004) observe that the delimitation of culture is a relational process, by which it is meant that part of the solution to defining who we are depends upon the existence of 'others' with whom we may make comparisons and draw distinctions. Without the 'other' the definition of the 'self' becomes problematic. The important point for this discussion is that the process of recognising the 'other' – who, in the context of tourism, will generally comprise the communities that are visited – implicitly shapes a power relationship in which the 'other' is accorded lower status (Aitchison, 2001).

However, the notion that tourism relations are shaped simply by the power of global capital and the economic force of tourists as consumers of commodified experiences that overwrite the narratives of local people who are powerless to resist, evidently provides an inadequate basis for describing what have become recognised as complex, negotiated relations. Foucauldian perspectives are particularly useful in this context since Foucauld envisages power not in terms of the relative strength of one over another, but as a relational process in which power flows in multiple directions. Most importantly, the relations between tourists and the communities they visit are mediated by third parties who act as brokers or middlemen (Cheong and Miller, 2000). Brokers may include the travel companies who sell tours at the point of departure, but will also include local entrepreneurs and elites who operate tourist businesses in destination areas; local politicians and planners who may regulate development; or local police who regulate tourist activity. This system is characteristically dynamic such that, for example, tourists may become local if they take up residence in areas that they have visited (through retirement or second home ownership) or local people may become brokers if they start to engage with tourism activity. Brokers, periodically, will also become tourists. But the essence of Cheong and Miller's (2000) argument is that tourists are more readily characterised as targets of power relations rather than agents, that is, enduring the consequences of power rather than exercising it. They operate from insecure positions since they will often be found in unfamiliar environments, possibly disadvantaged by an inability to speak local languages and exposed to different cultural norms and expectations that they may not comprehend. The tourist may possess economic power, but as Ateljevic and Doorne (2003) recognise, cultural power generally resides with the local communities.

The tourist encounter

At the heart of the processes of socio-cultural exchange between tourists and the communities they visit is the encounter and out of this encounter may emerge a range of social or cultural effects. Wall and Mathieson (2006) identify several important characteristics of the tourist encounter with local communities, especially in the context of mass tourism. Thus the encounter is typically seen to be:

- transitory;
- constrained in space and time, that is, seasonal in effect and focused on particular tourist sites;

- staged rather than spontaneous and often commercial in character;
- differentiated (or unequal) in terms of attributes such as wealth, but also less tangible qualities such as expectations.

To these may be added other attributes, for example, that the encounter is generally impersonal (Hunter, 2001).

However, although tourist encounters may be fleeting, impersonal and outwardly superficial, some form of socio-cultural effect will often emerge from the process of contact. This is particularly evident where tourism brings together regions and societies that are characterised by varying degrees of difference. International tourists, in particular, will tend to originate in a developed, urbanised and industrialised society and will carry with them the beliefs, values and expectations that such societies promulgate. But as the spatial range over which tourists roam is extended (and given the predisposition of many tourists to seek out places that are different), so the likelihood increases that encounters between local people and visitors will bring together opposing tendencies and experiences: development and underdevelopment; pre-industrial, industrial and post-industrial; traditional and (post)-modern; urban and rural; affluence and poverty; and all in a context in which the tourist is at leisure and probably enjoying novel situations, whilst the local people pursue the familiar routines of work.

Of course, the encounter is not always shaped in such simplistic ways (and we will return to this point shortly) and nor is the character of the encounter and its related effects consistent across the spaces of tourism. Spatial variation in the nature and consequences of the tourist encounter with local communities will occur for a number of reasons, but we may identify a set of key variables that will help to explain differences in effect. These are the situation of the encounter and visitor type; the nature of the location; the spatial proximity of tourists and local populations and the levels of involvement of the latter in tourism; cultural similarity; and the stage of tourism development.

Situation of the encounter and visitor type

It has been suggested by de Kadt (1979) that visitors and local people encounter one another in three basic situations:

- when tourists make purchases of goods and services from local people in shops, bars, hotels or restaurants;
- when tourists and hosts share the same facilities, such as local beaches and entertainment;
- when they meet purposely to converse and to exchange ideas, experiences or information.

The extent and the nature of socio-cultural impacts will clearly be influenced by whichever of these forms of contact prevails, but their incidence will also tend to reflect the type of visitor (see Chapter 1) and the periodicity and duration of their visits. When tourism is centred upon mass markets, contacts are most likely to be in either (or both) of the first two categories, but because of the limited seasons associated with many package holidays and the short duration of individual trips, the contacts are typically casual and brief. However, although contact may be limited, the scale of mass forms of tourism is still quite capable of producing a range of problems and changes through the demonstration effect or accul-turation. Purposeful engagement between the two groups is much rarer in modern tourism and is more typically a feature of independent traveller and explorer types of tourist. Since they are less numerous, these tourists are generally held to have lesser impacts upon local societies and cultures, although, strictly speaking, any form of contact is likely to produce some degree of social and cultural change, and if 'explorer' types of tourist spend extended periods in a host community, scope for cultural interchange will be significantly increased.

Nature of the location

Geographic elements are also important, both in the nature of destinations and in the effect of spatial proximity between hosts and visitors. The tolerance of tourism by local communities will be affected by the capacity of a locality to absorb tourism and, more simply, the degree to which tourists form identifiable groups and/or create visible problems. In metropolitan centres such as London or Paris, thousands of tourists may be accommodated with few discernible impacts because urban infrastructures are designed to cope with heavy use and, in many situations, the tourist simply blends with the crowds. In contrast, small rural communities that are not adapted to handling crowds may struggle to cope with more than a few hundred visitors, and because those visitors are far more conspicuous, scope for induced change through demonstration effects and acculturation will be enhanced.

Spatial proximity and levels of involvement

The nature and the intensity of exchange will be influenced by spatial and sectoral proximity of local populations and tourists. As we have seen in Chapter 4, tourism development has a marked tendency to spatial concentration at favoured locations, so patterns of development are uneven. Whilst some diffusion of impacts from centres of tourism into surrounding areas may be expected (e.g., through the employment of people who travel daily to work in tourism from a hinterland), the capacity of tourism to affect host societies and cultures will decline as distance from the tourist centres increases. Even within tourism areas, some locations remain untouched by tourists and their routine movements, whilst certain forms of development – especially enclaves – will purposely segregate locals and visitors, thereby minimising the social and cultural effects of each group upon the other. For similar reasons, different sectors within a local community will react variably to the presence of tourists. Business sectors and government are more likely to adopt favourable views of tourism because of the economic benefits that the industry can bring, whereas ordinary local residents who do not benefit directly from the tourism economy but whose lives are affected by the noise, overcrowding, congestion and overuse of facilities that tourism often creates will tend to form negative views. Thus attitudes and behavioural responses towards tourism are normally differentiated by the direct or indirect ways in which the various groups within communities experience tourism.

Cultural similarity

However, perhaps the most important factors in shaping the socio-cultural effects that arise from the encounter are the levels of cultural similarity or dissimilarity and the stage of tourism development that has been attained. Cultural 'distance' (which often tallies closely with spatial distance) between the visitor and the local community will be crucial in determining the level of effect that is likely to be felt. The maximum social effects tend to occur when a host community is relatively small, unsophisticated and isolated, and where affluence levels are markedly different. When local people and the visitors have similar levels of socio-economic and technological development, socio-cultural differences will tend to be less pronounced and tourism effects upon society and cultures are reduced in consequence.

Although international tourism does bring differing groups together, in many locations tourism also brings together culturally similar people. In North America, for example, interchange between Canadian and American tourists, whose lifestyles have much in common, produces comparatively few socio-cultural repercussions (although impacts upon Native American communities in the USA and Canada may be more pronounced). Even in the rapidly expanding markets of South-East Asia, a region in which cultural impacts might be expected to be an issue, over 75 per cent of international visitors originate within the region. Thus, although there are important differences between the major ethnic groups in this area, there remains a sufficient breadth of shared socio-cultural experiences to produce fewer effects than might have been anticipated. Unsurprisingly perhaps, studies of domestic tourism suggest that many of the socio-cultural effects that are linked to international travel largely disappear in situations where the visitor and local populations derive from the same socio-cultural milieu (see, e.g., Brunt and Courtney, 1999).

Stage of development

In Chapter 2, Butler's model of destination area development was used to illustrate how tourism places evolve through time (Butler, 1980). An important theme that is implicit in Butler's conception is that the effects of tourism on destination areas will also evolve through time, the natural tendency being for the scale of impact to grow as the destination progresses from exploratory stages (where effects are slight) to the stages of saturation (in which effects may be significant).

One of the most familiar articulations of this idea has been provided by Doxey's 'Irridex' – a contraction of 'irritation index' – which attempts to show how attitudes to tourism in a host area might change as the industry develops (Figure 6.1). The model suggests that initially the tourists are welcomed, both as a novelty and because of their scope for creating economic prosperity. As developments become more structured and commercialised, local interest in the visitors becomes sectionalised (i.e., some local people become involved with the tourists, others do not) and signs of apathy emerge, especially amongst the uncommitted. If growth continues, physical problems of congestion and spiralling development sow seeds of annoyance on the part of local people, whose lives are now increasingly affected and inconvenienced by tourism. In the final stage of Doxey's model, annoyance has turned to open antagonism and hostility towards the tourists, who are now blamed, fairly or unfairly, for perceived detrimental changes to local lifestyles and society.

Although it maps a pathway that may well be encountered in some tourism destinations (and the model was grounded in observations conducted in both Canada and the Caribbean),

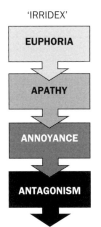

'IRRIDEX'

EUPHORIA — Initial phase of development, visitors and investors welcome, little planning or control mechanism.

APATHY — Visitors taken for granted, contacts between residents and outsiders more formal (commercial), planning concerned mostly with marketing.

ANNOYANCE — Saturation points approached, residents have misgivings about tourist industry, policy-makers attempt solutions via increasing infrastructure rather than limiting growth.

ANTAGONISM — Irritations openly expressed, visitors seen as cause of all the problems, planning now remedial but promotion increased to offset deteriorating reputation of destination.

Figure 6.1 Doxey's 'Irridex'

Doxey's model has drawn a number of criticisms. The most significant are that the concept is essentially a negative reading that permits little recognition of positive benefits, whilst the unidirectional quality of the model suggests only an inevitable sequence of decline in the relationship between local people and tourists (Murphy, 1985). The model also fails to acknowledge that the attitudes of people who are directly involved with, and benefiting from, the industry will differ from the attitudes of those who are not, and there is no recognition of the capacity for the trajectories of change to be altered by the application of effective planning systems.

In short, Doxey's model is too simple to capture the many nuances in the relations between tourists and local communities. This equally simple observation leads to some important, wider criticisms of some of the conventional interpretations of the relationship, especially those grounded in Smith's (1977) concept of 'hosts and guests'. The fundamental problem is that Smith's concept presents a basic dichotomy of just two cultural forms – the host and the guest – that whilst convenient, fails to replicate the complexities of the tourist encounter in real-world settings. The host–guest dichotomy may be critiqued from a number of perspectives.

First, it should be remembered that tourist flows to many destinations will be composed of tourists from a variety of sources with differing cultural backgrounds and contrasting levels of cultural difference from the communities they visit. The United Kingdom, for example, receives significant numbers of tourists from Europe, North America and Japan – some of whom are socially and culturally closer to the British than are others. The nature of the encounter and its effects will therefore be contingent on the degree of cultural proximity between host and guest and will be variable rather than consistent.

Second, it is also a mistake to assume that destinations are themselves culturally and socially homogeneous. Sherlock (2001: 285) makes the point that in contemporary social settings that are increasingly shaped by mobilities, local communities frequently comprise 'a fragmented and continually changing network of social ties'. Furthermore, many tourist destinations attract part-time residents – essentially second homers – who clearly occupy ambiguous and ill-defined positions in relation to a simple concept of hosts and guests. These people may appear as locals to the tourist, but as visitors in the eyes of longer-term residents.

Third, as an extension of the previous point, in many situations so-called hosts and guests share in the process of producing and consuming the places that are toured, such that the host may often be endowed with many of the attributes of the guest (Sherlock, 2001). Local leisure practices may merge imperceptibly into the practices of visitors – in restaurants, in entertainment venues, in retail environments or in the shared enjoyment of the spectacle of different types of public spaces, such as street markets (see Williams, 2003). Equally, the tourist encounter is not merely an encounter with host communities, but is also an encounter with other tourists whose actions and behaviour become integral to the production of the tourist experience (Crouch et al., 2001).

Fourth, and most importantly, however, we should recognise that the behaviour patterns of the visitors are often a diversion from their socio-cultural norms and do not, therefore, accurately represent the host societies from which they originate. Graburn's (1983a) concept of behavioural 'inversions' (see Chapter 1) indicates how tourist behaviours often display a significant degree of departure from normal patterns, with conspicuous increases in levels of expenditure and consumption, or adoption of activities that might be on the margins of social acceptability at home – for example, drinking and overeating, gambling, atypical dress codes, and nudity or semi-nudity. In other words, there exists within the visitors' normal culture what we might label as a 'tourist culture' – a subset of behavioural patterns and values that tend to emerge only when the visitors are travelling but which, when viewed by local people in receiving areas, projects a false and misleading image of the visitors and the societies they represent. This idea is illustrated diagrammatically in Figure 6.2 in which

Figure 6.2
Cultural 'distance'
and the socio-cultural
impact of tourism

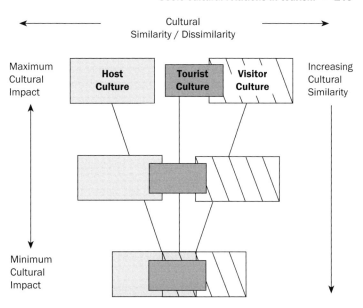

each box represents a culture with the tourist culture nesting within, but also extending outside, the normal visitor's culture, as a representation of the tendency to atypical forms of behaviour. The greater the extent of overlap between the cultures, the greater the socio-cultural similarities and the fewer the resulting tourism impacts. Conversely, the less overlap that exists between the three, the greater the cultural distance between host and visitor and, consequently, the greater the chance of the tourist encounter producing observable socio-cultural effects.

However, it is not just the culture of the tourist that may be misrepresented by the manner in which tourists present themselves to host communities. As we have already seen, the commodification of local culture and the frequency with which it is presented to tourists in staged, but inauthentic, performances means that in many situations host communities are also represented in ways that do not reflect their true nature. Consequently, the tourist encounter – although sometimes characterised as a meeting ground, with all the connotations of convergence and accord – is also fertile ground for mutual misunderstanding (Hunter, 2001).

Effects of tourism upon host communities

In this final section we review and illustrate some of the primary effects that have been recognised as potential outcomes from the tourist encounter with local cultures and societies. Much of the academic literature on the socio-cultural impacts of tourism tends to emphasise negative perspectives – such as increased incidence of crime, lowering of moral standards, breakdown of family structures or the commercialisation of tradition. Such emphases can, however, be misleading, as tourism development has also been shown to foster a range of positive effects – such as a reduction of cultural barriers, increased understanding and the promotion of pride in cultural heritage (Jafari, 2001). Popular views of tourism as the 'destroyer' of societies and their cultures are thus too simplistic. Tourism is not a monolithic force, nor does it stand apart from wider processes of development and change. It is both a cause and a consequence of socio-cultural development, and since it comprises a diversity of participants, agencies and institutions with differing motives and goals, its effects are diverse and often unpredictable. This leads to considerable spatial and temporal variation in the nature of relationships between tourism, society and culture and the effects that it creates. However, what is also clear is that, in general, the presence of visitors changes the object of their attention. This is a paradox that is common to several dimensions of modern tourism but it underpins many of the concerns that have been voiced over the socio-cultural impacts of tourism.

Empirical studies of the socio-cultural effects of tourism have highlighted a diversity of possible effects, and these are broadly summarised in Table 6.1. Very full discussions of these issues may be found within the general tourism literature (see, e.g., Mathieson and Wall, 1982; Murphy, 1985; Ryan, 1991; Richards, 1996; Meethan, 2001; Smith, 2003; Wall and Mathieson, 2006), so for the purposes of this book it is proposed to examine and illustrate a cross-section of effects which, for convenience, will be grouped under three broad headings: authenticity and commodification; moral drift and changing social values; and new social structures and empowerment.

Authenticity and commodification

Issues of authenticity and commodification generally reflect concerns over the way in which indigenous cultures are used to promote and sustain international tourism. The success of modern tourist destinations will often depend upon the ease with which distinctive images of a place may be formed and marketed, and although images may be constructed around a variety of natural and non-natural elements, socio-cultural characteristics are especially important. There is, though, a tendency for processes of image-building by marketing agencies to misrepresent societies and cultures or to simplify them by characterisation. Herein lies the seed of a major problem, since the image obliges local people to present their traditional rituals and events, folk handicrafts, music and dance, religious ceremonies or sporting contests – all of which are capable of attracting tourists and forming a central element in their experience of the destination – in ways that accord with the image, rather than reality. The attributes that tourist see, therefore, risk becoming projections of their own imagination rather than true representations of the object of their gaze (Cohen, 1988).

This is not to argue that tourist interests in culture are automatically detrimental as there is a solid base of evidence to show how cultural places, artefacts and performances have been sustained (and in some instances reacquired) through the interest and support of visitors. In

Table 6.1 *Primary positive and negative impacts upon host society and culture*

Main positive impacts
- Increased knowledge and understanding of host societies and cultures
- Promotion of the cultural reputation of the hosts in the world community
- Introduction of new (and by implication more modern) values and practices
- Revitalisation of traditional crafts, performing arts and rituals

Main negative impacts
- Debasement and commercialisation of cultures
- Removal of meanings and values associated with traditional customs and practices through commodification for tourist consumption
- Increased tensions between imported and traditional lifestyles
- Erosion in the strength of local language
- New patterns of local consumption
- Risks of promotion of antisocial activities such as gambling and prostitution and increased incidence of crime

Sources: Adapted from Mathieson and Wall (1982); Ryan (1991); and Wall and Mathieson (2006)

many tourist destinations, for example, the tourist souvenir trade has the potential – when correctly managed – to contribute significantly to local economies and to sustain material elements of traditional culture. The studies by Hall et al. (1992) and Ryan and Crotts (1997) of Maori culture in New Zealand illustrate several aspects of this process. Elsewhere, tourist demand has provided a basis for what Wall and Mathieson (2006: 274) describe as 'a renaissance in the production of traditional art forms'. Interestingly, in a number of situations the process of reasserting traditional practices illustrates Olsen's (2001) argument that rather than constituting a fixed attribute, authenticity is often negotiated through social processes. Smith's (1997) study of the Inuit provides one example in which tourist demand for authentic souvenirs stimulated novel practices – especially in stone carving – that provided new ways to express Inuit cultural identity, but in a form that both the Inuit and the tourists were prepared to accept as representative of native cultures. Medina's (2003) study of a community in Belize (which is summarised in Case Study 6.1) illustrates similar themes around the re-acquisition of identity linked to the rediscovery of ancient traditions through local involvement in tourism.

CASE STUDY 6.1

Tourism and Mayan identity

This study explores the proposition that the process of commodifying culture for tourist consumption may, when linked with new sources of knowledge, provide pathways to re-acquire forms of cultural identity. The study was focused on the village of San Jose Succotz in Belize which lies close to the ancient Mayan city of Xunantunich, one of the most-visited archaeological sites in the country. Claims to a Mayan heritage in San Jose do not, however, rest solely on proximity to a key site since the village also possesses a Mayan past of its own, with ethnographic studies conducted in the early twentieth century reporting significant populations of Mopan and Yucatec Maya, integral use of Mayan ritual in the agricultural cycle and some use of the Mayan language. However, the subsequent decline in the role of agriculture in the village has seen a reduction in traditional ritual practices, whilst population change has led to a marked reduction in the number of villages who identify themselves as Mayan or possess knowledge of Mayan language. Indeed, at the time of the study (1999), over 80 per cent of the villages identified themselves as *Mestizo* – a blend of European and indigenous ancestry. The key attributes of Mayan identity – ancestry, language and ritual – had tended therefore to become highly marginalised.

The key to the local revival of interest in Mayan identity has been the archaeological excavation of Xunantunich. This work has made the site available to tourists whose interest in the Mayan civilisation and their active demand for local souvenirs that reflect Mayan culture has stimulated new levels of local interest in, and respect for, Mayan identity. This has entailed several local developments that enable villagers to satisfy the tourist demands for authentic experiences of the Maya. For example, some of the local guides have acquired a basic understanding of the Mayan language which they deploy as part of their interpretations of the sites. More typically, many guides have spent time in acquiring knowledge – from books and from courses of study based at Xunantunich – about Mayan cosmology, history, customs and practice, to impart to the visitors. Beyond the framework of the guided tours, local production of souvenirs – especially carved stone and pottery – has also developed, with the artisan craftsmen using images and designs that have been uncovered at Xunantunich or are presented in written texts on the Maya as the basis for their work.

continued

Thus, by resorting to materials generated by modern scholarship, villagers have developed new ways of acquiring knowledge that can no longer be transmitted through the traditional pathways of social contact between generations, but which allow at least some elements of lost cultures to be recaptured. The authenticity of this re-invented form of Mayan identity may, of course, be less than perfect and the study showed that local people were themselves unsure of the extent to which a real Mayan identity had been reasserted in their community. The author of the study describes the process as producing 'a shared vision of "Maya culture" [that] is continually produced and consumed through the actions of archaeologists, tourism promoters, tourists, tour guides, artisans and vendors of artisanal products' (Medina, 2003: 357). But if we accept the notion that authenticity can be a negotiated attribute, then the experience at sites such as San Jose Succotz suggest that authenticity and commodification are not necessarily incompatible concepts.

Source: Medina (2003)

However, as levels of demand increase and the tourist experience of other cultures becomes progressively commodified, there are very real risks of negative repercussions as cultural artefacts and performances become absorbed into what Boorstin (1961) has described as 'pseudo-events' – elaborately contrived, artificial representations designed to meet (in Boorstin's rather elitist view) the undiscerning gaze of mass tourists. Pseudo-events generally share several characteristics:

- They are planned rather than spontaneous.
- They are designed to be performed or reproduced to order, at times or in locations that are convenient for the tourist.
- They hold an ambiguous relationship to real elements upon which they are based.
- Through time, they become authentic and therefore may replace the original event, practice or element that they purport to represent.

It is easy to see how tourist demand encourages these processes, and, on the positive side, it may be argued that by focusing tourist attentions upon staged representations of local culture – often within the comfort of the hotel lounge – the pseudo-event serves a valuable function in relieving pressures upon local communities and helps to protect their real cultural basis from the tourist gaze. But as it does so, it creates artificiality which detaches cultural elements from their true context. Tourists observing native ceremonies will usually lack the knowledge to comprehend the symbolism and true meaning of events, but there is a greater risk that through time the performers also lose sight of the original significance of a practice, and this alters its basis within the host culture. Likewise, the successful marketing of traditional objects as tourist souvenirs may alter their meaning or value, and where tourism markets develop, a tendency towards increased dependence upon mass forms of production will marginalise the true craft worker. Mass production typically takes control over the development and sale of craft goods out of local communities and into the hands of non-native producers. Blundell's (1993) study of the problems of the Canadian Inuit souvenir trade illustrates this issue very well.

The extent to which authenticity issues and the commodification of culture by tourism is a real concern is very much a matter of opinion. The natural temptation is to decry the manner in which commercial tourism erodes and alters the cultural basis of host societies, but it is important to reiterate the point that culture is not static, it is dynamic and adaptive, and a vibrant society will constantly re-create and reconstruct its cultural basis. It is also a mistake to characterise native populations as passive objects of the tourist gaze since it is evident that host communities may actively construct and promote representations of their culture to attract the visitor, and many practices have thereby acquired new meanings and values. In this way, tourism should perhaps be conceived not as a force that exists outside local cultures but rather as an integral part of ongoing processes of cultural formation (Crang, 2003; Shaw and Williams, 2004).

Moral drift and changing social values

A second area of general concern focuses upon the potential for contact between visitors and hosts to alter value systems and the moral basis to local societies, generally producing a drift towards adoption of more permissive or relaxed moral standards. However, because cultural attributes such as values are more deep-seated, it has proven rather harder to distil specific effects that may be linked with tourism (Wall and Mathieson, 2006).

To the local observer, the casual lifestyle of many tourists, their conspicuous consumption, their rejection (albeit temporarily) of normal strictures of dress and some elements of etiquette can create very diverse reactions amongst local people, although the strength of that reaction will depend upon the cultural distance between the host and the visitor (see Figure 6.2). Where differences are clear, the demonstration effect will tend to draw some elements in a society towards the alluring lifestyle that the tourists project (especially the young), whilst others (particularly older groups) will resist what are perhaps perceived as immoral forms of behaviour. (Such divisive tendencies have, for instance, been noted in several Mediterranean destinations, where the imposition of largely agnostic or atheistic North European tourists onto predominantly Catholic or Greek Orthodox communities with quite restrictive moral and social codes has been problematic – see, for example, Tsartas, 1992.) In time, however, the natural processes of succession within the community will ensure that the value systems instilled in the young today are likely to become the norm for the society at large in the future. As a result, the effect of exposure to tourists may be to produce a drift towards changed moral and social values.

Once again, the extent to which this constitutes a 'problem' will depend upon the positions from which such changes are viewed, and whilst the temptation is to paint the tourist as the moral polluter of simpler, traditional societies, there are cases where roles are reversed and impacts are greater amongst the tourists than amongst the hosts. Visitors to Scandinavian countries or the Netherlands, for example, may find prevailing moral codes that adopt rather more tolerant attitudes to social issues such as sexuality and bisexuality, drug use or prostitution than those that prevail at home. Under these circumstances it is the tourists, not the hosts, who are likely to experience a challenge to their traditional moral codes and behaviours.

An outline review of the literature will reveal a core of common social issues that are routinely linked with tourism. These include tendencies for tourism to be associated with increased incidence of gambling, prostitution and certain types of crime, together with a rather more subtle impact upon religions within host countries (Wall and Mathieson, 2006).

Gambling, prostitution and crime are frequently inter linked, both through organisational structures in which the ownership of casinos and brothels is often vested in the same hands and sometimes financed by profits from criminal activity, and through spatial proximity in which clubs, casinos, bars and brothels cluster to produce 'red light' or 'entertainment' districts. London's Soho and Amsterdam's Warmoesstraat are examples. The links to tourism are, however, much less clear.

The hedonistic character of many forms of tourism inevitably fosters some interest in activities such as gambling or prostitution, although interest will occur selectively and, with the occasional exception of resorts such as Las Vegas or Monte Carlo (see Leiper, 1989), only small minorities of visitors will actually indulge in these activities. However, gambling is a growth area in tourism (Fadington, 1999) and has shown a demonstrable capacity both to sustain the expansion of established resorts such as Las Vegas (which has emerged as the fastest growing city in the American West over the last decade (Douglass and Raento, 2004)) and to revive places that were previously in decline – such as Atlantic City. Timothy's (2001) study of border tourism has also shown how the establishment of casinos on the borders between states that permit gambling and those that do not has become a common form of development, reflecting significant levels of suppressed demand for gambling in many local communities, as well as amongst visitors. The effects of gambling tend to be variable although it is probably fair to characterise the benefits as being broadly economic, whereas the problems tend to be social in nature, especially when addictive forms of behaviour develop that lead to breakdown of family or other social ties (Pizam and Pokela, 1985; Wall and Mathieson, 2006).

The sensitivities that surround prostitution have resulted in there having been – until fairly recently – comparatively few empirical studies of the role of tourism in its development, although the essay by Graburn (1983b) is an exception. More recent studies (e.g., Hall, 1992, 1996; Cohen, 1993; Muroi and Sasaki, 1997; and Opperman, 1999) have tended to suggest that whilst tourism may create an environment that is conducive to the development of prostitution and may promote existing practices, it seldom introduces activity in a direct sense. Thailand, for example, has developed a dubious reputation for sex tourism, yet it is clear that prostitution was an established element in local urban subcultures in Thailand long before the arrival of large-scale tourism. The primary effect of tourism seems to have been to encourage the addition of a tier of expensive, elite young women to meet the new demands of the tourist market (Cohen, 1993), although there is also some evidence to link tourism with newer forms of sexual exploitation, especially involving children.

Similarly, links between tourism and local crime are not always clear and consistent. Visible differences in levels of affluence between visitor and host are sometimes held to account for increases in the incidence of robbery and muggings, especially as tourists moving around strange locations, and often unable to distinguish 'safe' from 'unsafe' areas, are easy targets for streetwise criminals (Ryan, 1993; Prideaux, 1996; Sheibler et al., 1996; Harper, 2001). Tourism development has also been linked to increased rates of burglary, vandalism, drunk and disorderly behaviour, sexual and drug-related offences and soliciting by prostitutes (which is a criminal activity in many countries). However, statistical linkages do not necessarily mean that tourism causes such activity. The normal practices of tourists create conditions and environments in which many forms of crime will flourish, but except in situations where the tourists themselves are perpetrators of crime (as, for example, in the rising incidence of drunken and violent behaviour by young British tourists in Spanish Mediterranean resorts or, less commonly, the smuggling of drugs or other illegal goods by visitors), tourism is unlikely to introduce crime to a host society. The tendency must already exist, albeit, perhaps, in a latent form.

The moral value systems in many societies are rooted (if only distantly) in religious beliefs and practices, so the capacity of local communities to resist changes to moral codes may be partly dependent upon the strength of the religious basis to daily life. The links between tourism and religion have changed through time in some interesting ways. Religion was, and still is, a basis for particular forms of tourism based around pilgrimage, but whilst in many societies – especially in the developed world – belief in religion has been eroded in the face of growing agnosticism and atheism, religious sites have become an increasingly popular object of the tourist gaze, even if people do not subscribe to the beliefs that such places represent (Vukovic, 2002). There is no doubt that worshippers at the great Anglican cathedrals of England are greatly outnumbered by the millions of tourists who come simply to view the buildings.

This is a potential source of conflict when the practices of the devout are directly compromised by the idle curiosity of the masses. For most tourists, religion has become entertainment, typically in the form of casual inspection of religious sites or the observation of religious ceremony. For the worshipper, or the participant in a religious ceremony, the place or the event has quite different meanings and may be a source of profound spiritual, moral and psychological support. Any devaluation of the experience, therefore, whether it be through the commodified performance of religious spectacles for tourist consumption or irreverent behaviour on the part of tourists towards religious places or practices, may be deeply disturbing (see Case Study 6.2). Yet as before, the effect of such encounters will be unpredictable. On the one hand, it may serve to reinforce local adherence to religiously based practices and values, strengthening a sense of local cultural identity. Equally, it may erode the position of religion within societies, altering the meaning and symbolism of ceremonial events and opening the way towards wider processes of social and cultural change.

New social structures and empowerment

The composite effect of many of the socio-cultural changes that have been associated with tourism may eventually lead to significant shifts in local social structures and new patterns of social empowerment. As before, effects are likely to be most pronounced where tourism brings together hosts and visitors from contrasting socio-economic traditions, but where such differences are marked, repercussions could be significant.

Changes result through a number of pathways, but three are worth emphasis. First, tourism creates new patterns of employment and opportunities for work amongst groups who, in traditional societies, may not normally work for remuneration, for example women. As we have seen in Chapter 4, tourism creates particular opportunities for employment for women, and it has been argued that one of the beneficial effects of tourism has been to help the liberation of women from traditional social structures, to provide the independence that comes with a personal income, and to promote, through time, more egalitarian social forms and practices. Ateljevic and Doorne's (2003) study of the production of 'tie-dyed' fabrics as tourist commodities in the Chinese prefecture of Dali showed how local women engaged in the production of fabric acquired a greater level of independent control over their lives whilst the income they earned raised the prospect of improved educational and employment opportunities for their children. In many traditional agrarian societies, the arrival of tourism may also be beneficial to young people who gain employment in the industry. This enables new levels of financial independence, partial or total release from the traditional social controls of their elders – especially within extended families – and new choices in matters such as place of residence or selection of marriage partners.

Such social empowerment may also arise through the second key process – language change – although this is also an area that has received only modest levels of attention with respect to the role of tourism (Cohen and Cooper, 1986). Language is a significant defining feature of a society. It provides identity and acts as a cultural marker (Wall and Mathieson, 2006) but, more significantly, it underpins social relations by defining who talks to whom. However, because international tourism is generally conducted through one of a very few languages that have worldwide usage (most typically English and, to a lesser extent, French), tourists that originate in countries where these 'global' languages are based will often harbour expectations that the hosts will have at least some grasp of their language. Expectation is often reinforced by practice, at least to a basic level of communication. Foreign ownership of tourism developments may impose a new language as the norm for business purposes, whilst training in the hospitality industries will also strive to give personnel some grasp of languages that they are likely to encounter. But as with employment, so the acquisition of new language skills empowers people in several significant ways. It provides wider access to globalised media and the influences that the media convey; it makes easier the possibility of migration in search of employment or improved prospects; and it alters the status of the individual within their home society through the acquisition of a powerful skill that others may lack. However, as skills in international languages are acquired, so there is a risk of displacement of local languages (Huisman and Moore, 1999), although there are few empirical studies that have demonstrated this tendency with clarity, White's (1974) study of the decline in the use of Romansch in areas of tourism development in Switzerland being an exception.

The empowerment that comes with employment or the adoption of new languages is best envisaged as operating at the individual or group level. But occasionally, whole communities and cultures become empowered through the development of tourism and its integration into local socio-cultural development. Picard's (1993, 1995) studies of tourism development on Bali showed how the appeal of the distinctive Balinese culture to international tourists provided powerful political and economic 'levers' that could be deployed to advantage by the local Balinese authorities in their dealings with the central Indonesian government. The desire of the Indonesian authorities to showcase Balinese culture, in order to project positive images of the state to the international community, enabled a reassertion of local identity and a political elevation of Bali in ways that might not otherwise have been possible.

This links to the third pathway to change through the creation of local resistance. Pitchford (1995) has noted that where social groups possess a distinctive culture that forms the basis of an attraction to tourists, this becomes a resource of both material and cultural significance in any local assertion of identity and resistance. Such resistance may be directed against the homogenising effects of globalisation and the mass marketing of international travel, but may also be deployed to counter colonialism, whether as an external or an internal process. Pitchford's (1995) study examines how tourism development in Wales has been used to promote and protect Welsh culture in the face of a protracted and systematic erosion of Welsh identity through internal colonisation by the English. Picard's (1993, 1995) work on Bali demonstrates a comparable process in terms of Balinese resistance to the colonialism of the Indonesians.

However, resistance is not just a process of engagement between the communities that receive tourists and the wider world, it also works *within* communities as a process through which the cultural acceptance of tourism is negotiated. This notion of mediated resistance is examined in the final case study in this chapter which serves not just to illustrate how the local relations between tourism and cultural practices might be rationalised, but also illustrates several of the wider issues that have been discussed in this chapter.

CASE STUDY 6.2

Mediated resistance to tourism in a Hindu pilgrimage town

The Hindu town of Pushkar lies in the Indian province of Rajasthan and has a well-established tradition as a centre of religious pilgrimage. The town stands on the shores of a lake that has mythological links to the god Brahma and a large number of temples and shrines, together with *ghats* (step-like embankments that are ritual locations for worship, bathing and cremation of the dead) provide focal points for *puja* (ritual patterns of devotion). The practice of *puja* draws pilgrims from all over India and provides an important livelihood for Brahman *pandas* – religious men who conduct the various rituals but who also arrange for the more routine requirements of pilgrims for food and accommodation and who receive payments for their services.

Tourism to Pushkar originated in the late 1960s through small-scale visiting by hippies, but, under the guidance of state and regional governments that are anxious to bring new sources of wealth to a particularly impoverished area, it has evolved into a much larger-scale activity involving large numbers of wealthy, white, Western tourists. This has impacted, first, on the physical landscape of the town as hotels, guest houses, restaurants and tourist shops have been developed amongst the religious sites, and, second, upon the cultural basis to the community. These cultural impacts have included:

- commodification of *puja* through its adaptation (and abbreviation) as a staged performance that is conducted in return for money (as donations) and delivered in a blend of pidgin English;
- the relinquishment by some *pandas* of their traditional role as mediators of pilgrimage and who, instead, apply their knowledge to meeting the needs of tourists from whom much higher earnings may be derived;
- signs of acculturation, especially amongst young adult groups who have begun to adopt Western fashions (such as jeans and baseball caps) and who also aspire to ownership of luxury goods such as cameras;
- a realignment of local production of crafts to meet the demands of tourists for souvenirs rather than catering for the needs of pilgrims;
- replacement of local cuisine in many local restaurants with alternative dishes that are held to be more suited to the tourist palate.

Such change has inevitably created widespread local concern amongst those not directly involved in tourism about the loss of tradition and the debasement of the religious significance of the site. Many tourist practices – such as dress codes and the tendency to photograph sacred sites such as the temples and the activity of the *ghats* – are seen as an affront to Hindu beliefs and customs, yet at the same time a significant portion of the local population have come to be dependent upon tourism's economic benefits. Many local people have therefore been forced to adopt an ambivalent position in relation to tourism – on the one hand condemning or expressing concerns over the damaging impacts of the activity on local culture; whilst on the other, finding ways to rationalise (or mediate) that resistance in ways that permit continued participation in a valuable activity.

The authors of the study propose that a key part of the process of mediating resistance has been to deflect accountability for tourism's impacts onto outsiders, but not the tourists themselves. This draws upon some traditional local divisions between the Brahman identity of long-term residents (the insiders) and those from the wider India (the outsiders) who have moved to Pushkar – perhaps as administrators or to take advantage of commercial opportunities – and on whom the 'unwelcome' development of tourism may be blamed. Similarly, government is also held to account for the way in which the religious and cultural identity of the town has been promoted to international tourists. This has helped to sustain a political rhetoric in which local politicians and pressure groups have often been able to strengthen their local positions by leading periodic attacks on state and national government on issues relating to tourism development, but, crucially, conducted in ways that do not directly attack the tourist industry itself.

However, perhaps the most interesting form of mediation of local resistance has been to invoke local religious beliefs – in particular, the Hindu belief in *kalyuga* – as a means of

continued

rationalising the damaging presence of tourists. *Kalyuga* refers to Hindu understanding of how cosmic ages change through a cycle of progressive decline in physical and moral standards that are rejuvenated at the start of the next age. Fortuitiously, perhaps, *kalyuga* constructs the present age as one of discord and disintegration and for the devout in Pushkar, this helps to explain – and perhaps render more acceptable – the deviation from the traditional modes of social conduct and behaviour that is evident in the declining status of pilgrims and the increasing presence of tourists in holy places.

This study provides some interesting insights into ways in which a local community that has both vested interests in, and genuine concerns over, tourism development can come to terms with the challenges that tourism creates. In some respects, of course, the recourse to a rhetoric that deflects accountability onto external forces (the role of outsiders, government, or even cosmic cycles) will appear to an outside audience as a convenient way of evading awkward local dilemmas. However, the more important issue is how the local community views the process and if such forms of mediated resistance make what the authors of this study call the disjunction between tourism and local culture palatable and which functions to create conditions for the acceptance of cultural change, then there is surely merit in that process.

Source: Joseph and Kavoori (2001)

Conclusion

The impacts that tourism brings to host societies and cultures are remarkably diverse and often inconsistent in their effect, reflecting the many different ways under which people travel and the variations in local conditions that they encounter. In some situations, where cultural distances between hosts and visitors are slight, socio-cultural effects of tourism are minimal. Elsewhere, changes are more significant and although the tendency in many of the discussions of socio-cultural relationships between tourists and the communities they visit is to emphasise the negative, the preceding discussions have attempted to show that there are often significant and tangible benefits from encounters between tourists and local people too. Thus, for example, tourism can actually become an agent for empowerment, a means for sustaining cultural identities and a way of asserting distinctive local identities in a world that is increasingly shaped by global processes.

It is also important to re-emphasise the point that societies and cultures are not fixed entities, nor are hosts the passive receivers of the stimuli to change that the visitor may bring. Society and cultures evolve constantly, in response to a wide range of external and internal influences – one of which is clearly international tourism. But it must be remembered that tourism is one of many such influences and disentangling the effects of tourism from those of, *inter alia*, multinational corporations, international political organisations, global media, NGOs aid and charitable groups, and cultural exchange and educational programmes is probably an impossible task.

Summary

The relations between tourism, societies and cultures are often reflected in imprecise ways, but few doubt that through contact between tourists and the societies and culture that are toured, tourism has the power to alter socio-cultural structures in destination areas even though the precise forms of such effects are often uncertain and spatially variable. A diversity of factors may be held to account for such variation, including the nature and scale of tourist encounters, the cultural 'distance' between the different groups and the stages of tourism development that have been attained. There is also a range of possible effects, including issues of cultural commodification and (mis)representation, the introduction of new moral codes, or the promotion of new social value systems. However, whilst the tendency is to represent tourism as a form of socio-cultural 'pollution', there is a growing body of evidence to show that processes of cultural influence are often two-way, and, further, that positive socio-cultural impacts around local empowerment and the maintenance of cultural identities and their associated practices may be initiated through contact between tourists and local communities.

Discussion questions

1   If tourist encounters with host communities are characteristically transitory and superficial, why is tourism considered to be a threat to local societies and cultures?
2   What are the primary social and cultural benefits that tourism might bring to the places that tourists visit?
3   Evaluate, with reference to case studies, the evidence that tourism produces moral drift in host communities.
4   Discuss the validity of MacCannell's concept of authenticity as a basis from which to explain contemporary tourist interest in foreign societies and cultures.
5   To what extent is it fair to characterise tourism relations with host communities as asymmetric and processes of cultural change through tourism as being unidirectional?
6   Is there still a place for traditional concepts such as the demonstration effect and acculturation theory in developing understanding of how tourism produces social and cultural change?

Further reading

Students with interests in the conceptual basis to the relations between tourism, society and culture might usefully commence their further reading by reference to a number of classic texts and papers. These will include:

Cohen, E. (1988) 'Authenticity and commodification in tourism', *Annals of Tourism Research*, Vol. 15 (2): 371–86.
MacCannell, D. (1973) 'Staged authenticity: arrangements of social space in tourist settings', *American Journal of Sociology*, Vol. 79 (3): 589–603.
Smith, V.L. (1977) *Hosts and Guests: The Anthropology of Tourism*, Philadelphia: University of Pennsylvania Press.
Urry, J. (1990) *The Tourist Gaze: Leisure and Travel in Contemporary Societies*, London: Sage.

More recent writings that either develop themes that are explored in these sources or which introduce valuable new perspectives include:

Crouch, D. (ed.) (1999) *Leisure/Tourism Geographies*, London: Routledge.
Crouch, D., Aronsson, L. and Wahlstrom, L. (2001) 'Tourist encounters', *Tourist Studies*, Vol. 1 (3): 253–70.
Fisher, D. (2004) 'The demonstration effect revisited', *Annals of Tourism Research*, Vol. 31 (2): 428–46.
Rojek, C. and Urry, J. (eds) (1997) *Touring Cultures: Transformations in Travel Theory*, London: Routledge.

Smith, V.L. and Brent, M. (eds) (2001) *Hosts and Guests Revisited*, New York: Cognizant Communications.

An excellent general discussion of the impact of tourism upon societies and cultures may be found in:
Wall, G. and Mathieson, A. (2006) *Tourism: Change, Impacts and Opportunities*, Harlow: Prentice Hall.

Similarly useful discussions may also be found in the following:
Mowforth, M. and Munt, I. (2003) *Tourism and Sustainability: Development and Tourism in the Third World*, London: Routledge.
Murphy, P.E. (1985) *Tourism: A Community Approach*, London: Routledge, pp. 117–51.
Ryan, C. (1991) *Recreational Tourism: A Social Science Perspective*, London: Routledge, pp. 130–66.

A very readable collection of case studies of tourism impacts and cultural conflicts is provided in:
Robinson, M. and Boniface, P. (eds) (1999) *Tourism and Cultural Conflicts*, Wallingford: CAB International.

For a comprehensive examination of culture as a tourist attraction in Europe see:
Richards, G. (ed.) (1996) *Cultural Tourism in Europe*, Wallingford: CAB International.

7 ▸ Strategies for development: the role of planning in tourism

Implicit in many perspectives upon sustainable tourism – and, indeed, on tourism development in general – is the view that planning has a key role to play in assuring orderly and appropriate patterns of development and, within this process, resolving many of the conflicts that such development may generate (Gunn, 1994; Inskeep, 1991). Tourism planning provides a primary mechanism through which government policies for tourism may be implemented (Hall, 2000) and, in its different forms, can be a mechanism for delivering a range of more specific outcomes. These will include:

- the integration of tourism alongside other economic sectors;
- the direction and control of physical patterns of development;
- the conservation of scarce or important resources;
- the active promotion and marketing of destinations;
- the creation of harmonious social and cultural relations between tourists and local people.

Hall (2000) argues that tourism planning has the potential to minimise the negative effects, maximise economic returns to the destination and build positive attitudes towards tourism in the host community. Conversely, where effective planning of tourism is absent, there are evident risks that tourism development will become unregulated, formless or haphazard, inefficient and likely to lead directly to a range of negative economic, social and environmental impacts.

This chapter attempts three tasks. The first section aims to explore the basic nature of planning processes and some of the types of planning approach that have been applied to tourism development. Second, the importance of planning tourism is explained and some of the main strengths and limitations in both conception and implementation of tourism plans are highlighted, especially in relation to issues of sustainability. Finally, the differences in approach to tourism planning at national, regional and local levels are examined and illustrated.

Planning and planning processes

'Planning' has been defined in various ways, but a common perspective recognises it as an ordered sequence of operations and actions that are designed to realise either a single goal or a set of interrelated goals and objectives. Murphy (1985: 156) writes that 'planning is concerned with anticipating and regulating change in a system, to promote orderly development so as to increase the social, economic and environmental benefits of the development process'. This conceptualisation implies that planning is (or should be) a process:

- for anticipating and ordering change;
- that is forward looking;
- that seeks optimal solutions to perceived problems;
- that is designed to increase and (ideally) maximise possible developmental benefits, whether they be physical, economic, social or environmental in character;
- that will produce predictable outcomes.

From this broad definition it follows that planning may take on a variety of forms and may be deployed in a great diversity of situations. These will include physical and economic development, social policy, service provision, infrastructure improvement, marketing or business operations and environmental management.

A general model of the planning process

Although there are a diversity of potential applications for planning, the basic nature of the planning process is, in theory, remarkably uniform, even allowing for the variation in detail that will reflect the specific applications in which planning is being exercised. Figure 7.1 sets out a normative model of the planning process in which the principal elements in devising and implementing a plan are envisaged as a series of key stages or steps (WTO, 1993). These key stages are essentially self-explanatory but there are several features of the model to emphasise:

1 There is a progression within the planning process from the general to the specific. The process begins with broad goals and refines these to produce specific policies for implementation.
2 There is an evident circularity in the process by which objectives and the options for realising those objectives are open to review and amendment in the light of either background analysis or the performance of the plan in practice.

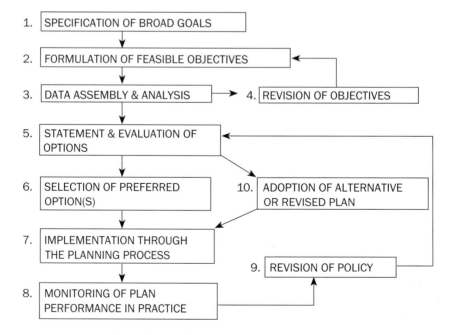

Figure 7.1 Normative model of the planning process

3　The process is dynamic. The general model maps out a set of procedures that allow planning to be adaptive to changing circumstances. This is a quality that is especially important to tourism planning where patterns of demand and supply are often volatile. Flexibility should be a key concept for tourism planners.

It should, of course, be conceded that normative models of this type have attracted some criticism. Hall and Jenkins (1995) make the point that planning, whilst theoretically informed, actually takes place in real-world settings that are complex and in which there are a host of embedded values, powers and interests that tend to modify or subvert ideal patterns. However, whilst acknowledging the validity of this view, the model is still held to retain a utility in outlining an ideal sequence, even if ideals are not always attained in practice.

Types of plan and planning approaches

The general model defines a typical process that may be applied in a wide range of planning contexts. However, within this diversity of contexts, the actual planning approach commonly falls into one of four categories (although these are not mutually exclusive):

- boosterism;
- industry oriented;
- physical–spatial;
- community oriented.

This classification is derived from work by Getz (1986) and has been widely deployed in the subsequent literature on tourism planning.

According to Hall (2000), boosterism is arguably the dominant planning approach in tourism development in many destination areas, but whether it actually constitutes planning in a true sense is debatable. Under a boosterist approach the process of planning is generally aligned with exploitation of the natural or cultural resources of a destination to maximise economic returns. Local participation in the planning process is characteristically minimal, with decisions primarily deferred to government or industry 'experts'. Wider impacts are thus accorded little attention as planning is focused upon the capacity of tourism to foster (or boost) economic development.

Industry-oriented approaches perhaps represent a more regulated form of boosterism (in so far as they still constitute a planning approach that is shaped by economic imperatives), but there is a stronger focus upon the use of planning to achieve efficient, sustainable use of resources to deliver increased employment and regional development. Activities such as marketing and promotion tend to become conspicuous elements in tourist plans under an industry-oriented approach, whilst socio-cultural or environmental concerns, or issues of how tourism wealth may be distributed across communities, are often accorded much lower priority (Burns, 1999).

Physical–spatial approaches are grounded in traditional forms of urban planning and its primary concerns for regulating physical development and the proper ordering of land use and associated infrastructures. This form of planning approach is strongly represented in several influential texts on tourism planning – especially Gunn (1994) and Inskeep (1991) – and reflects an emphasis that is widely encountered in the work of statutory planning authorities relating to the physical design and layout of tourist areas and the zoning of activity. As a response to the rising tide of concerns around issues of sustainability, physical–spatial approaches have become more attuned to managing the environmental impacts of tourism, but the integration of socio-cultural concerns into this approach is much less evident.

Community-oriented approaches – which are now widely associated with Murphy's (1985) groundbreaking work – have become much more conspicuous in recent years as the emphasis upon finding sustainable forms of development has become more widely reflected in both policy and practice. This reflects the growing realisation that local participation is often essential to securing sustainable development of tourism and effective management of tourism's environmental impacts, and also that community involvement is generally the most effective way of resolving socio-cultural tensions between tourists and local people. Thus, in theory, planning that is community oriented can help not only to provide essential frameworks through which local living standards may be raised and infrastructures improved in ways that benefit local people as well as visitors, but also enable development that is aligned with the cultural, social and economic agendas of local communities (McIntosh and Goeldner, 1986).

Within these contrasting approaches, different types of plan may then be developed, including master plans, incremental plans and – occasionally – systematic plans. The master plan approach is arguably the most traditional and also the least suited to the particular requirements of tourism. Master plans centre on the production of a definitive statement that provides a framework for guiding development. The plan defines an end-state (or set of targets) towards which public and/or private agencies are encouraged (or required) to work. Targets are normally expected to be attainable within set time periods – typically a five-year time horizon – and once set in motion, a master plan is normally left to run its course until its time has elapsed. At the end of the plan period, a new master is produced. The master plan approach has the advantage of adopting a comprehensive view of development processes but has also been widely criticised. Burns (2004), for example, suggests that master plans suffer from being:

- too complex and too demanding of government resources;
- likely to encourage a reductionist, homogenising view of tourism that only reflects the patterns of known market segments;
- undemocratic and over-dependent upon expert knowledge at the expense of local participation;
- limited by national political boundaries and therefore increasingly at odds with globalising tendencies and the erosion of national space.

However, such deficiencies have not prevented widespread use of master planning approaches in tourism (see, e.g., Gant and Smith (1992) on national development planning in Tunisia).

As an alternative approach, the natural dynamism in tourism (whereby new tourists and new tourism products and destinations tend to redefine patterns more or less continuously) has encouraged some tourism planners to move away from a master plan approach and towards the more adaptable forms of incremental (or continuous) planning. The key difference between incremental plans and master plans is that whereas the master plan is a periodic exercise, incremental planning recognises a need for constant adjustment of development process to reflect changing conditions. So whereas the master plan approach, in defining a blueprint for development, would place an emphasis upon Stages 1 and 2 of the general model (specification of broad goals and objectives), the incremental approach shows a much greater concern for Stages 8–10 (monitoring, revision of policy and objectives, and adoption of revised plans). Since one of the primary objectives of tourism planning is to match levels of demand to supply, this capacity to adjust planning programmes as required is a particular advantage.

One of the recurring themes in the tourism planning literature is the need to plan such a diffuse activity comprehensively and in a manner that integrates the planning of tourism with the other sectors with which it has linkages. Given the breadth of those linkages and the diverse impacts that tourism tends to generate, a planning approach that is comprehensive yet allows for the need for regular readjustment in physical development, service delivery and visitor management is clearly advantageous. Such an approach is provided by systems planning.

The systems approach (which originated in the science of cybernetics but is now applied widely in a range of investigative, managerial and planning contexts) is founded on the recognition of interconnections between elements within the system, such that change in one factor will produce consequential and predictable change elsewhere within the system. Thus in order to anticipate (or plan for) change, the structure and workings of the system need to be fully understood and taken into account in any decision making. In a planning context, systems approaches attempt to draw together four key elements – activity, communications, spaces and time – and map the interdependence between these in producing patterns of development.

The advantages of a systems approach to planning are that it is comprehensive, flexible, integrative and realistic, as well as being amenable to implementation at a range of geographic scales. On the negative side, however, a systems approach requires a great deal of information in order to comprehend how the system actually works (Stage 3 of the general model); it is dependent upon high levels of expertise on the part of the planners and is, therefore, an expensive option to implement that will often exclude local people from effective participation in planning processes because of its complexity. For these reasons it remains the least widely applied of the three methods described, although as planning techniques become more developed, it is an approach that may become more prominent through time.

Tourism and planning

Planning is important in tourism for a wide range of reasons. First, through the capacity of physical planning processes to control development, it provides a mechanism for a structured provision of tourist facilities and associated infrastructure over quite large geographic areas. This geographic dimension has become a more significant aspect as tourism has developed. Initially, most forms of tourism planning were localised and site-specific, reflecting the rather limited horizons that originally characterised most patterns of tourism. But as the spatial range of tourists has become more extensive as mobility levels have increased, planning systems that are capable of coordinating development over regional, national and even international spaces have become more necessary.

In view of the natural patterns of fragmentation within tourism (with its multiplicity of providers and tourist segments), any systems that permit coordination of activity are likely to become essential to the development of the industry's potential. This fragmentation is mirrored in the many different elements that are often required to come together within a tourism plan, including accommodation, attractions, transportation, marketing and a range of human resources (see Figure 7.2). Given the diverse patterns of ownership and control of these factors in most destinations, a planning system that provides both integration and structure to these disparate elements is clearly of value (Inskeep, 1991). Planning systems (when applied in a marketing context) will also enable the promotion and management of tourism places and their products, once they are formed.

Planning can also be a mechanism for the distribution and redistribution of tourism-related investment and economic benefits. This is a particularly important role for planning given that tourism is becoming an industry of global significance but one where activity does not fall evenly across different regions and where the spatial patterns of tourist preference are

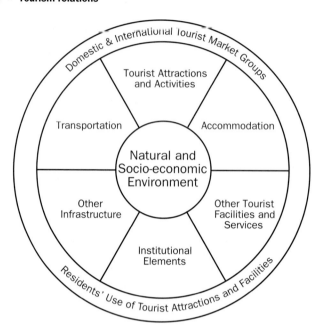

Figure 7.2
Principal components
of a tourism plan

also prone to variation through time. Planning may assist both the development of new tourist places and, where necessary, the economic realignment of established places that tourists have begun to desert.

The integration of tourism into planning systems gives the industry a political significance (since most planning systems are subject to political influence and control) and therefore provides a measure of status and legitimacy for an activity that has not always been taken too seriously as a force for economic and social change.

A common goal of planning is to anticipate likely demand patterns and to attempt to match supply to those demands. Furthermore, through the exercise of proper controls over physical development and service delivery, planning will aim to maximise visitor satisfaction. There is now ample evidence from around the world that the unplanned tourist destinations are the ones that are most likely to be associated with negative impacts and low levels of visitor satisfaction, whereas the application of effective planning has often enhanced the tourism product, to the benefit of both host and visitor alike. Baidal's (2004) study of the evolution of tourism planning in Spain highlights both the damaging consequences of unplanned tourism development in the 1960s, as well as the improvements that have accompanied the establishment of a strong base in tourism planning by the Spanish autonomous regions after 1994.

Finally, as noted in the introduction to this chapter, there are clear links between planning and principles of sustainability. Implicit in the concept of sustainable tourism are a range of interventions aimed not only at conserving resources upon which the industry depends, but also at maximising the benefits to local populations that may accrue through proper management of those resources. Sustainable tourism strives to balance economic growth, environmental protection and social justice (Coccossis, 1996) and this requires integrative planning (Hall, 2000).

Links between planning and sustainability are especially important at a local scale. This is partly because implementation of the principles of sustainability has become focused at a local level by the widespread adoption of Local Agenda 21 as a framework in which to develop sustainable practices, but also because the requirements for strategic sustainable tourism planning will vary significantly from place to place and are therefore best understood in a local context. Local level planning is also more easily informed through public participation and through the involvement of stakeholders – both of which are seen as essential ingredients in delivering sustainable forms of development (Simpson, 2001; Ruhanen, 2004).

Hall (2000: 38) proposes five key areas in which local consultation can help to shape a sustainable planning programme by establishing:

- the primary values that both local people and visitors attach to the area;
- the aspirations that residents may hold for tourism in their area;
- the fears that local people may harbour around possible tourism impacts and effects;
- the special characteristics that locals may wish to share;
- the aspects of the local area that might detract.

By considering these criteria, Hall asserts, local destinations are 'better placed to determine their positioning in the tourism market, product development, infrastructure requirements, development constraints, local needs and preferred futures' (Hall, 2000: 38).

The diversity of roles and functions that is set out above does, however, lead to problems in defining the essential dimensions of tourism planning. In fact, tourism planning, as a concept, is characterised by a range of meanings, applications and uses. It encompasses many activities; it addresses (but does not necessarily blend) physical, social, economic, business and environmental concerns and in so doing involves different groups, agencies and institutions with their own particular agendas. Tourism planning may be exercised by both the public and the private sectors and be subject to varying degrees of legal enforcement. It also works at local, regional, national and (occasionally) at international scales. So to talk of 'tourism planning' as if it were a single entity is, therefore, highly misleading, and Table 7.1 attempts to reinforce this point by summarising a cross-section of applications that are located within the broad realms of tourism planning.

Apart from ambiguities over what may actually constitute tourism planning, there are further constraints and weaknesses to be taken into account. These include a tendency towards the adoption of short-term perspectives, organisational deficiencies and problems of implementation.

The adoption of short-term perspectives is a common characteristic in tourism, and, in the view of some authors, has limited the development of longer-term, strategic planning in the sector (see, for example, Cooper, 1995). The primacy of short-term responses arises for several reasons. It is a reflection of the natural rhythm of annual cycles within tourism whereby the industry tends to adopt a season-by-season perspective on its performance. But it is also a consequence of the structure of the industry at most destinations and the dominance of small- or medium-scale enterprises – a sector that adheres strongly to short-term, tactical views of tourism and is difficult to integrate into wider, longer-term planning frameworks.

Those frameworks may themselves be subject to a range of organisational shortcomings. In many destination areas, the speed with which the need for tourism planning has grown has outstripped the development of organisations, expertise and knowledge to undertake the task. Studies of tourism planning in some of the newer global destinations such as New Zealand and the micro-states of the South Pacific, for example, reveal common problems of inconsistencies in the development of tourism strategies both between and within states and regions; fragmentation and division of responsibility between different public and private agencies; lack of knowledge of patterns of tourism in localities; and an absence of planners with specialist knowledge of the industry (Craig-Smith and Fagence, 1994; Page and Thorne, 1997). Yet even destinations with well-developed planning structures and a

Table 7.1 *Diversity of tourism planning*

Planning sector	Typical tourism planning concerns/issues
Physical (land)	Control over land development by both public and private sectors Location and design of facilities Zoning of land uses Development of tourist transportation systems Development of public utilities (power, water, etc.)
Economic	Shaping spatial and sectoral patterns of investment Creation of employment Labour training Redistribution of wealth Distribution of subsidies and incentives
Social/cultural	Social integration/segregation of hosts and visitors Hospitality Authenticity Presentation of heritage/culture Language planning Maintenance of local custom and practice
Environmental	Designation of conservation areas Protection of flora and fauna Protection of historic sites/buildings/environments Regulation of air/water/ground quality Control over pollution Assessment of hazards
Business and marketing	Formation of business plans and associated products Promotional strategies Advertising Sponsorship Quality testing and product grading Provision of tourist information services

good understanding of the tourism markets are not immune from these difficulties. In the UK responsibility for 'planning' tourism falls to a range of agencies, including regional tourist boards, national park authorities and local government planning departments – the last of which rarely contain tourism experts. As a result, the emergence of what has been termed an 'implementation gap' – that is, a divergence between what is intended by a tourism plan and what is actually delivered – has been a problem in many localities (see Kun et al. (2006) for a discussion of implementation problems in Chinese tourism).

Tourism planning at contrasting geographical scales

The use of geographical scale is a particularly valuable device for drawing out key differences in emphasis and application within tourism planning, and to illustrate the point, the chapter now addresses tourism planning across a spectrum of scales from the national to the local. (Some writers, for example Hall (2000), also recognise an international or supranational scale of intervention. However, this is at best an emerging area that is evident – for instance – in some aspects of planning within the EU and in some of the regulatory frameworks that now shape international conservation or the preservation of designated World Heritage Sites, but perhaps lacks the unity and holism that characterises the forms of planning on which this chapter is focused. For this reason the following discussion is confined to an examination of national, region and local approaches.)

Before we examine these three scales of tourism planning, three general points are worth noting. First, although we may distinguish various geographic scales of planning intervention in tourism, these should be seen as interconnected rather than separate spheres of development. In a model framework, such a relationship might be viewed hierarchically so that within a nation state, national policies set a broad agenda for development that directly shapes regional-level policies, whilst these in turn form a framework for locally implemented plans. As the scale of intervention diminishes, so the level of detail in planning proposals is likely to increase, but the overall aims of planning at each level remain complementary and consistent in direction (Figure 7.3).

In practice, however, neat hierarchical arrangements are rarely found, sometimes because one of the tiers is missing or only partially developed, or, where all tiers are in place, differences in political or institutional attitudes at the different levels may frustrate implementation. In the UK there is no clearly defined regional level of planning so that the regional tourism strategies devised by the tourist boards have been limited in their effect by the absence of legal frameworks for their implementation. In contrast, in New Zealand, concerted attempts to produce regional tourism strategies have been frustrated by the absence of clear policy at the national level (Page and Thorne, 1997). Geographic area will also be a factor, with the absence of a regional tier being especially typical in small nations where regional subdivisions of the national space offer no particular logic or advantage.

In view of the interconnectivity between the different scales of planning, it follows that some areas of concern will form strands that run across all three levels, albeit with varying degrees of emphasis. Economic considerations are one element that may provide a focus of interest at all three scales, as are concerns for infrastructure improvements such as transportation and public utilities.

It is also inevitable that, given the widely differing developmental situations in which tourism planning is applied, there will be marked differences within – as well as between – levels, and this will also vary from place to place. In some situations strong national level policies have been criticised for concealing or failing to address significant regional disparities (Baidal, 2004), whilst difficulty around what is sometimes termed as the 'articulation' (or coordination) of policy across planning levels has been a recurring theme in many destination areas (see, e.g., Pigram, 1993; Kun et al., 2006). The following discussion should therefore be treated primarily as a generalisation of planning at the different spatial levels, with allowance made for the capacity of individual states, regions or localities to vary substantially from the patterns described.

Tourism planning at the national level

The significance accorded to national-level planning of tourism may vary quite markedly between destinations but is typically conceptual in character and normally seeks to define primary goals for tourism development and identify policies and broad strategies for their implementation. Within this framework, however, several more specific emphases may emerge and, in particular, we should note a common concern in national tourism plans with economic issues. This reflects the perceived capacity of international tourism to affect positively a country's balance of payments account and to create employment. Consequently, a growing number of nations, especially in the developing world, have positioned tourism centrally within their national economic development plans.

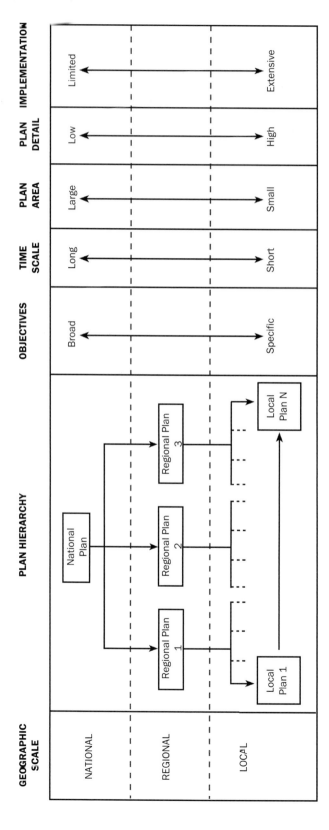

Figure 7.3 Model planning hierarchy

A second common role for national tourism plans is the designation of tourism development regions (see, e.g., Alipour (1996) and Tosun and Jenkins (1996) on the case of Turkey). This may be done for any of several reasons: to help structure programmes for the redistribution of wealth and to narrow interregional disparities; to create employment in areas where unemployment is an issue; or to channel tourism development into zones that possess appropriate attractions and infrastructure and are therefore considered suitable for tourism. As well as reflecting economic concerns, regional designation may also be guided by environmental factors, in particular a need to protect fragile regions from potentially adverse effects of tourism development.

National-level tourism planning is also often directed towards strategic marketing. This is especially prominent amongst developed destinations that possess the expertise and the resources to form and promote a distinctive set of national tourism products. For example, the strategic planning of British tourism development at the national level is largely absent and the primary role of national agencies such as the British Tourism Authority (BTA) is the marketing of British destinations to domestic and, especially, foreign travellers.

These economic and marketing roles of national level plans are reflected across the globe. Table 7.2 summarises findings from a study by Baum (1994) of national tourism policies in some forty-nine countries worldwide and places in rank order the eight most important determinants shaping national-level tourism planning in those countries. This study emphasises the economic and marketing functions already mentioned, but Table 7.2 also draws attention to national issues that occur more selectively. For example, some national tourism plans reflect needs to improve and develop infrastructure, especially transport; others include provision for educational and employment training schemes; whilst a smaller number recognise the potential for tourism to forge international linkages and to maintain positive images of a country within the international community.

However, the approaches to delivering the objectives set out in Table 7.2 vary considerably between nations. Some destinations have adhered to quite rigid programmes of national tourism master plans of the kind adopted in countries such as Indonesia and Tunisia (Gant and Smith, 1992; Pearce, 1994), whilst others, such as the UK, prefer a more low-key, flexible approach of policy guidance. The physical planning of tourism in England and Wales is only loosely shaped by government by means of Planning Policy Guidance (PPGs). These documents, which are effectively memoranda to local government planning departments, set out key issues to be addressed and preferred pathways for development, but allow considerable leeway for local interpretation of the guidance.

Institutional contexts of national tourism planning are also variable. In the study of national tourism planning referred to in Table 7.2 (Baum, 1994), only half of the countries surveyed had established a government department with sole responsibility for national tourism planning and nearly 15 per cent apparently had no governmental-level interests in

Table 7.2 *Main determinants of national tourism plans and policies in forty-nine countries (in rank order)*

- To generate foreign revenue and assist balance of payments
- To provide employment
- To improve regional and local economies
- To create awareness of the destination/country
- To support environmental conservation
- To contribute to and guide infrastructure development
- To promote international contact and goodwill

Source: Baum (1994)

the sector at all. Elsewhere, tourism was accorded only secondary interest and this is often reflected in the movement (or division) of tourism planning briefs between government departments through time. In the United Kingdom in the fifteen years between 1983 and 1998, tourism development was first the responsibility of the Department of Trade and Industry, then the Department of Employment and finally the Department of Culture, Media and Sport (which was originally called the Department of National Heritage). This rather uncertain status reflects the secondary position that tourism holds in many national planning frameworks and is a weakness that, in the longer term, may well need to be addressed.

To illustrate a current approach to tourism planning at the national level, Case Study 7.1 describes the planning of tourism in one emerging destination – Morocco.

CASE STUDY 7.1

National-level tourism planning in Morocco

The kingdom of Morocco, which lies on the north-west Atlantic coast of Africa, embodies many of the planning challenges of an emerging destination area. Although Morocco is separated by less than 10 miles of sea from Europe and its near-neighbour Spain (see Figure 7.4), its geographic and cultural positioning as a Muslim state on the continent of Africa had limited tourism development prior to 2001. Although the country is now recognised as possessing significant tourism assets that include:

- a high-quality, undeveloped coastline;
- a rich cultural heritage that blends Roman, Moorish, French and Arab influences and includes seven World Heritage Sites;
- important ecological areas;
- foreign visitor levels in 2001 stood at only 2.2 million. Most of these visitors were drawn to the main beach resort of Agadir and the inland city of Marrakech.

Somewhat unusually, the first Moroccan tourism development plan – titled 'Vision 2010' – is a personal initiative of the king, Mohammed VI. However, the approach adopted is essentially a partnership – enshrined in a framework agreement – between the Moroccan government and the General Confederation of Moroccan Enterprises (CGEM). The framework agreement makes explicit the intention of the government to position tourism as a national economic priority and to harness the potential for tourism expansion to stimulate economic development, attract inward investment, create jobs and increase the level of foreign earnings. In the process it is anticipated that local problems of unemployment, poverty and a lack of educational attainment will all be ameliorated.

To achieve these broad goals, some ambitious planning targets have been set:

- Levels of foreign visiting are to be increased to 7 million annually by 2010.
- 600,000 new jobs are to be created to support the tourism sector.
- 80,000 new hotel rooms are to be constructed, which will increase bedspaces from 70,000 in 2001 to 230,000 by 2010.
- Six major coastal resort areas are to be developed.
- Existing resorts, especially Agadir and Tangier, are to be refurbished.
- Cultural tourism to heritage sites such as Fes, Meknes, Marrakech and the Roman site at Volubilis is to be encouraged.

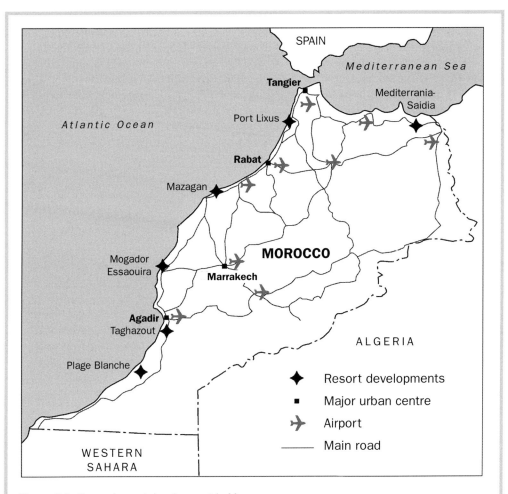

Figure 7.4 Planned resort development in Morocco

To enable these targets to be achieved, government and private sector investment is being channelled towards infrastructure improvement – especially the further development of several regional airports and a programme of new road construction; new promotional work by the Moroccan Tourist Board; and new tourism training programmes for Moroccans seeking employment in tourism. Plans to increase the number of incoming flights from Europe and associated deregulation through the adoption of an 'open-skies' policy (signed in December 2005) is also an essential part of the plan.

However, the key element in 'Vision 2010' is the creation of new coastal resorts, known as the 'Plan Azur'. These resorts are envisaged as up-market destinations, designed to raise the market profile of Morocco and attract high-spending tourists from Europe and north America to the luxury hotels, condominiums, golf courses and leisure complexes that form the basis to each development. The locations of the Plan Azur resorts (of which one is on the Mediterranean coast and the remainder on the Atlantic coast) are shown in Figure 7.4. These sites have been consciously selected to achieve a better balance of economic development across the country as a whole and divert growth away from the traditional economic heartland in the area between Casablanca and Rabat, the Moroccan capital. Each

continued

resort is based upon beach tourism but is intended to offer a slightly different product. So, for example, Plage Blanche (which is in the least-developed location and possesses important ecological sites) is expected to appeal to an eco-tourist market; Taghazout and Port Lixus will offer a range of sport-related activities; whilst Mogador has potential to attract heritage tourism because of its proximity to the World Heritage Sites of the Medinas at Essouira and, more distantly, Marrakech.

But whilst Vision 2010 clearly offers significant opportunities, there are also evident challenges to its full implementation. For example:

- Although foreign investment in the Plan Azur has been attracted from Belgium, Holland, Spain, South Africa and the USA, a general lack of national government funding for heritage sites has been identified as a problem.
- Outside investment has been affected by a lack of detailed information on market trends in tourism in Morocco and by bureaucracy and fiscal complexity at several levels of government.
- The wider development of rural tourism (which offers scope for trekking and adventure tourism) is severely constrained by poor conditions and a lack of basic amenities (such as clean water) in many rural areas.
- There is an urgent need to develop local programmes of training in tourism and hospitality management if local people are to benefit from the jobs that the plan will create.

It is clear, therefore, that these – and other – issues will need to be addressed if the ambitious agenda of Vision 2010 is to be fully realised.

Sources: Ambrose (2002); Morocco Ministry of Tourism (2005)

Tourism planning at the regional level

In comparison with national forms of tourism planning, regional tourism plans are usually distinguished by a marked increase in the level of detail and a sharper focus upon particular developmental issues. National plans tend to be broad statements of intent or, as in the case of Vision 2010 in Morocco, provide a template for development, but at the regional level the implications of those intents can be mapped far more precisely and planning can reflect specific requirements. Since the implications of development proposals for individual localities also become more apparent, some degree of public interest or participation within the tourism planning process may also be evident. There is, therefore, a sense in which regional levels of planning provide a means for balancing (or connecting) national and local interests, as well as enabling integration of rural and urban development (Church, 2004).

Several themes are likely to be carried through from the national to regional levels. In particular:

- Concerns for the impact of tourism upon regional economies and employment patterns.
- Development of infrastructure, including transport systems to assist in the circulation of visitors within the region, as well as provision of public utilities such as power and water supplies, both of which are frequently organised at regional levels.
- Further spatial structuring in which tourism localities within regions are identified.
- Regional-level marketing and promotion, especially where the region possesses a particular identity and/or set of tourism products.

However, there will also be distinctive features of a regional tourism plan that may not be found at a national level.

Regional plans commonly show greater levels of concern over environmental impacts. Except in the case of small nations, the uneven spatial patterns that are associated with tourism mean that environmental impacts are seldom felt at a national level but are manifest within regions and localities. The tendency of environmental effects to spread through natural systems and across wider geographic spaces (see Chapter 5) also means that a regional scale of planning is often the appropriate level for intervention with planned solutions to such impacts.

Second, regional plans will often contain consideration of the type and location of visitor attractions, together with supporting services such as accommodation. Such matters are rarely articulated in detail within a national tourism plan, but in regions the reduction in geographic scale makes it easier to define locations that will support tourism, establish how far existing capacities match expected demand and, thus, plan developments of new attractions and services that are required to meet identified deficiencies.

Regional plans may also reflect needs associated with the management of visitors. Distinctions between management and planning in tourism are often blurred, but unless the nation is small, the regional level is normally the first point at which tourist management issues begin to emerge clearly, albeit still at a macro scale. Regional zoning strategies aimed at either concentration or dispersal of visitors, the planning of tourist information services, designation of tourist routes and strategic placement of key attractions may all form part of regional tourism management strategies.

However, whilst an identifiable role for regional planning may be clearly discerned, experience suggests that the implementation of tourism planning at a regional scale has often proven to be problematic. Several factors may be held to account for these difficulties. For example:

- The concept of a 'region', as applied to tourism, is often rendered ambiguous – or even meaningless – by the manner in which highly mobile tourists shape tourist space.
- Regions generally contain local areas that may wish – for local reasons – to pursue a pattern of development that is not necessarily aligned with national objectives and which therefore sets up tensions between different levels of planning.
- There may be difficulties in connecting the wider agenda of tourism planning (which will often include issues such as marketing and promotion) with the more specific processes for land planning and development control that often form the focus of statutory planning procedures.

To illustrate some of these themes, Case Study 7.2 examines some of the difficulties faced in the development of tourism in a regional planning framework in Spain.

CASE STUDY 7.2

Regional tourism planning in Spain

Despite its position as a primary international destination and the critical significance of tourism to the Spanish economy, the general absence until very recently of effective tourism planning has been remarkable (Garcia et al., 2003). The expansion of Spanish tourism in the period between 1960 and 1975 took place under the dictatorship of General Franco and

continued

in the framework of an authoritarian, centralised state (Baidal, 2003). Developments were shaped by macro-economic policies and boosterist approaches, with high levels of speculation and little or no regulation of physical development and its impacts. Plans, where they existed, were usually considered as study documents rather than instruments for direct implementation (Baidal, 2004) and the consequences in terms of rapid deterioration of many tourism areas was, in hindsight, entirely predictable. However, following the death of Franco in 1975 and particularly since about 1990, the Spanish government has recognised the need for a more informed, planned and sustainable approach to tourism (Burns and Sancho, 2003). This has been pursued as part of a wider policy of progressive decentralisation of power in which the regions that comprise the Spanish state have acquired an autonomous status in a wide range of policy areas, including tourism (Pearce, 1997). This regional approach has been reinforced following the entry of Spain into the EU in 1986 and subsequent access to the various support mechanisms within European regional policy.

Initially the regional plans were expected to reflect national priorities to support, in a more structured way, the expansion of tourism and, in particular, the diffusion of tourism from heavily used coastal areas to other regions that were receiving comparatively low levels of use. Part of this process has also been directed at diversification of tourism products, to capitalise on undeveloped sectors, such as rural tourism (Mintel, 2003b). However, as the regional approach has matured, regional plans have also come to reflect – as one would expect – specific regional issues. Thus in the Balearics, for example, the regional plan is focused by the extreme pressures of growth that have been exerted on destinations such as Mallorca and emphasises policies such as limitations on the growth of urban resorts; increases in the extent of protected areas; stricter controls on building in rural areas; and restructuring of mature tourism regions on the coast (Baidal, 2004).

In the process of strengthening regional approaches to tourism development, the impact of EU regional policy – particularly the European Regional Development Fund – has been especially important. In the 1990s almost three-quarters of Spanish territory qualified as Objective 1 regions (in which GDP per person was below 75 per cent of the EU average) whilst smaller areas qualified as Objective 2 regions (areas affected by industrial decline) or Objective 5b (areas of rural development). These designations gave a significant boost to regional policy and planning in the autonomous communities, as well as a considerable increase in the availability of economic resources (Baidal, 2003).

The application of ERD funding has provided further impetus to the identification of specific approaches in different regions. Baidal (2003) notes that whilst there are several strategies that are common to tourism planning across all 17 autonomous communities (for example, a general commitment to developing tourism's growth potential and addressing territorial imbalances), different issues now provide a focus for planning in communities with established or emerging tourism industries. Hence in the communities with established tourism industries (such as Catalonia, Valencia and Andalusia), regional programmes are typically focused round issues such as strengthening of competitiveness, modernisation and overcoming structural defects – especially in older resorts. In contrast, in areas of emerging tourism development (such as Aragon, Galicia and Castilla-la-Mancha), issues such as economic diversification, capacity building and the development of new forms of tourism around nature, culture and heritage tend to shape the regional agenda.

Overall assessments of regional tourism policy in Spain (e.g., Pearce, 1997; Baidal, 2003, 2004) suggest some significant areas of benefit around quality improvement in traditional coastal resorts; the spatial diffusion of tourism into emerging tourism regions and an associated diversification of regional economies; and the development of new tourism products in sectors such as urban, rural, events, sports and heritage tourism. However, although regional planning has played an essential role in establishing the basis for all of these areas of gain, a number of evident weaknesses around the processes of regional planning have been recognised as impediments to progress. Several areas of concern have been noted.

The devolution of power has not, in practice, produced entirely clear demarcations of responsibility between national, regional and local levels of governance. Although all the autonomous regions have developed regional tourism strategies that are intended to guide local implementation, the powers invested in the local municipalities within the regions allow local areas considerable discretion in defining their own strategic directions. At the same time, the Spanish government continues to influence strategic planning directions at a national level. It retains a responsibility for the promotion of tourism abroad (Pearce, 1997) and, as a consequence of the Law of the Coasts (1988), the state also retains authority over the development of the Spanish coastline (Burns and Sancho, 2003). So whilst regional policy is envisaged as a 'bridge' between state objectives and the local agendas of the municipalities, articulation across these planning levels is, in practice, less than perfect.

Second, regional tourism plans and land planning processes have not been integrated in a consistent fashion and, indeed, in several regions – for example, Cantabria, Castilla y Leon, Murcia, Nararrc and Valencia – tourism and land planning are quite independent. This weakness has been reinforced in many areas by a marked reluctance to develop and apply legally enforced planning regulations (Baidal, 2004). There is still a sense in which the hesitancy and lack of urgency that characterised Spanish planning in the early phases of tourism development is still present and Pearce (1997) has expressed doubts over the capacity of some of the autonomous communities to deal effectively with the rising number of EU directives that require translation into law and implementation.

Regional planning approaches have also suffered from familiar problems that occur in many institutionalised systems of being complex, slow and lacking in flexibility. Moreover, the coordination that is essential in planning a complex area such as tourism is often missing. Baidal (2004: 329) notes that 'tourism plans often contain proposals that are hardly viable because they involve various administration departments which, on many occasions, have not even participated in the preparation of the plan'. This problem is compounded by the way that tourism planning responsibilities have been incorporated into the administrative systems of the autonomous communities. The normal pattern has been for tourism to become a part of a larger department and only one region – the Balearics – saw the need to establish at an early stage a special Department of Tourism in its administrative framework (Pearce, 1997).

Thus whilst the Spanish example demonstrates quite well the transformative potential of a regional approach to tourism policies and the ability to shape planning directions to suit contrasting regional and local needs, it also highlights the importance of effective coordination and integration of policy between the different levels of governance and within the organisational frameworks that are responsible for delivering planning on the ground. Without the latter, the potential of the former will only be partly realised.

Tourism planning at the local level

Local-level planning of tourism is a highly variable activity, reflecting the diversity of local situations in which tourism is developed. Yet at the same time, this level of application is also the easiest at which to identify a core of common planning concerns.

Local tourism planning will frequently reveal a focus upon the physical organisation of tourism facilities (accommodation, local transport, catering and local attractions) and the control of physical development (such as hotel construction). Local plans are typically short term and regulatory in nature (rather than being longer-term, strategic statements) with particular concerns for reducing development conflicts and harmonising activities that use the same spaces and/or resources. Local plans will show some similarities to regional-level plans in their attention to the logistics of provision of supporting infrastructure – such as power supplies, water and sanitation, local accessibility – but will be distinctly more detailed in their approach. Unlike regional plans, however, local planning of tourism will pay much greater levels of attention to the physical design and layout of developments – something that is rarely encountered at the larger geographic scales of intervention.

The local level is often the scale at which physical land use plans and associated tasks such as the spatial zoning of land uses are implemented. This is for two reasons. First, it is the planning level at which there is most likely to be a legally enforceable system of planning control. In England and Wales, for instance, although the county structure plans map the broad planning strategies at a regional or sub-regional scale, implementation is mainly via the development control process, which is operated through the medium of local plans and local decisions on development. Second, in most cases the appropriateness of a proposed development is most effectively judged in a local context, since this is the level at which impacts are to be most clearly felt. For this reason, it is also the level at which questions of public reaction to development are best considered, as the implications of proposed developments become prominent and measurable.

Although controlling development is an important and distinctive function of local plans, they may also reflect issues that are addressed at regional and national levels, especially economic and environmental effects of tourism. Economic impacts are likely to be considered in terms of scope for local employment, new-firm formation and potential multiplier effects of tourism incomes. Environmental and conservation issues will also be addressed, especially since the existence of legal controls and the increased use in many local planning procedures of EIA (see Chapter 5) reinforce the capacity of local planning systems to protect conservation areas and fragile environments from potentially harmful physical developments.

The local level is often an appropriate scale at which to devise detailed plans for visitor management. Regional scale strategies – such as the national park plans created by the individual national parks in England and Wales – will usually articulate a broad strategy for managing visitors by, for example, a macro-scale zoning policy (see Figure 5.2). However, within zones – and especially those that are planned to accommodate significant concentrations of tourists – there is generally a need for a detailed and place-specific articulation of planned visitor management.

Finally, local planning has emerged as the level at which public or community participation in planning processes is now most clearly developed. This is evident not just in a progressive shift towards more holistic approaches that encourage communities to comment on planning proposals (Church, 2004) but also in the rising prominence of partnership approaches in policy and planning that are aimed both at the mediation of tourism's cultural and social impacts upon communities (Bramwell, 2004) and at addressing issues of sustainability through Local Agenda 21. The Rio Earth Summit in 1992 adopted Local Agenda 21 as a framework for requiring local governments to develop policies that were sustainable and which would draw communities into participatory forms of planning, although in practice the impact of Local Agenda 21 on tourism policy and planning has been highly variable (Jackson and Morpeth, 2000).

The rising significance of participatory planning and partnership approaches at the local level invites further consideration. The value of local engagement in tourism planning processes has been recognised for some time (see, e.g., Murphy, 1985; Pearce, 1989; Inskeep, 1991) and it is generally held to deliver more positive outcomes than might otherwise occur in its absence. Timothy (1999) suggests that participation in planning can be seen in two basic ways: in decision-making processes and in derived benefits from tourism development, but the benefits are especially significant when planning approaches are based around partnerships that engage local stakeholder groups. Bramwell and Sharman (1999) suggest that the benefits of including local stakeholders and the knowledge that they bring will include:

● reduced levels of conflict;
● increased political legitimacy;
● improved coordination of policy across physical, economic, social and environmental sectors;
● increased likelihood of sustainable solutions.

However, the same authors also acknowledge that participatory local planning through stakeholder partnerships can be affected adversely by several areas of potential difficulty. These will include:

● the extent to which collaborating stakeholders represent all sections of the local community;
● the nature and frequency with which stakeholders are involved in the process;
● inequalities in power and influence between stakeholders;
● the level of understanding – both of the process and, importantly, of other stakeholder views;
● a willingness to accept consensus and implement resulting policies (Bramwell and Sharman, 1999).

Mitchell and Reid (2001) have provided a useful conceptual framework for examining how community perspectives on tourism are formed and integrated into planning processes (Figure 7.5). The framework envisages a three-stage process of which the most critical is arguably the first stage, in which the level of integration within the community is established and – by extension – its capacity to influence the subsequent planning process. Three critical parameters are seen as shaping the endogenous environment (i.e., the environment within the community): the community's awareness and understanding of tourism; the degree of unity within the community (and hence the range of opinions that the community may hold); and the nature of power relationships both within the community and between the community and the outside world. The latter is seen as providing an exogenous environment that shapes considerations such as tourism demand and which exerts a moderating effect upon the parameters that determine the views and actions of the community itself. In other words, the community cannot take decisions in isolation from wider influences.

Clearly, differences in the key parameters of awareness, unity and power will tend to regulate the capacity of the community to engage as participants in the planning process and thus shape the impacts that arise from the outcomes (Stages 2 and 3 of the model). This helps us to understand why the success of participatory local planning approaches evidently differs from place to place. For example, Timothy (1999) notes that local participation appears to be much harder to achieve in Third World areas than in developed nations such as the UK or Australia. Timothy's study of participatory local planning in Indonesia found that local involvement was hampered by a range of factors including a lack of expertise and understanding of tourism development issues on the part of local people, as well as cultural and political traditions that encouraged locals to defer to others whom they perceived to possess power and status. In other words, although the Indonesian communities revealed quite high levels of unity in a social sense, their lack of awareness and power reduced their

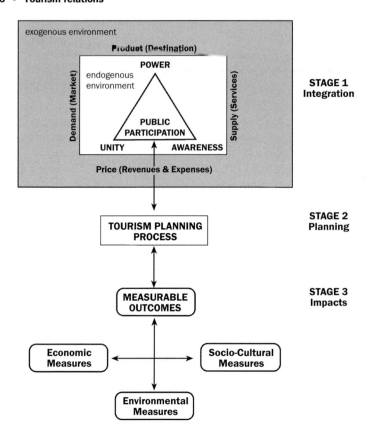

Figure 7.5
Integration of community perspectives in tourism planning

capacity to influence planning decisions which tended, as a consequence, to be imposed from outside by government.

However, low levels of participation are not inherently shaped around a First World–Third World dichotomy. A recent study of local planning in thirty tourism destinations in Queensland (Australia) found that despite the rhetoric around public participation, effective involvement by local people in tourism planning was evident in only a minority of cases (Ruhanen, 2004). Here the author concluded that lack of awareness and understanding – particularly of concepts of sustainable tourism – on the part of those who were empowered to make local planning decisions was a key constraint. In contrast, the study by Bramwell and Sharman (1999) of the Hope Valley area of the Peak District National Park (UK) that is described in Case Study 7.3 reveals a much more successful pattern of community involvement in developing a tourist management plan (although it must be conceded that this case does not typify the British experience as a whole, where community participation is often very poorly developed). Here, Mitchell and Reid's three parameters for integration were largely met by a community that was able to marshal quite high levels of unity through a clear framework for organising and expressing local opinion; displayed high levels of understanding of the issues through the inclusion of articulate and informed local residents; and was empowered – both by formal processes of inclusion and by the 'power' that comes with understanding of the issues and an ability to articulate that understanding. As a result, some genuine local participation in developing planning policies for tourism in the local area was revealed.

CASE STUDY 7.3

....

Community integration in tourism planning: Hope Valley, UK

Hope Valley and Edale occupy a central position within the Peak District National Park in northern England. Centred on the rural communities of Edale, Hope and Castleton and within easy reach of major industrial cities such as Manchester and Sheffield, the area is an established tourism destination that receives in excess of 2.5 million visitors annually. The majority are day visitors who are drawn by the appeal of attractive rural settlements such as Castleton and local scenic attractions such as Winnats Pass (a natural gorge), by the moorland and dale landscapes that provide opportunity for a wide range of outdoor activities, and by a range of tourist attractions that include spectacular underground caverns and historical sites such as Peveril Castle. Edale is also the southern starting point of the Pennine Way, the premier long-distance footpath in the UK (Figure 7.6). The local population numbers around

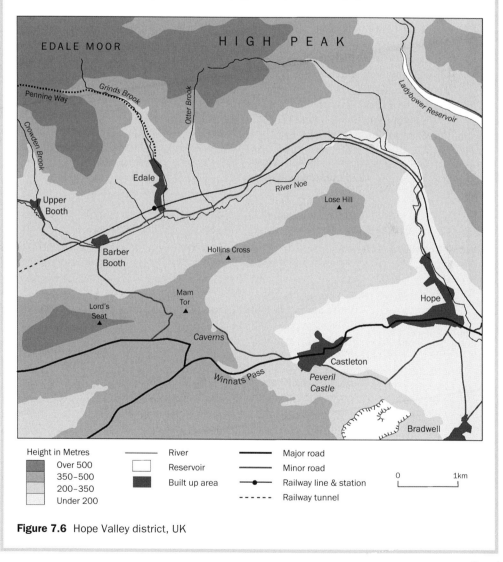

Figure 7.6 Hope Valley district, UK

continued

2,000 people and although tourism supports a large number of local businesses and jobs, significant local concerns have arisen in relation to issues such as traffic congestion, overcrowding of village centres, pressures on local attractions and a general loss of local amenity.

In the mid-1990s the Peak Tourism Partnership (a public–private sector organisation with a remit to promote sustainable tourism and develop visitor management) initiated the production of a new visitor management plan for the area. The broad objectives of the plan were to address issues of overcrowding by finding ways to limit visitor numbers at peak times, reduce the number of cars at key sites, divert usage from sensitive locations, and improve both the local benefits from tourism and the quality of life for local residents.

A prominent feature of the approach to developing the management plan was a conscious attempt to involve local expertise and opinion in its development. The framework for consultation is shown graphically in Figure 7.7 and which identifies several key stages:

- Stage 1 comprised an initial exercise in community mapping (by consultants) that identified key local stakeholders (with knowledge and understanding of the local community and tourism issues) and who came together in a tourism workshop. The focus of this workshop was on finding ways to develop a community agenda for tourism based on local need rather than tourist demands.
- Arising from the workshop, Stage 2 saw a smaller body of thirty-one representatives of local government, community groups, landowners, environmental agencies and groups, tourism businesses and recreational users formed as a Visitor Management Plan Working Group that was charged with developing the plan and building a consensus on the way forward.

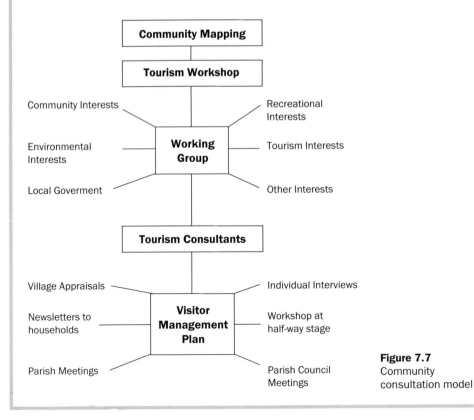

Figure 7.7
Community consultation model

- Following the drafting of plan proposals by a second group of tourism consultants, Stage 3 saw an extended exercise in public consultation through devices such as parish meetings, interviews and discussions with key individuals, further workshops, village appraisals and newsletters to all households that informed the production of the final plan.

However, although this approach was comprehensive and genuine in its desire to build a plan through community engagement, in their evaluation the authors of the study noted several areas of difficulty. To some extent some of the problems that were noted were perhaps an inevitable outcome of a democratic process and should not, therefore, devalue the principle of local engagement in tourism planning. So, for example, not everyone felt that they had had the opportunity to register their views, whilst in other instances there were questions over whether individuals who acted as representatives of larger stakeholder groups were, in fact, truly representative of group opinion. The issue of power was evident too, with some voices being more audible and influential than others. Perhaps, for this reason, consensus proved elusive in some key areas, although lack of agreement was also clearly a reflection of how people perceived the impact of the plan on their own interests. For example, strongly opposing views emerged relating to proposals to regulate visitor parking, with traders fearing loss of business whilst residents with no direct economic interest in tourism welcoming the chance to free some village areas of tourist cars.

Set against these areas of difficulty, however, were a number of positive benefits from the participation of the community in the planning process. On balance the project was broadly successful in building an inclusive approach and initiating dialogue between stakeholder groups that many participants recognised as being both open and beneficial in helping to build understanding of other points of view. The fact that the plan that was finally produced commanded a broad basis of support within the Hope and Edale communities was an additional indicator of success but perhaps more importantly signalled a collective sense of 'ownership' of the plan by local people that in turn endowed the project with a political legitimacy when the process moved to the important phases of implementation.

Source: Bramwell and Sharman (1999; 2000)

These examples and case studies are intended to illustrate the range of applications of planning in tourism development. It is important to realise that the preceding text provides only an outline review of a truly extensive topic. Readers with a desire to probe the subject more deeply are therefore encouraged to explore the literature cited below and in the overall Bibliography.

Summary

By focusing on the role of planning in shaping physical developments, the chapter highlights those aspects of tourism planning in which geographical perspectives are most useful in delivering an understanding of processes of change, although, clearly, this is not the only way in which tourism is influenced by planning processes. The chapter also illustrates that tourism planning is an overtly geographic phenomenon, not just because it is an essential mechanism for ordering tourism space, but also because the way in which planning is applied to tourism varies across space and between different locations. Planning at national, regional and local scales is now widely encountered and whilst there are common themes and issues that link the different scales of intervention, there are also distinctive dimensions that typify planning for tourism at these different spatial levels.

Discussion questions

1   To what extent do boosterism, industry-oriented, physical–spatial and community-oriented traditions in tourism planning represent complementary rather than contrasting approaches?
2   With reference to a selection of national tourism plans, examine the extent to which common themes and issues emerge at this level of intervention.
3   To what extent is the effectiveness of tourism planning inhibited by the nature of tourism itself?
4   What are the strengths and weaknesses of adopting a regional-level approach to tourism planning and why might these vary between different destination areas?
5   Critically assess the potential and the difficulty of developing community-level participation in local planning for tourism?

Further reading

There are several texts on tourism planning that have acquired a classic status and which still provide informed discussion of key planning themes, in particular:
Gunn, C.A. (1994) *Tourism Planning*, New York: Taylor and Francis.
Inskeep, E. (1991) *Tourism Planning: An Integrated and Sustainable Development Approach*, New York: Van Nostrand Reinhold.
Murphy, P.E. (1985) *Tourism: A Community Approach*, London: Routledge.
Pearce, D.G. (1989) *Tourist Development*, Harlow: Longman.

In addition a more recent and very detailed consideration of the links between policy and tourism has been provided by:
Hall. C.M. (2000) *Tourism Planning: Policies, Processes and Relationships*, Harlow: Prentice Hall.

A shorter essay on links between local and regional policy and planning is provided by:
Church, A. (2004) 'Local and regional tourism policy and power', in Lew, A.A. et al. (eds) *A Companion to Tourism Geography*, Blackwell: Oxford, pp. 555–68.

An excellent discussion of the links between tourism planning and sustainability is provided by:
Hunter, G. and Green, G. (1995) *Tourism and the Environment: A Sustainable Relationship?*, London: Routledge.

Useful recent case studies of tourism planning in practice are to be found in:
Baidal, J.A.I. (2004) 'Tourism planning in Spain: evolution and perspectives', *Annals of Tourism Research*, Vol. 31 (2): 313–33.
Burns, P. (2004) 'Tourism planning: a third way', *Annals of Tourism Research*, Vol. 31 (1): 24–43.
Garcia, G.M., Pollard, J. and Rodriguez, R.D. (2003) 'The planning and practice of coastal zone management in southern Spain', *Journal of Sustainable Tourism*, Vol. 11 (2/3): 204–23.
Ruhanen, L. (2004) 'Strategic planning for local tourism destinations: an analysis of tourism', *Tourism and Hospitality Planning and Development*, Vol. 1 (3): 1–15.

Similarly, useful discussions of community participation in tourism planning are provided by:
Bramwell, B. and Sharman, A. (1999) 'Collaboration in local tourism policymaking', *Annals of Tourism Research*, Vol. 26 (2): 392–415.
Simpson, K. (2001) 'Strategic planning and community involvement as contributors to sustainable tourism development', *Current Issues in Tourism*, Vol. 4 (1): 3–41.
Timothy, D.J. (1999) 'Participatory planning: a view of tourism in Indonesia' *Annals of Tourism Research*, Vol. 26 (2): 371–91.

Part III
Understanding the spaces of tourism

To some extent the sub-title to this final collection of chapters – understanding the spaces of tourism – might also serve as a sub-title for the text as a whole, since much of the preceding discussion has pursued precisely that goal. However, most of the content in Parts I and II has been largely shaped by a traditional view of the scope and remit of tourism geography and of the ways of understanding or interpreting the spatial patterns of tourism or the relationships between people – as tourists – and places. But it is clear, both from the shifting nature of the tourist gaze, as well as from the development of new critical perspectives in human geography, that any approach that confines itself to the range of material that we have explored in Parts I and II will reveal only a partial view of what tourism geography now constitutes. We need, therefore, to begin to think about tourism – to *understand* it – in some new ways and it is in this sense that Part III addresses this subject.

There are two key questions that, although not formally stated, actually shape the approach in these final chapters. First, how should we understand the position of tourism in post-industrial (or, if we prefer, postmodern) society and the new spaces of tourism that have emerged in association with the post-industrial/postmodern shift? Second, how has the so-called 'cultural turn' in human geography and the related development of postmodern perspectives altered how we understand the changing place of tourism in contemporary life and the different geographies that it creates?

In practice, of course, these two ideas are intimately connected and part of the object of the discussion in Part III is to explore these connections and show how they relate to the evolving character of tourism geography. Tourism has always been more than just the simple practice of travelling for pleasure or for the rewards of rest and relaxation. It has always been an activity encoded with layers of meaning, some subtle, some much more overt, but until quite recently those meanings and the practices that they generate have seldom been seen as constitutive of daily life to any degree of significance.

All that is changing as tourism moves from being a marginal activity pursued in what Turner and Ash (1975) once described as 'pleasure peripheries' to an activity that is often central to shaping the spaces of post-industrial life in the twenty-first century. Thus, for example, as a more embedded facet of post-industrial life, tourism is helping to reshape space and, more importantly perhaps, to redefine images of places (particularly in urban settings). Tourism has become much more central in the construction of identity – both of places and, critically, of individuals whose patterns of consumption in tourism are often consciously managed to confer upon themselves particular identities and status. Tourism has also been recognised as an important arena in which people explore the self and their understanding of who they are (through embodied forms of tourism such as adventure travel), whilst in other contexts (such as heritage) tourism may become a form of resistance to the anonymous, de-personalised world that it often felt to characterise the post-industrial, postmodern era.

This notion of tourism as a form of resistance provides an important connection to a wider theme. One of the essential differences that the cultural turn and the wider adoption of postmodern perspectives in human geography has made has been to reposition the tourist as *subject* – recognising that people have agency and that rather than being necessarily the passive recipients of managed tourist experiences, they are active in shaping experience for themselves. Postmodern discourse instinctively rejects the notion of overarching understandings or universal explanations of phenomena such as tourism in favour of a multiplicity of positions that reflect the fact that people make sense of the world they inhabit as individuals and on their own terms. So whilst many sectors of tourism are still shaped by practices of mass consumption and the geographies that those practices support, an important message to take from these chapters is that other forms of tourism are emerging that are far more reflective of individual tastes and preferences. Consequently, the spaces and places that locate tourism are becoming more diverse, more numerous and harder to differentiate from the other spaces that we occupy and we need to recognise and appreciate the implications of these trends for the wider understanding of tourism geography.

8 Inventing places: cultural constructions and tourism geographies

Places, and images of places, are fundamental to the practice of tourism. The demand for tourism commonly emanates from individual or collective perceptions of tourist experiences that are usually firmly rooted in associations with particular places, whilst the promotion and marketing of tourism depends heavily upon the formation and dissemination of positive and attractive images of destinations as places. Tourism therefore maps the globe in a distinctive, though highly subjective, manner and one of the ways in which we may view the geography of tourism is as a visible manifestation of perceptions and images of what constitute tourism places. However, as those perceptions and images are recast and re-formed – in response to changing public expectations, tastes, fashions, levels of awareness, mobility and affluence – new tourism geographies emerge, modifying or often replacing previous patterns as different forms of tourism promote new areas of interest.

This chapter explores some of the ways through which such tourism geographies are formed and, in so doing, aims to introduce Part III of this book – understanding the spaces of tourism. In particular, the discussion aims to show that although part of the process of inventing tourism is centred on the physical development of tourist space – and much of the preceding discussion in this book has been explicitly concerned with this aspect of tourism – the making of tourist places is not simply a physical process. When we define a location as a tourist place, we apply to that place an additional layer of distinction and whilst part of that distinction may indeed be grounded in the physical attributes of a place, it is – fundamentally – a culturally informed process.

The cultural distinction of tourist places is evident in several ways, but two are worth noting at this juncture. The first is in the *roles* that we ascribe to tourist places. Tourist places need to serve a purpose – whether as places of fun, or as places of excitement and challenge, or as places of spectacle, or as places of memory. Yet none of these attributes exists in isolation, they are cultural constructs that reflect the values, beliefs, customs and behaviours by which we define ourselves as individuals and as members of a society. Second, tourist places are generally made distinct by the incidence of recognisable tourist *practices*. A number of writers (e.g., Crouch et al., 2001; Edensor, 2000a, 2001) have drawn attention to the ritualised, performative nature of tourism with its shared sets of conventions and assumptions about appropriate behaviours. Tourist places are therefore actively produced through the performances of the tourists who congregate at favoured sites and whose presence and actions, in turn, reinforces the nature and character of those sites as tourist places.

It is also useful to note how the evolution of tourist places through time is shaped by underlying socio-cultural processes and, in particular, change in cultural markers such as taste and fashion. The initial recognition of places as tourist destinations generally reflects an appraisal of resources that is located in cultural evaluations and the subsequent physical development of tourism places typically depends as much upon social and institutional structures and organisations as it does upon the more tangible impacts of (say) innovations in transport technology. Hence, the original growth of sea bathing resorts in eighteenth-century England mirrored key shifts in health practices and beliefs, whilst the later development of mountain tourism in Alpine Europe owed its impetus to the alternative views of landscape that grew out of the new taste for the romantic picturesque that was popularised in the first decades of the nineteenth century. Later still, the growth of mass forms of Mediterranean tourism only became really established with the fashion for sunbathing from about 1920 (Turner and Ash, 1975). Railways (and later aeroplanes) may have provided a physical mechanism for moving tourists in large number to new destinations, but it required the transformations in the social organisation of tourism (e.g., the guided tour and later the packaged holiday) and the importation of holidaymaking into popular culture to realise that potential to the full.

By recognising the significance of culture as a primary influence upon the tourist's identification of (and with) places, it allows us to start to comprehend the bewildering range of locations that now present themselves as destinations for the (post)modern traveller. Contemporary societies are shaped by mobilities and we are constantly confronted by choices around where to visit and what to do (Franklin, 2004). Part of the sheer diversity of tourism places in the contemporary world arises, therefore, from the simple fact that different people will apply different criteria in resolving the choices that are at their disposal. However, it is important to recognise that the decisions that we make – whilst often formed at the level of the individual and thus reflective of personal inclinations and dispositions – are mediated in some important ways. Once again, part of that mediation is derived from the cultures in which we reside and which help to shape preferences and inform the codes of behaviour that we exhibit as tourists. But in addition – and most importantly – our recognition of, and identification with, tourist places is also mediated by the agency of others whose role it is to influence perceptions and promote places as appropriate objects of tourist attention.

In simple terms, therefore, we may view the identification of tourism places as arising from the interplay of:

- the agency that we exercise as individuals and the performative nature of our behaviours as tourists;
- the cultures in which we reside and which help to determine those individual and performative characteristics;
- the agency of others whose role it is to shape perceptions and promote tourist places.

To try to develop a clearer understanding of these important ideas and the ways in which they intersect, the chapter examines four related themes: the construction of tourist places through the 'gaze'; the performative nature of tourism; the role of place promotion; and the theming of tourist environments. However, before we move to the main discussion, it is necessary to explore in a little more detail, geographical understandings of the concept of 'place' and how this relates to the invention of tourism places.

The concept of place

Over the last thirty years, 'place' has become one of the central organising concepts in human geography, yet it remains an elusive and, at times, intangible idea. By tradition, the examination of place provided a focus of geographical investigation in the early part of the twentieth century and was widely reflected in the work of geographers such as de la Blache, Hartshorne and Fleure (Castree, 2003). However, the understanding of place that was deployed at this time was essentially of places as physical locations – distinctive points on the earth's surface at which characteristic physical or human patterns could be isolated and described. More recent understandings of place, especially following the reassertion of humanistic approaches in the 1970s by writers such as Relph (1976) and Tuan (1977), have extended and deepened the concept in some important ways, in particular by seeking better understandings of how people relate to places (Crang, 1998).

A key facet of modern understandings of place has been the recognition that places are essentially social constructs rather than just physical entities and whilst in their simplest guise places constitute points on a map, they are more importantly a locus of institutions, social relations, material practices and forms of power and discourse (Harvey, 1996). Places are not merely bounded spaces or *locations*, but are also settings (or *locales*) in which social relations and identities are constituted and in which is developed a *sense of place* (Agnew, 1987). The sense of place – which essentially relates to the unique qualities that places acquire in people's minds – is formed in complex ways. In part it is a product of the physical attributes of the setting that mark the place as being distinctive – such as local landscapes or styles of building. But it is also a product of the personal attachments to place that people will develop and the consequent ways in which they endow places with subtle symbolic or metonymic qualities (i.e., the place comes to represent more complex emotions and feelings). Places therefore provide individuals with a sense of belonging that is progressively reinforced by memories (both collective and individual) that become associated with the places in question and which together help to reinforce people's sense of identity. There are, therefore, very powerful imaginative and affective dimensions that cause people to identify with particular places (Castree, 2003) that, whilst unseen, are hugely influential on attitudes and behaviour. 'Places say something about not only where you live or come from, but who you are' (Crang, 1998: 103).

However, it is very important to acknowledge that places are dynamic rather than fixed entities. As a Marxist, Harvey (1996) is keen to emphasise the political–economic basis to place and the ways in which places evolve in response to change in the (global) systems of production and consumption. Thus, for example, places that were once defined by productive industry and the communities that were forged in associations with those industries (such as urban docklands) are being progressively remade as new places of consumption with new place identities shaped by different social dynamics (such as gentrification) or new activities (such as tourism). Harvey (1996) also emphasises the role of places as symbols of power and notes several of the ways in which institutions such as the church and the state routinely identify and revere a range of places as symbolic expressions of institutional power and related social meaning. Yet these are seldom fixed entities either.

One of the most powerful forces of change that is now widely believed to affect the distinction of places is globalisation. Harvey (1996: 297) observes that places are no longer protected by the 'friction of distance' whilst Castells (1996) asserts that the flows of people, information and goods that lie at the heart of globalisation is breaking down the barriers that once made places different. Relph (1976, 1987) has also provided a detailed dissection of the ways in which modern urban development has rendered a growing number of places as 'placeless', that is, indistinct from each other because of the homogeneity of their built environments and their associated styles of living. Ironically, perhaps, tourism – which by tradition has been widely represented as a quest for difference – has become one of the most influential agents in promoting placelessness and homogeneity in some of its main destination areas, such as the Mediterranean coastline.

From this brief exploration of the concept, it may be deduced that tourism intersects with place in a number of important ways:

● Many forms of tourism are firmly grounded in a sense of place, and without the distinction that tourist places are able to generate, much of the rationale for modern travel would be undermined.

- Tourist perceptions and motivation (and hence behaviours) are directly shaped by the ways in which we imagine places and are encouraged to imagine places by the travel industry.
- Tourist places often possess strong symbolic and representational qualities that form a primary basis to their attraction.
- Tourism is a primary means through which we construct and maintain identity – where we visit says much about who we believe we are and about the images and identity that we wish to project to others.
- Tourism can be a medium through which people develop personal attachments to places and for whom the place becomes a site of meaning.
- Tourism places provide people with important sites of memory – we tend to recall tourist experiences long after more routine aspects of our daily lives are forgotten and we commonly engage in actions (such as photography or gathering of souvenirs) that enable us to store these memories of tourist places for future recall.
- Tourist places help to provide some people with a sense of belonging – particularly when places become sites of habitual visiting to which people return on annual (or more frequent) 'pilgrimages'.

The tourist gaze

In developing a closer understanding of how tourists relate to the places that are toured and how their actions shape their experience of place, one of the most influential ideas to emerge in the last twenty years has been Urry's (1990) notion of the 'tourist gaze'. Urry's book sets out to answer a question that is fundamental to tourism, namely, why do people leave their normal places of work and residence to travel to other places to which they may have no evident attachment and where they consume goods and services that are in some senses unnecessary? The answers that Urry proposes are shaped by two fundamental assumptions: first, that we visit other places to consume the sights and experiences that they offer because we anticipate that we will derive pleasure from the process; and, second, those experiences will in some way be different from our everyday routines and, therefore, out of the ordinary. Urry (1990: 12) further explains that the extraordinary may be distinguished in several ways. For example:

- in seeing a unique object or place – such as the Eiffel Tower or the Grand Canyon;
- in seeing unfamiliar aspects of what is otherwise familiar – such as touring other people's workplaces or visiting museums or other tourist sites that allow us to glimpse how other people live (or lived) – such as the stately home or the restored miner's cottage in an industrial museum;
- in conducting familiar routines in unfamiliar settings – such as shopping in a north African bazaar.

In these types of ways, Urry argues, our gaze – as tourists – becomes directed to features in landscapes and townscapes which separate them from everyday experience and whenever places are unable to offer locations or objects that are out of the ordinary then, almost by definition, there is 'nothing to see'. And seeing is a central component in the concept of the gaze – indeed the term itself prioritises the visual forms of consumption of tourist places as the means by which most tourists relate to the places they visit. 'When we "go away" we look at the environment with interest and curiosity . . . we gaze at what we encounter' (Urry, 1990: 1).

The concept of the gaze is valuable because it posits an understanding of both the construction and the consumption of tourist places that is grounded in observed tourist practices and a common-sense rationale, whilst also providing a useful point of entry to understanding the selective ways that tourism maps space and defines tourist places. Most importantly, it emphasises the subjective nature of tourism and the position of the tourist as subject (MacCannell, 2001), and in so doing, the concept of the gaze points to two important consequences. First, it locates tourists as consumers in a central role within the process of making tourist places; whilst second, in acknowledging that different groups will construct their gaze in differing ways, it provides an explanatory rationale for the diversity that is evident across the range of tourist places that we commonly encounter.

The metaphor of visualisation that is explicit in the term 'gaze' is also a key to comprehending many modern tourism practices and their associated meanings. Tourism is a strongly visual practice. We spend time in advance of a tourism trip attempting to visualise the places we will visit by examining guide books and brochures, or in anticipatory day-dreams; we often spend significant parts of the trip itself engaged in the act of sightseeing in which we gaze upon places, people and their artefacts; and we relive our experiences as memories and recollections, aided by photographs or home video footage that we have consciously taken to act as visible reminders of the trip (see Figure 1.5). For Urry (1990: 138ff.) photography is intimately bound up with the tourist gaze. It provides a means of both appropriating the objects of our gaze as we 'capture' interesting scenes or actions in our cameras and verifying to others that we have really witnessed the places that our photos represent. Photography also idealises places by the way that we select scenes, frame and compose our images and – in the digital age – manipulate the output to enhance further the qualities of the settings we have recorded if the true image fails to satisfy. The postcards that we habitually buy and send to others similarly act as a surrogate means of representing and signalling the genuineness of the tourist experience (Yuksel and Akgul, 2007). Many aspects of tourism therefore become what Urry terms 'the search for the photogenic' – a quest for the visual experience that directly shapes the way we tour places as we move from one 'photo opportunity' to the next (see also Crang, 1997; Crawshaw and Urry, 1997).

However, the entire process of visualisation, experience and recall of tourist places is, of course, socially constructed and strongly mediated by 'cultural filters'. We gaze and record places in a highly selective fashion, disregarding some places altogether and, from the remainder, removing the unappealing or the uninteresting. In the process, we are inventing (or reinventing) places to suit our purposes. The gaze is also a detached and superficial process, as the term itself suggests. This superficiality increases the role of cultural signs within the invention and consumption of tourist places – not signs in the literal sense of directional indicators, but figurative signs: places or actions that represent, through simplification, much more complex ideas and practices. So for the tourist, a prospect of a rose-decked, thatched cottage may come to represent or embody a much wider and more complex image of 'olde England' and the lifestyles and practices that mythologies associate with the rural past. There is a real sense in which some forms of tourism have become an exercise in the collection of such 'signs' – the postcards and the holiday photographs from the great tourism sites of the world conferring a status upon the individual, the true mark of the modern (or postmodern) tourist.

The emphasis within Urry's conception of the gaze on places that present the extraordinary also helps to explain the clear tendency for tourism geographies to change through time. As places become unacceptably familiar, there is an evident need for at least a part of the tourist gaze to be refocused upon new destinations or perhaps upon elements in existing destinations that had not previously been a part of the tourist circuit. So, Brighton and Torbay are replaced by Biarritz and St Tropez, whilst the seasoned tourist to Paris, no longer simply content with views of the Eiffel Tower, may now sign up for guided visits to the city's nineteenth-century sewers (Pearce, 1998). The tourist gaze is seldom fixed, but, rather, it shifts in response to changes in fashion, taste, accessibility and – not least – change in the character of places under the active agency of tourism development and, indeed, of the gaze itself.

There is no doubt, as MacCannell (2001) concedes, that Urry has accurately described a prominent form of tourist travel and a characteristic mode of engagement of tourists with place. However, and perhaps unsurprisingly, dissenting voices have been raised, for although the concept of the gaze provides a valid and convincing explanation of some important areas of tourist behaviour, it provides a rather less convincing basis for understanding the full range of those behaviours. As Franklin (2004: 106) observes, 'one would not want to dispute the foundational and influential nature of the tourist gaze, but we might say it is only one among many types of touristic relationship with objects'.

Two areas of concern are worth noting, both of which arise from the basic assumptions that informed Urry's original ideas. The first concern relates to whether tourism necessarily engages with the extraordinary. Urry (1990: 12) proposes that 'tourism results from a basic binary division between the ordinary/everyday and the extraordinary', which implies that the objects of the tourist gaze should be exceptional. This is problematic because it assumes the existence of 'an ordinary' against which comparisons may be drawn and that tourism retains a level of distinction that enables meaningful differentiation from other socio-cultural practices. Yet one of the evident impacts of postmodern change has been a progressive dissolution of assumed boundaries – what Lash and Urry (1994) have termed 'de-differentiation' – so that tourism becomes harder to separate from other social and cultural practices, whilst the 'extraordinary' has become infused into daily life. In a robust challenge to the tourist gaze, Franklin (2004: 5) drives the point home by noting that 'the everyday world is increasingly indistinguishable from the touristic world. Most places are now on some tourist trail or another . . . [and] . . . most of the things we like to do in our usual leisure time double up as touristic activities and are shared spaces'. In a related line of argument, MacCannell (2001) also criticises Urry for assuming that the everyday cannot be extraordinary and that modern life is intrinsically boring, thereby creating the need for periodic escape to extraordinary places through tourist travel. Franklin (2004: 23) is of a similar mind, asserting – with confidence – that 'with modernity there is never a dull moment'. This bold claim risks over-stating its case as it seems an evident truism that most people would be quite capable of identifying many aspects of their modern lives that are grindingly dull and routinised. However, the point that many facets of modern life (and many modern places) are strongly synthetic of tourist experience remains an important perspective.

To some extent this argument turns on whether the exceptional is necessarily unfamiliar. Notions of 'exceptional' and 'ordinary' are, after all, relative terms that are normally defined at the level of the individual. Franklin's (2004) thesis is persuasive around the theme of the dissolution of boundaries and the embedding of many of the experiences that we acquire through tourism into daily life, but he perhaps loses sight of the fact that people – as reflexive individuals – will still accord 'extraordinary' status to many of their tourist trips (even when they are made to familiar places) and that these trips will tend to remain as distinctive, memorable events within their wider lifestyles. As a frequent tourist to France, this writer has become familiar with large swathes of French territory and with many aspects of French life, some of which is also routinely encountered during home life. However, that familiarity does little to diminish the *frisson* of anticipation that always accompanies a trip to France, nor to dilute the sense of engagement with foreign, extraordinary, and even exotic places that such trips tend to provide. Perhaps, as Urry suggests, it may be the *scale* of difference that is important here, rather than difference as an absolute condition.

The second significant critique of Urry's concept is that the tourist gaze proposes an essentially detached engagement of tourists with places and the experiences that places provide. Urry's gazing tourist is characteristically an observer, a collector of views and someone for whom sightseeing is a primary *modus operandi*. Yet the increasingly diverse nature of contemporary tourism reveals many areas of engagement in which the gaze is marginalised or even irrelevant. In a later chapter we will consider the embodied nature of tourist experience in which, far from being a detached observer, the tourist becomes an active participant in local customs, practices and activities in which places (and their occupants) are contacted through embodied experience. The burgeoning interest in adventuresome and active forms of tourism (such as climbing, trekking, surfing, hang-gliding or bungee-jumping) and the more selective, but, locally important, engagement of tourists in sectors such as wine and food tourism, sex tourism and naturism, tells us that many people are not content simply to look, but must also feel, taste, touch, smell and hear (see, e.g., Franklin, 2004; Hall et al., 2000; Inglis, 2000; Macnaghten and Urry, 2001; Veijola and Jokinen, 1994). The inability of the concept of the gaze to account for these forms of experience does not negate its wider value as a theoretical perspective on what is probably the dominant form of relation between tourists and place, although clearly it is a weakness that should be noted.

Tourism places as places of performance

The manner in which tourists direct their gaze is an important aspect in the making of tourist places but we should recognise that it forms a part of a wider process of engagement that is sometimes described as the tourist 'performance', that is, the actions, behaviours, codes and preferences that tourists exhibit whilst visiting a destination. Interest in the performance of tourists is a relatively new critical position that has developed through new cultural perspectives in geography (and other social sciences) and reflects recognition of a very basic observation, namely, that 'tourism cannot exist independently of the tourists that perform it' (Franklin, 2004: 205). In other words, whilst the tourist industry may produce and promote any number of tourist spaces, these remain as inert entities until such time as they become populated with people – as tourists – whose engagement with the sites and with each other produces the institutions, relations and practices that define the site as a place of tourism. Edensor (2001: 59) writes that 'tourism is a process which involves the on-going (re)construction of praxis and space in shared contexts' and tourists thus possess a dynamic agency that continually produces and reproduces diverse forms of tourism and tourist places through the actions that comprise their performance.

The performative nature of tourism is interesting because there is both a circularity (through which performance reinforces particular codes and practices) and an opportunity for resistance to those codes and expectations. In terms of circularity, it is immediately evident that some aspects of tourist performance reflects what Bourdieu (1984) defines as *habitus* and which establishes habits and responses that are shaped by – and inform – the everyday lives that we live. These are normative codes of behaviour that help us to organise our lives and engage effectively with others within wider communities. Tourism generates or acquires its own shared sets of conventions with regard to behaviour and expected actions (i.e., the tourist performance) by drawing on our wider habitus (which provides unreflexive dispositions which we seldom abandon completely) but supplemented by particular expectations of how we should behave *as tourists* at given locations. In this light, tourism may be seen as constituting a collection of embodied practices and meanings that are commonly understood and which are reproduced by tourists through their performances. The circularity within this process should therefore be evident in so far as part of the performances we give as tourists reflect a shared understanding of what we are *expected* to do as tourists. Moreover, as Edensor (2001) notes, as performers we are generally subject to the disciplinary gaze of other tourists – our co-performers – and the surveillance of others helps to reinforce conventions around behaviours that are appropriate to ways of being a tourist in a particular place.

However, as Edensor (2001) makes clear, tourism also provides sites of resistance. He describes how tourism offers the opportunity to discard our everyday 'masks' and explore temporarily new roles and identities. In so doing we may choose not only to confront routine habits and pursue the inverted forms of behaviour that Graburn (1983a) has suggested are characteristic of tourism, but we may also challenge established codes of tourist behaviour. It is through these processes that tourist performances reveal their dynamic qualities. Although it is often strongly mediated (e.g., by tour guides), the tourist performance is not fixed but may be subverted, adapted or contested in order to meet the particular purposes of the performers. In certain circumstances it may be used as a means of asserting identity (especially when performance diverges from the norm) and it may be consciously non-conformist; for example in resisting the conventions of organised forms of tourism in favour of more personalised explorations of places and experience. More simply, perhaps, it is also true that different types of tourist will encounter the same places with contrasting perspectives and expectations, so that whilst there may be normative codes that shape tourist practice across most places, alternative styles of tourist performance will emanate not only from studied modes of resistance, but also through the plain fact that people are different.

Case Study 8.1 illustrates the divergent ways in which tourist performance reflects – and is reflected in – place, through a study of tourists at the Taj Mahal in India.

CASE STUDY 8.1

Performing tourism at the Taj Mahal

This case study centres on an original enquiry by Edensor (1998, 2000a, 2001) into the nature of tourist performance at the Taj Mahal in India. Through the adoption of an ethnographic approach that was based upon extended observation and engagement with tourists at this world-famous site, Edensor provides some perceptive insights into the contrasting ways in which tourists construct an understandng of place and how the differing ways in which tourists strive to make sense of the places they visit are reflected in the 'performances' they give (i.e., their actions, dispositions and behaviours).

The Taj Mahal – sometimes described as 'the world's most famous tomb' – is located at the city of Agra in Uttar Pradesh, India. It was constructed between 1631 and 1653 on the orders of the Muslim Moghul emperor Shahjahan as a mausoleum for his wife Mumtaz Mahal, but it has become the most renowned icon of modern India and a powerful magnet for global tourism.

However, the Taj is essentially a symbolic site and as Edensor (1998: 7) notes, symbolic sites tend to be diversely represented, with contested notions about what they mean being articulated by differing groups of people. Edensor proposes several narratives that he suggests inform the understandings of the place that most tourists exhibit, including:

- A 'colonial' narrative that is largely Western in origin and which reflects romantic constructions of an imaginary India that are shaped by an aesthetic appreciation of the grandeur and sublimity of the building and its representation of the mystical East.

- A 'secular, nationalist' narrative that has emerged (following Indian independence in 1947) as a form of resistance to colonial views of India as inferior and backward and which emphasises the Taj as an Indian achievement of world importance – a pinnacle of Indian cultural synthesis and a symbol of Indian identity and unity.
- A 'Muslim narrative' that continues to view the Taj as a sacred site and which enables Indian Muslims to express both identity and a pride in the achievements of what they view as a golden age of Moghul rule and a cultural high point in Indian history.

These narratives, in turn, shape the performances of the dominant groups of visitors (package tourists of largely Western origin; backpackers and independent travellers – also of largely Western origin; domestic Indian tourists; and religious visitors) in both subtle and evident ways. This produces contrasting forms of engagement with the place that are expressed particularly through the embodied actions of walking, gazing and photographing.

For example, with regard to their performative characteristics, the actions of most package tourists are disciplined and closely choreographed. Their visit is often mediated by initial images that they have formed in advance of their visit (and which they seek to affirm), as well as the way in which tour guides regulate the use of both time and space. Itineraries around the site follow a limited number of prescribed routes to a select number of key locations, at which scripted information is conveyed and photographs that capture 'classic' views of the Taj are taken (Plate 8.1). The tight time schedules that are often applied to guided tours means that although the Taj is widely perceived as the highlight of the holiday, the actual experience of the place is often abbreviated. These tourists are especially susceptible to the colonial narratives and the mythologies of an exotic 'other', yet the scope for reflective, contemplative gazing is limited by the regulated brevity of their visit. For most tourists in this category, photography is a dominant part of the performance – in Edensor's (1998: 128) words, 'materializing the tourist gaze' and capturing images that both verify the experience and provide an essential basis for subsequent acts of remembering.

In contrast, the performance of independent travellers and backpackers provides important points of difference. These tourists spend much more extended periods of time at the site and indulge in a much wider range of actions: gazing, strolling, people-watching, socialising, writing and resting. They tend to explore the whole site, displaying improvised rather than choreographed movements. Although strongly susceptible to the same romantic and aestheticised gaze that informs the actions of many of the package tourists, these people are more likely to adopt a reflexive approach to understanding the Taj, rather than being content to acquire the reams of scripted knowledge that is imparted to packaged tourists. Photography is a less conspicuous part of their engagement with the Taj, although other forms of visualisation are often deployed (such as sketching or contemplative reflection).

For many domestic Indian tourists, their visit is commonly motivated by a desire to witness a national monument and see for themselves a potent symbol of Indian national identity. Their walking is usually purposeful and direct whilst their photography is generally directed at capturing portraits of their family group (which is how many visit the Taj) with the monument itself as background rather than subject. Their gazing is normally unreflexive, mirroring – as it tends to do – the secular national narratives and is a much less intense gaze than either the backpackers or the packaged tourists from overseas. Domestic tourists tended to spend the shortest periods of time at the site and often subjected the building and the grounds to only superficial attention, so much so that one of Edensor's (English) subjects

continued

Plate 8.1
'Materialising the gaze': a classic tourist image of the Taj Mahal, India

was moved to comment that Indians 'just don't know how to be tourists' (Edensor, 1998: 126) – a subconscious but nevertheless revealing admission that tourism is indeed often about performing a role.

However, not all Indian visitors performed in this way. For Indian Muslims, the Taj is a sacred site and a powerful symbol of what many still consider as a golden age of stability and progress under Muslim rule. In consequence, their visits are generally of an extended nature and the gaze is reverential. Edensor (1998) observed how many Muslims progressed slowly around the site, would spend time reading the Quranic inscriptions on the buildings, pray in the adjacent mosque (where few from other groups chose to venture) and sit in silent contemplation.

These *vignettes* of tourist behaviour (which are explained extensively and in fascinating detail in Edensor's original work) demonstrate very clearly how tourists make sense of places in diverse ways. The symbolic and emotional importance of the Taj Mahal is different for domestic and foreign tourists and, within those groups, is different for secular and religious visitors or for organised and independent travellers. Key parts of the performance – such as photographing and gazing – are also informed in varying ways by the contrasting narratives of the Taj and this is further revealed in the differences between the romantic, reflexive and reverential gazes of – respectively – packaged tourists, backpackers or Indian Muslims. The detail that informs these broad observations also encourages Edensor to an important overall conclusion – that is, the diverse character of tourist performance at the Taj tends to 'disavow the idea that places have some essential identity. Rather, places are continually reconstituted by the activities that centre upon them' (Edensor, 1998: 200).

Tourism place promotion

The essential argument in the preceding two sections is that the manner in which we gaze upon tourist sites (and sights) and the performances we impart as tourists to those sites contribute directly to the creation of a location as a tourism place. However it is important to recognise that whilst the gaze and its associated forms of performance are a product of our own social, educational and cultural backgrounds, they are also a reflection of the systematic production and presentation of tourism places within the media in general, and the travel industry in particular. Urry (1990) has characterised this as a form of 'professional gaze' through which media such as film, television, magazines, travel books and advertisements constantly produce and reproduce objects for tourist consumption. This is an enormously powerful influence that infiltrates the subconscious of everyday life, creating new patterns of awareness, fuelling desires to see the places portrayed and instilling within the travelling public new ways of seeing tourism destinations. Most visitors' perceptions of tourism places are often vague and ill-formed, unless those perceptions have been sharpened through previous experiences. Hence there is clear potential (through the process of place promotion) for marketing and promotional strategies to shape both the character and the direction of the tourist gaze, and in the process, invent new tourism places.

Place promotion is defined by Ward and Gold (1994: 2) as 'the conscious use of publicity and marketing to communicate selective images of specific geographical localities or areas to a target audience'. This is a useful starting point as it clearly positions the use of images as a critical element in the process of choosing a destination (Molina and Esteban, 2006). Morgan (2004: 177) comments that 'place promotion presents the world as image, inviting the viewer to become an imaginary traveller to an imagined place', and as we have already seen, the exercise of our imaginative faculties at the early stages of planning tourist excursions is an important precursor to the actual business of travel (see Chapter 1). However, such images are – almost by definition – selective representations of the places in question, and as a consequence, all place promotion campaigns are founded on fragmentary assemblies of place elements that promoters judge will resonate with potential visitors (Ward and Gold, 1994). Hughes (1992) further explains that the construction of such imaginary geographies succeeds by linking the promoted images to the store of perceptions and experiences that are already embedded (from other media and knowledge pathways) in what he terms 'commonsense understandings' of the destinations in question. However, because such images form a 'text' that is used to represent the tourist destination – and because such texts will be read in variable ways by different 'readers', that is the tourists (Jenkins, 2003) – some important practical advantages are derived from this approach. In particular, by drawing selectively on alternative sets of images, it enables the same place to be sold simultaneously to different customers (Ashworth and Voogd, 1994), although it may also create tensions around how a particular place should be represented.

This is an important point because it encourages us to recognise that place promotion is more than simply an exercise in marketing. Morgan (2004: 174) writes that whilst place promotion 'has a clear business function and marketing rationale', the discourse of place promotion actually 'reveals underlying narratives of place'. This connects directly to the wider understandings of place as loci of social relations, material practices, power and resistance (see, e.g., Ringer, 1998; Aitchison et al., 2001) that, in an ideal world, would directly inform the promulgation of promotional images. However, to date, much of the research on place promotion has tended to mirror industry perspectives and focus on visual representations of place.

According to Dann (1996) surprisingly little analytical work has been conducted upon the role of advertising in the cultural representation of tourist places, although work by Dilley (1986) – amongst others – has established the importance of cultural themes in shaping the presentation of destinations in tourism brochures. Some of the first promoters of tourism places were the railway companies, which, in their efforts to secure a commercial market, produced some enduring images of places. Visitors to contemporary Torquay, for example, are still welcomed to the 'English Riviera' – a conception (originally in the more spatially specific guise of the 'Cornish Riviera') of the Great Western Railway in the first decades of the twentieth century. Under the distinctly patriotic slogan 'See your own country first', it exhorted potential travellers to explore the delights of distant and exotic Cornwall in preference to Italy, with which it drew direct parallels in terms of the mildness of climate, the natural attractiveness of its (female) peasantry and even the shape of the two lands on the map, albeit with some cartographic licence in the case of Cornwall (Plate 8.2). In so

Plate 8.2 'See Your Own Country First': an early example of place promotion by the Great Western Railway, UK

doing, the railway promoters fed off such limited perceptions of Cornwall as may have existed at the time, but primarily they invented an image that was then reinforced through associated guidebooks and literature that presented Cornwall as some form of distant, yet still accessible, Arcadia (Thomas, 1997). Comparable strategies were also evident at the same time in the USA, where railroad companies were active promoters of tourist travel encouraging, for example, citizens from the eastern USA to visit the 'old west' before it disappeared into history (Zube and Galante, 1994).

This tradition of creative promotion of tourist places has continued to the present and content analysis of contemporary tourist brochures reveals texts that are often unashamedly escapist in their tone and which, when combined with photographic representations, emphasise difference, excitement, timelessness or the unspoilt, tradition or romance – according to the perceived market at which the publicity is aimed.

Such creative constructions of tourist places are most prevalent, of course, in the representation of foreign destinations, where fewer people will have had the direct experience needed to balance the claims of the brochures and the guidebooks. Messages are often subtly encoded. A study by Dann (1996) of a cross-section of British travel brochures promoting foreign places found, for example, that 25 per cent of illustrations showed only empty landscapes and, especially, beachscapes (reinforcing ideas of escape); that pictures showing tourists were nine times more common than pictures of local people (reinforcing notions of exclusivity and segregation); and that written text placed overwhelming emphasis upon qualities of naturalness (as an antithesis to the presumed artificiality of the tourists' routine lives) and the opportunities for self-(re)discovery. Only occasionally were senses of the exotic conveyed by use of images of local people, whilst reassurance that the experience of difference would not be so great as to be disorientating and unpleasant was provided by pictures showing familiar (though culturally displaced) items – typically as background elements. Examples of the latter might include 'English-style' pubs in the Spanish package tour resorts or, most ubiquitously of all, glimpses of the familiar red emblem of the Coca-Cola Company.

Promotional material that presents selective representations of realities is, of course, to be expected. What is more interesting, perhaps, is the emerging trend in some sectors of tourism towards promotion of places on the basis of historic rather than contemporary associations or, especially, through largely imagined reconstructions of a locality (see, e.g., Prentice, 1994). Hughes (1992: 33) comments that 'the past is being reworked by naming, designating and historicizing landscapes to enhance their tourism appeal', whilst the modern fascination with visual media such as television and film has also become widely embedded in tourism place promotion (Butler, 1990).

These trends have been nicely exemplified in England and Wales by the growing practice within regional and local tourism boards of appropriating legendary, literary or popular television characters or events to provide a form of spatial identity to which tourists will then be drawn (Figure 8.1). Some are well established. The term 'Shakespeare Country' to designate the area around Stratford-upon-Avon dates back to railway advertising of the 1930s and, along with similar descriptions of the Lake District as 'Wordsworth Country' or Haworth as 'Bronte Country', possesses some grounding in the real lives of individuals. 'Robin Hood Country' is more problematic given the uncertainties surrounding the actual existence of Robin Hood. However, descriptions of parts of Tyneside as 'Catherine Cookson Country' (after the books of the popular novelist) or the Yorkshire Dales as 'Emmerdale Farm Country' (after the TV soap opera) or Exmoor as 'Lorna Doone Country' (after the eponymous fictional heroine) take the process one stage further removed. They confuse reality with fictional literary and television characters or locales, and tourists are thereby confronted by a representation of what is already a representation. It is then only a short step to the totally artificial worlds of Disney in which cartoon characters – albeit in the form of

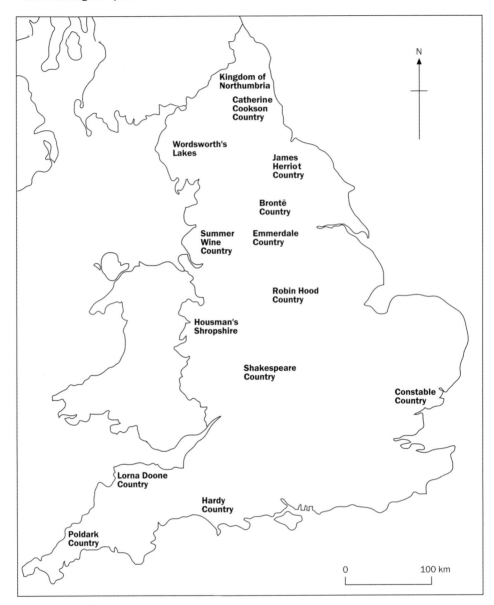

Figure 8.1 Imagined tourism 'countries' in England

employees in costume – step into the sunlight of Anaheim or Orlando to be photographed with the tourist.

Such practices represent commodification of tourism in one of its most overt forms. This is the tourism industry constructing a product and marketing it as an inclusive and convenient experience of another place. It draws selectively upon the real nature of places and presents only those elements that will appeal to the market segments at which the holidays are directed. But given the alacrity with which tourists consume such commodified and invented places, a question is raised over the significance, or otherwise, of 'real' experiences of place.

The problem for providers of tourism is that having constructed specific images of peoples and places in order to draw the visitor, it is obligatory for destinations to match the images that are projected. The tourist must confirm his or her expectations, otherwise the return visits, or the visits of others made on the basis of personal recommendation (one of the most important methods for disseminating knowledge of tourism places), will not occur. In this way, tourist images tend to become self-perpetuating and self-reinforcing (see Jenkins, 2003) with the attendant risk that, through time, tourist experiences become increasingly artificial.

The argument that tourist experience is founded on artificial rather than real situations is an idea that has been debated since the early 1960s, when a number of scholars – most notably Boorstin (1961) – attempted to argue that the traveller does not experience reality but thrives instead upon 'pseudo-events', that is, commodified, managed and contrived forms of provision that present a flavour of foreign places in a selective and controlled manner. This is evidenced in several distinct directions, including isolation of tourists from host environments, forms of cultural imposition and the staging of events.

The physical isolation of the tourist from the host environment receives its most obvious expression in the forms of enclave resort development in the Third World (see Chapter 4), where visitors are provided with the familiar creature comforts that may literally have been imported from their place of origin, set within a physical environment that has been deliberately contrived to reflect popular images of what an exotic location should be like. But many forms of tourism tend to locate visitors in what has been termed an 'environmental bubble': a protective cocoon of Western-style hotels, international cuisine, satellite television, guidebooks and helpful, multilingual couriers – 'surrogate parents' that cushion and, as necessary, protect the tourist from harsher realities and unnecessary contacts. As a result, the tourist gaze is often akin to gazing into a mirror. We construct tourism places to reflect ourselves, rather than the places we are visiting.

As we have seen in Chapter 6, such expectations on the part of tourists often lead to forms of cultural imposition of particular forms of development and provision upon host communities. Many tourists, quite illogically, expect a home-from-home experience, even in foreign lands, and the necessity for local providers to match those expectations inevitably changes the nature of the places that we visit. In the most extreme forms of this phenomenon, places actually begin to lose their sense of identity – they become placeless (Relph, 1976) – quite indistinct from other tourist places and quite unrepresentative of the realities of indigenous places. The mass tourist resorts of the Spanish coast, for example, commonly present a bland, placeless uniformity that says little about the 'real' Spain that exists often only a few miles inland.

Artificiality in the tourist experience of place may also be a consequence of staged events. One of the many ironies in international travel is that a primary motive for touring is exposure to foreign culture and custom, yet this is often met through contrived purveyance of supposed custom, whether via the sale of inauthentic souvenirs or via staged events or places (MacCannell, 1973, 1989). Sanitised, simplified and staged representations of places, histories, cultures and societies match the superficiality of the tourist gaze and meet tourist demands for entertaining and digestible experiences, yet they generally provide only partial representations of the places that are toured and their societies and cultures.

But do the rising levels of artificiality in tourist experience of place really matter in the business of place promotion? The answer, probably, is 'no'. Writers such as Poon (1989), Urry (1994b; 2000), Ritzer and Liska (1997) and Franklin (2004) have clearly signalled the importance of postmodern forms of tourism as a force shaping the development of new tourist places and that the artificiality of experience of place is no barrier to success. The postmodern tourist is widely recognised as embodying a new spirit of playfulness as a dominant mode of experience. These people are not deceived by the pseudo-realities of contemporary tourism but are happy to accept such constructions at face value as an expected and, indeed, a *valued* aspect of new forms of experience. No one is fooled, for example, by the staged representations of other places or other epochs that are assembled in almost perfect detail in the themed hotels of Las Vegas, but that does not stop these places being hugely popular – and hugely enjoyable – as tourist places. Indeed, as has previously been noted, the *in*authentic is, today, often preferred to the authentic as being a much more reliable, comfortable and pleasurable experience.

The theming of tourist environments

In their discussion under the heading of 'landscapes of pleasure', Shaw and Williams (2004: 242) remind us that 'tourism spaces are dynamic in that they are constantly being created, abandoned and re-created'. They also note how industrialisation and modernity contributed directly to the creation of tourist places such as sea-bathing resorts and that in turn, post-industrialisation and postmodernity are creating new and rather different tourism places. An important facet of this process of invention of new tourism places (and, indeed, the reinvention of existing ones) has been the trend towards 'theming' the environment.

Theming is a planned process that strives to impart a sense of both identity and order to a given place through a combination of physical design (or redesign) of space and the development of an associated set of cultural narratives, all of which connect to a common theme or a set of related themes. The chosen theme(s) may relate to distinctions of both time and/or place and is generally overlain by complex associations of economic, social or cultural practices. Within the environment in question, themes are actively reinforced through material and symbolic means to encourage both direct and subconscious engagement on the part of the user with the themes in question. These devices might include:

- development of landscape elements (or, in a historic environment, retention of selected elements from the previous landscape) that are suggestive of the theme;
- naming of places, streets, public spaces or premises with titles that connect to the theme;
- development of attractions (such as museums, tourist trails or places of entertainment) that reflect the selected theme;
- adoption of street furniture and signage that identifies with the theme by its styling or by carrying logos;
- inclusion of styles and mixes of retailing (or other business) that relate to the theme;
- marketing of souvenirs in which the theme is mirrored;
- incorporation of the theme into place promotion media (such as brochures).

The increasing prominence of themed environments undoubtedly reflects the reorganisation of space – especially in cities – as centres of consumption and the redesign of space around visual images and scripted themes or narratives is a key part of this process (Paradis, 2004). Paradis further suggests that themed environments are predominantly designed to appeal to tourists or visitors and that the emergence of theming of built environments, in particular, has coincided with the growth of tourism as a major form of urban economic development. There is, therefore, an important linkage between theming, place promotion and the commercial success of places as tourist destinations and whilst themes may often reflect the spectacular, they may also be attached to otherwise mundane products or locations as a way of enhancing their competitive position and attracting attention (Paradis, 2004). However, theming does not simply deliver commercial benefits, there are also important opportunities to project varying forms of cultural identity through the development of themed spaces and in some situations complex mosaics of themes will become evident as different interest groups seize opportunities to develop themes that reflect their differing purposes and agendas.

Unsuprisingly, there is a wide range of potential tourist spaces to which theming may be applied and Figure 8.2 presents a theoretical framework to illustrate this point. Several aspects of the typology are worth noting.

Theming may be applied at a range of geographical scales that span the spectrum from the micro-level of individual premises to the macro-level of regional or even national space. At the micro-level, theming of individual premises – such as bars or restaurants – has become a very familiar aspect of contemporary retail change, as was evidenced, for example, in the widespread development of 'Irish style' pubs in British cities during the 1990s. At a higher spatial level, different types of urban zone have become themed areas (such as waterfront regeneration zones or, more selectively, major retail malls of which ventures such as West Edmonton Mall (Canada) and the Mall of America at Minnesota (USA) are ultimate examples (Goss, 1999). Theming may sometimes engulf entire townships – such as the 'Bavarian' themed town of Leavenworth in Washington (USA) (Frenkel and Walton, 2000) or the town of Ironbridge in Shropshire (UK) (which styles itself as 'the birthplace of industry' because of its role in the British Industrial Revolution). Theming at the scale of cities is not unknown – as is evident in Las Vegas (USA) – whilst as Figure 8.1 has already illustrated, theming at a regional level as a form of place promotion has become a widespread practice in many destination areas. Finally, theming at a national level may sometimes be encountered when a destination area possesses a single, very strong product that forms the dominant basis to its place promotion activity. The case of Egypt and the role of its ancient past in defining its contemporary identity as a world tourism destination is perhaps an example.

Second, it is also evident that theming can be applied to differing *forms* of space. Some themed environments comprise enclosed or contained spaces that operate as commercial attractions – such as theme parks – but theming may also be used to imbue public spaces with particular qualities or identities. The popular current trend in some British cities to designate cultural or cafe 'quarters' within redeveloped city centres is an example of this

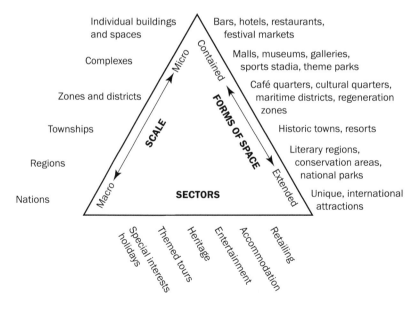

Figure 8.2 Typology of themed tourism spaces

practice, as is the wider process of regional designations that was discussed earlier. The regional examples make a further interesting point since it is evident in these cases that themed space need not necessarily be contiguous, but may comprise an assemblage of spatially disconnected sites around which tourists must literally tour in order to engage fully with the chosen theme. Equally, themed spaces may be disaggregated into smaller areas, each of which projects its own theme. Disney theme parks exemplify this process in the subdivision of park space into contrasting zones such as 'Frontierland' and 'Fantasyland', whilst similar approaches are used in themed shopping malls. Shaw and Williams (2002) provide the example of the Trafford Park retail mall near Manchester (UK) which makes a virtue (in its publicity) of the way in which the various spaces of the mall each reflect different parts of the world.

Third, the typology suggests that theming may be applied to different *sectors* of tourism. The sector to which it has perhaps been applied most widely is the expanding realm of heritage attractions, but from the preceding paragraphs it is clear that theming is also a popular approach in sectors such as entertainment and retailing.

Theme parks

We will revisit several of these themed environments in subsequent discussions of heritage and of tourism in urban places (see Chapters 9 and 10). To conclude this chapter, however, it is proposed to consider the development of theme parks in greater detail. Theme parks commend themselves for further consideration from a number of perspectives, but three key ideas are worth pursuing, in particular:

● the role of theme parks in the globalisation of culture;
● the capacity of theme parks to shape new geographies of tourism and invent tourist places;
● the postmodernity of theme parks and the influence that the concept has exerted on other postmodern tourist spaces.

The specific character of theme parks varies from place to place but for purposes of this discussion a theme park is viewed as a self-contained family entertainment complex designed around landscapes, settings, rides, performances and exhibitions that reflect a common theme or set of themes.

Modern theme parks have become strongly implicated in the development of globalised tourism cultures, both through their spatial extension from their origins in North America to other world regions such as Europe and the Pacific Rim of Asia and through the important role that major corporations – especially Disney – now play in global media and communications. As Davis (1996) explains, the ancestry of the modern theme park may be traced to the American fairground-style amusement parks (such as Coney Island) that had been established towards the end of the nineteenth century and which were believed to number in excess of 1,500 by the early 1920s. A smaller number of similar ventures had also been created in Britain and Europe prior to 1950, such as the Efteling Park in the Netherlands.

However, it was largely through the entry into the amusement park industry of the cartoon and movie-maker Walt Disney that the modern theme park evolved. Disney opened his first park at Anaheim (Los Angeles) in 1955, and although he drew several ideas from the Efteling Park in shaping the initial designs, Disneyland – as his first park is named – was a genuinely innovative development of a new type of tourism space and one which anticipated – but also influenced – important shifts in taste and preference amongst people at leisure. In particular, Disney realised the potential to initiate wider connections between the theme park, the media of TV and film, and the advertising and marketing of these media and their associated products. The immediate success of Disneyland (which attracted 3.5 million visitors in its first year of operation (Bryman, 1995)) partly reflected, therefore, the fact that the Disney company's film and media outputs had become accessible as physical attractions that could be visited and experienced at first hand (Plate 8.3).

To achieve such synergies of entertainment and corporate promotion, major theme parks are produced as a carefully scripted and intricately designed set of physical spaces that blend the attractions and other forms of entertainment with an appropriate range of commercial opportunities, all set within a highly regulated and controlled environment. Davis (1996: 402) describes the landscape of places such as Disneyland as 'exhaustively commercial . . . a virtual maze of public relations and entertainment . . . a site for the carefully controlled sale of goods (souvenirs) and experiences (architecture, rides and performances) "themed" to the corporate owner's proprietary images'. Indeed, within Disney parks, the overarching 'theme' is clearly Disney itself (rather than the narratives of history, fantasy, nature or exploration that outwardly shape the different 'lands' that comprise each park). Hence the rides, architecture, products and sub-themes all connect to the films, TV shows, comics and music of the Disney corporation in a self-reinforcing circle of promotion and cross-references. (For a highly detailed analysis of the design and operation of Disney parks, see Fjellman, 1992.)

The impact of Disney's initial venture and the popular allure that has been created around Disney's parks has had two important effects. First, the commercial success of the

Plate 8.3 The innovator and his innovations: Walt Disney and Mickey Mouse greet the visitors to Disneyland, Los Angeles

Disney theme park concept encouraged others to enter the field, in particular major entertainment corporations and film companies such as MCA and Time-Warner (which operates the highly successful Universal Studios theme park in Hollywood). The quest for new business on the part of these corporations has prompted a spatial extension of theme parks into what has become a global market (see below). Second, the process of globalising the theme park has been recognised as a highly influential medium of cultural exchange and – especially – influence. According to Wasko (2001) the image of Mickey Mouse is reputedly now the best-known cultural icon in the world, which clearly testifies to the global 'reach' of major entertainment corporations such as Disney.

However, whilst there is no doubt that corporations such as Disney deliver valued forms of entertainment to a global audience (see Wasko et al., 2001), there is also a significant critical discourse around perceived detrimental cultural impacts. Deconstruction of Disney's movies and theme parks (e.g., Byrne and McQuillan, 1999) tends to expose the partial and selective ideologies that shape the Disney product (e.g., around processes such as colonisation of the American West or of gender and race relations) or the extreme levels of rationalisation and control that govern the operation of the parks (Ritza and Liska, 1997). Buckingham (2001: 270) captures well a common view of Disney when he writes that Disney 'encapsulates everything that is wrong with contemporary capitalism: the destruction of authentic culture, the privatization of public space, the victory of consumerism over citizenship, the denial of cultural differences and of history'. Yet, as Byrne and McQuillan (1999: 2) observe, Disney retains a 'powerful hegemonic hold over children's literature, family entertainment, mainstream taste and Western popular culture' that not only remains intact, but is also continuing to reach new audiences.

The development of theme parks as tourist attractions illustrates several aspects of the contemporary redefinition of tourism practices and places, and they illustrate, very effectively, the idea of invented places. The notion of theme parks as invented tourism places operates at two levels. First, in many parks the visual and contextual fabric is often an invention since it portrays imaginary characters and places that circulate around cartoon figures, fairy stories, myths or legends. 'Magic Kingdoms' and 'Fantasy Lands' are popular constructions in theme parks across the world, whilst themes that have a stronger grounding in reality, such as Disney's 'Frontierland' and 'Main Street USA', present idealised and highly selective recollections. Second, theme parks are quite capable of inventing new tourism geographies by the way in which they are located. Whilst some ventures have gravitated towards established tourism areas, such as the theme park developments in Florida, others have been obliged (through their considerable land requirements) to take on greenfield sites in places where tourism was not previously present or conspicuous. The original Disney development at Anaheim, for example, was located in a nondescript zone on the city fringe where the local tourist stock at the time amounted to just seven rather modest motels. Similarly, the subsequent development of Euro-Disney on a 2,000ha site at Marne-la-Vallee, 32km east of Paris, although close to a tourism city of global significance, also introduced large-scale tourism to an area that had only been lightly affected previously.

Initially the development of American parks led to an expansion of European theme parks (from the late 1970s) and then of East Asian and Australian parks (during the 1980s) (Davis, 1996). The growth in theme parks on the Pacific Rim, notably in Japan and more recently South Korea and China, has been especially impressive. For example, a study of theme park development in Japan (Jones, 1994) showed that before 1983 there were only two theme parks in that country, but by 1991 that number had risen to twenty-seven (Figure 8.3). The catalyst for change in Japan was the opening of Tokyo Disney in 1983, which although not owned by the Disney Corporation, was designed by Disney staff and is, as Bryman (1995) observes, unashamedly American in its approach although it does offer some important concessions and adaptations to Japanese culture and taste. Annual

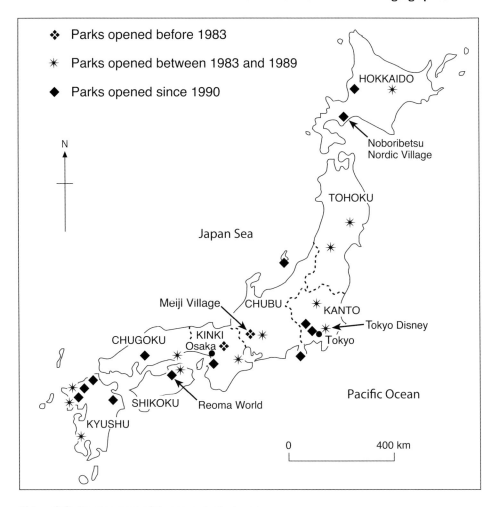

Figure 8.3 Development of theme parks in Japan

levels of visiting to Tokyo Disney soon exceeded 10 million, spawning a growing number of alternative theme parks, some of which mimic the Disney concept whilst others develop the Japanese fascination with the cultures of other places that embraces an eclectic mix of themes that include Nordic villages and medieval German towns. A recent study of theme park development in China (Zhang, 2007) has also demonstrated the rapidity with which theme parks are emerging as attractions to both domestic and international tourists in China. Zhang's study is interesting, not just for its account of the expansion of Chinese theme parks (which has seen hundreds of small parks established since 1980 and a number of major projects at cities such as Beijing, Shanghai and Shenzhen) but more importantly because of the challenge to the prevailing hegemony of the Disney model that Chinese theme parks appear to offer. Parks such as the 'Overseas Chinese Town' at Shenzhen are not just environments of leisure and tourism, but carefully designed spaces that project and reinforce key messages about the political, social, educational and cultural modernisation of China.

The spatial expansion of theme parks as tourist attractions is, of course, a reflection of the success of the concept and its almost universal appeal – something that is strongly reflected in their capacity to draw huge numbers of visitors. These are family attractions that, perhaps surprisingly, also appeal to older tourists. In Britain, Blackpool's Pleasure Beach attracted 6.2 million visits in 2002, with Alton Towers drawing 2.45 million visitors in the same year (BTA, 2003). This pales almost into insignificance, however, compared to the market leaders in North America and the Far East. Tokyo Disney attracted almost 13 million visitors in 2006 while in the same year the combined total attendance at the six Disney parks in the USA (Disneyland, California Adventure, Magic Kingdom, EPCOT, Animal Kingdom and Disney World/MGM – the last four all in Florida) exceeded a staggering 65 million people. Aggregate attendance at the top twenty theme parks in the USA and Canada in 2006 is estimated at exceeding 119 million (www.themeparks.about.com).

The spatial distribution of the major parks is interesting. As the Japanese case shows, there are clear advantages to being close to major urban markets and/or established tourism regions. In the USA (see Figure 8.4) the largest parks generally cluster in the warmer states such as Florida and California, since these represent the preferred destinations for American tourists in general. The more attractive climates in these locations clearly favour outdoor parks. (This was a requirement that Disney discovered to its cost in the near-disastrous opening of Euro-Disney in the damp and often cold outskirts of Paris.) But interestingly, parks can also be developed successfully in less propitious locations. Loverseed's (1994) analysis of patterns of visiting to American theme parks in the 1990s showed that, at that time, the most rapid rates of expansion in visits were being recorded in less popular tourism areas such as Illinois, Ohio, Missouri, Tennessee and Kentucky, although the actual levels of visiting were well below the market leaders in California and Florida. Similarly, in Britain, the most popular theme park, Alton Towers, is buried deep in the lanes of rural north-east Staffordshire, one of the less-visited counties in England – especially if visitors to the park are deducted from aggregate totals. The capacity of theme park tourism to define new places as objects of the tourist gaze is, therefore, quite significant.

Lastly, brief mention should be made of the postmodern character of theme parks and their influence on other themed tourist spaces. Because of a number of primary characteristics, theme parks have been widely viewed as the quintessential postmodern spaces. This postmodernity is evident in, for example:

- the overt and conscious mixing of architectural styles and spaces that produces places that are a collage or pastiche of otherwise incompatible genres;
- the deliberate confusion of the real with the artificial and a common reliance on what Eco (1986) terms the 'hyper-real' – objects or situations that are more real than reality itself (such as the chance to be photographed with Mickey Mouse);
- the extended use of simulacra (which are representations of originals that do not actually exist – such as Tom Sawyer's island which visitors to Disneyland tour by boat);
- the exaggeration of time–space compression in which park visitors cross simulated time and space with effortless ease or encounter juxtapositions of diverse epochs and cultures in ways that are otherwise impossible (Bryman, 1995);
- the widespread incidence of 'de-differentiation' (Lash, 1990) which, as noted above, refers to the dissolution of conventional boundaries between institutional orders and related distinctions. This is evident in aspects such as the seamless integration in Disney's latest generation of parks of retailing, entertainment and tourist accommodation, but, according to Rojek (1993b), extends to a more fundamental dissolution of the distinctions between theme park experience and the spectacle of daily life itself (see, also, Franklin (2004) for a similar argument).

These are environments that appeal strongly to the 'post-tourists' – the playful consumers of superficial signs and surfaces that some writers see as embodying the new age of tourism

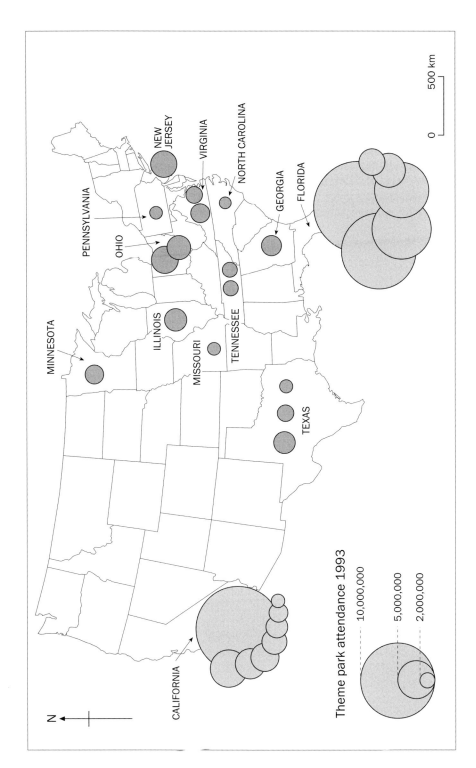

Figure 8.4 Major theme parks in the USA in the 1990s

Theme park attendance 1993

10,000,000

5,000,000

2,000,000

(e.g., Feifer, 1985; Rojek, 1993b). Bryman (1995: 178) captures this essence very well in noting how theme parks 'through their cultivation of excitement, their presentation of sound-bites of history (and the future) and their fabrication of simulacra which are better than their original referents and exhibit impossible juxtapositions, create an environment in which "the post-tourist emphasis on playfulness, variety and self-consciousness" (Urry, 1990) can be given full rein'.

Critically, however, these effects and influences are no longer confined to the spaces of theme parks, but – perhaps because of the enormous popularity of such places – have come to influence a much broader range of places that people encounter – either as tourists or simply as part of daily routines. Davis (1996) observes how theme parks have expanded from being stand-alone attractions to form complexes and resort regions (such as Orlando and Las Vegas) and become widely influential on the form and design of an extended range of other leisure environments: heritage centres, museums, hotels, casinos and shopping malls (Bryman, 1995). Whether we care for it or not, theming has become an embedded part of daily experience and it is in this wider influence on postmodern places that the real significance of theme parks probably lies.

Summary

This chapter has explored some of the ways in which new tourist geographies and associated tourism places are formed. Whilst many factors contribute to the formation of new patterns of tourism, particular emphasis has been placed upon cultural influences that not only shape our understanding of places and the way in which we select the different sites on which we gaze, but also shapes directly the performances we deliver as tourists. These performances, in turn, contribute further to the process of defining (or making) tourist places. However, although many aspects of our gaze are individuated, our actions are also mediated by others. This is evident especially through the practice of place promotion (which raises our awareness of potential destinations and actively influences the way in which we form images of those places), but is also revealed in the increasingly popular practice of 'theming' – a trend that is not only influential in the places that we visit as tourists, but is also becoming widely embedded in the places in which we live our daily lives.

Discussion questions

1   Identify and evaluate the strengths and the weaknesses of Urry's concept of the tourist gaze as a way of understanding how tourists relate to place.
2   What do you understand by the term 'tourist performance' and why might the performative characteristics of tourism be important?
3   In what ways does place promotion in tourism tend to shape only selective repre-sentations of tourism places?
4   Why has the practice of theming tourism places become so popular, both with developers and with tourists themselves?
5   What do you believe are the key components that have defined the success of Disney's theme parks?

Further reading

There is a substantial geographical literature on the concept of place, but the following essay provides a very convenient point of entry:

Castree, N. (2003) 'Place: connections and boundaries in an interdependent world', in Holloway, S.L. et al. (eds) *Key Concepts in Geography*, London: Sage, pp. 165–85.

Although they are not a recent publications, the most influential discussions of 'placelessness' and the manner in which modern developments have eroded the distinctive nature of places are probably:

Relph, E. (1976) *Place and Placelessness*, London: Pion.
—— (1987) *The Modern Urban Landscape*, London: Croom Helm.

The concept of the tourist gaze and an interesting collection of essays on cultural inventions of place are to be found in:

Urry, J. (1990) *The Tourist Gaze*, London: Sage.
—— (1995) *Consuming Places*, London: Routledge.

For recent critiques of Urry's concept of the tourist gaze, see:

Franklin, A. (2004) *Tourism: An Introduction*, London: Sage.
MacCannell, D. (2001) 'Tourist agency', *Tourist Studies*, Vol. 1 (1): 23–37.

For further insights into the performative character of tourism see:

Crouch, D., Aronsson, L. and Wahlstrom, L. (2001) 'Tourist encounters', *Tourist Studies*, Vol. 1 (3): 253–70.
Edensor, T. (1998) *Tourists at the Taj: Performance and Meaning at a Symbolic Site*, London: Routledge.
—— (2000) 'Staging tourism: tourist as performers', *Annals of Tourism Research*, Vol. 27 (2): 322–44.

A varied set of interesting essays on image formation in tourism may be found in:

Selwyn, T. (ed.) (1996) *The Tourist Image: Myths and Myth Making in Tourism*, Chichester: John Wiley.

Place promotion is considered in some detail in:

Gold, J.R. and Ward, S.V. (eds) (1994) *Place Promotion: The Use of Publicity and Marketing to Sell Towns*, Chichester: John Wiley.

Useful recent essays on the representation of tourist places can also be found in:

Jenkins, O.H. (2003) 'Photography and travel brochures: the circle of representation', *Tourism Geographies*, Vol. 5 (3): 305–28.
Yuksel, A. and Akgul, O. (2007) 'Postcards as affective image makers: an idle agent in destination marketing', *Tourism Management*, Vol. 28 (3): 714–25.

Theming of tourist places is discussed in:

Paradis, T.W. (2004) 'Theming, tourism and fantasy city', in Lew, A.A. et al. (eds) (2004) *A Companion to Tourism Geography*, Blackwell: Oxford: 195–209.
Shaw, G. and Williams, A.M. (2004) *Tourism and Tourism Spaces*, London: Sage, Ch. 10.

A concise analysis of the development of theme parks as features on the global tourism landscape is provided by:

Davis, S.G. (1996) 'The theme park: global industry and cultural form', *Media Culture & Society*, Vol. 18 (3): 399–422.

For a critical evaluation of the work of Walt Disney and the parks that he designed see:

Bryman, A. (1995) *Disney and His Worlds*, London: Routledge.
Fjellman, S.M. (1992) *Vinyl Leaves: Walt Disney World and America*, Boulder: Westview.

The impact of the Disney product on global culture is extensively explored through a set of national case studies in:

Wasko, J., Phillips, M. and Meehan, E.R. (eds) (2001) *Dazzled by Disney? The Global Disney Audiences Project*, London: Leicester University Press.

9 Urban tourism in a changing world

In the concluding essay to their edited work on *The Tourist City*, Fainstein and Judd (1999a: 261) comment that 'tourism has been a central component of the economic, social and cultural shift that has left its imprint on the world system of cities in the past two decades'. This simple statement captures the evident truism that cities in the twenty-first century represent important tourist destinations and so any attempt to develop an understanding of the spaces of tourism needs to examine these primary tourist locations. The contemporary significance of urban tourism derives in part from the scale of activity and its diversity – embracing, as it does, several forms of pleasure travel, business and conference tourism, visiting friends and relatives, educational travel and, selectively, religious travel. But more importantly, perhaps, urban tourism has acquired a level of significance through its new-found centrality in the processes of reinvention of cities under post-industrial, postmodern change and the related restructuring of urban economies and societies around consumption. Urban tourism has variously become an essential tool for physical redevelopment of urban space, for economic regeneration and employment creation, for place promotion, for re-imaging cities and helping to create identity in the new global systems. As a consequence of these processes, tourism (and its infrastructure) has become deeply embedded within both the urban fabric and the daily experience of people who live within these places.

However, although the importance of urban tourism is now widely acknowledged, it is a subject that has, until very recently, been an area of relative neglect within tourism studies. This is a recurring introductory theme in most recent texts on the subject (see, *inter alia*: Page, 1995; Law, 2002; Shaw and Williams, 2002; Page and Hall, 2003; Selby, 2004). Common explanations for the tendency to disregard urban tourism include difficulties entailed in isolating and enumerating tourists from the local urban population which, in turn, create practical problems in measuring urban tourism and identifying its economic, cultural and environmental impacts. In addition, the predisposition in tourism studies to focus upon the holiday sector has not encouraged research into key component areas of urban tourism, such as business tourism or the visiting of friends and relatives (Law, 1996, 2002). Neglect may also owe something to the fact that whilst some cities (such as Paris, Rome and Venice) have a lengthy tradition as destinations for tourists that dates back at least to the Grand Tour (see Chapter 3), the wider patterns of tourism development have often been shaped around a very clear desire on the part of urban populations to *escape* the environments of major industrial cities and conurbations rather than to visit them (Williams, 2003). It is only with the onset of post-industrialism that cities across the urban spectrum have become major objects of the tourist gaze and tourism geography is perhaps still coming to terms with the full implications of this highly contemporary process.

This chapter attempts to distil some of the more important current perspectives on urban tourism by exploring three related areas. The next section examines the changing urban context in order to isolate and explain the wider processes that are reshaping contemporary cities and which help us to appreciate more clearly why tourism has become a prominent component of the post-industrial, postmodern city. This is followed by a discussion of the tourist city which aims to explain how cities function as tourist destinations, whilst the final section examines how tourism intersects with the new urbanism that is revealed in cities of the twenty-first century.

The urban context

Any understanding of the changing significance of urban tourism must be grounded in a wider awareness of the contemporary urban context and, especially, the emergence of post-industrial and postmodern cities. This transition has exerted a fundamental influence upon not only the internal organisation of urban space and the structuring of economic and social relations *within* cities, but also the relationships *between* cities. Four broad, related themes are central to this process: globalisation; economic and social restructuring; the remaking of urban identity; and new political agendas.

Under globalisation, the contemporary mobility of capital, labour, materials and information has directly shaped the development of new global networks of exchange and new transnational systems of production and consumption in ways that are producing an essentially different narrative of urbanism at the start of the twenty-first century. Fainstein and Judd (1999a: 261) describe some of this dynamic in commenting that 'the present epoch involves a different, more flexible organization of production, higher mobility of both capital and people, heightened competition among places, and greater social and cultural fragmentation'. Social and cultural fragmentation arises particularly from the international migration of labour and helps to produce the characteristically heterogeneous urban populations in postmodern cities that are a particularly visible product of globalisation.

Closely linked to globalisation and highly influential on the development of post-industrial cities has been the process of economic restructuring (Dear and Flusty, 1998). Restructuring has been shaped by two primary phases: a period of significant *deindustrialisation* (from about 1970) in which the traditional forms of manufacturing that shaped the industrial, modern city of the Western world have largely collapsed, to be replaced – under a period of *reindustrialisation* – by new growth sectors in the urban economy shaped around the information economy, new technologies and a significant growth in service industries associated with these activities. This has been central to the transition from production to consumption as a defining logic of contemporary cities in Europe, North America and Australasia, but has also triggered major reworking in the productive use of space. In particular, the margins of major cities have emerged as dominant areas of new industrial activity. Here firms that new communications technologies have often rendered footloose may capitalise on the lower land cost, the environmental attraction and the enhanced levels of accessibility that peripheral sites often provide, to establish spatially fragmented but functionally linked zones of new (post-Fordist) production. At the same time, older areas of (Fordist) manufacturing fall derelict as traditional production is supplanted by new modes, creating significant opportunities – and indeed a need – for urban regeneration. Soja's (1989, 1995, 2000) detailed analyses of the growth of the new, peripheral developments of technology industries in Los Angeles and the progressive abandonment of older industrial districts in central Los Angeles provides an excellent example of the spatial reorganisation of the urban economy under postmodernity.

However, although the process is driven by an economic imperative, it also has significant social ramifications in the related development of new and similarly fragmented patterns of social space. Globalisation and the post-industrial shift are widely held to have sharpened distinctions and disparities – some of which are economic (e.g., in the gaps between wealth and poverty), whilst others are socio-cultural (e.g., in the identities of the many minority communities that congregate in modern cities). But these distinctions also become etched in space – in wealthy or poor neighbourhoods, or in the enclaves that are forged by ethnic minorities. This is not, of course, a new characteristic of urban social geography, but it has become much more extensive and more sharply drawn, especially in major cities. In many situations such spatial distinctions are further emphasised through the overt defence of space, either through social practices or active surveillance and policing. Davis's (1990) brilliant dissection of the social landscapes of Los Angeles paints a fascinating picture of this tendency as it is variously revealed in the gated and patrolled communities of the wealthy, the protected 'public' spaces of the municipal cores and even in the 'turfs' of the street gangs that flourish in poor areas such as Watts and South Central and who assert their own, distinctive hold on space.

The third key process in the post-industrial/postmodern transformation of cities has been the active remaking of cities and city identity. This has been partly a response to the decline of old centres of production which created a need to pursue regeneration policies as a means of addressing the economic and social malaise that followed the widespread loss of traditional areas of work, as well as the physical problems of derelict and 'brownfield' land that deindustrialisation usually created. At the same time, there are important links to globalisation as the need to forge new identities and images as a way of enabling post-industrial places to compete effectively in a global context has became a key driver of change. Part of the new relationship between cities under globalisation is that they now need to compete on a world scale for capital investment, labour – and, indeed, tourists – and their ability to compete shapes their scope for future development (Gospodini, 2001; Hall, 2005). In many cities the process of 'manufacturing' new sites of consumption – in regenerated waterfronts, in themed shopping malls, or in state-of-the-art museums, galleries and sports stadia – has significantly affected the visual character of cityscapes and the components that comprise the human setting in contemporary cities. Theming has become a *leitmotif* of postmodern cities but so has the progressive development of a new aesthetic around postmodern urban design that is evident in the eclectic, collage-like mixing of architectural styles and traditions and the rising significance of signs and signifiers as cultural markers in the malls, cafe quarters and reconstructed waterfronts of the new urban landscape. As we will see in a subsequent section, this trend has had some important implications for the development of urban tourism.

Finally, the impact of new political agendas after 1980 should be noted, in particular the rise of what has been termed 'New Right' politics under the leadership of Ronald Reagan in the USA and Margaret Thatcher in the UK. Judd (1999) and Law (2002) both comment on how the entrepreneurial approaches favoured by both these administrations moved the focus of urban policy away from the social-welfare agendas of the 1970s and placed new emphases on public–private partnerships and the active selling of cities as places for investment as the way forward. This helped to establish a new urban political climate that was highly conducive to the development of tourism (amongst other service sectors) as strategically important in the remaking of cities.

This introductory discussion – of necessity – paints only the briefest of outlines of central themes in the changing urban setting but it provides an essential context in which to develop an understanding of tourism in urban places. Table 9.1 provides a key-point summary of the primary characteristics of urban post-industrialism and postmodernity.

The tourist city

How are the spaces of tourism developed and arranged within cities? The preceding description of the urban context has perhaps tended to treat cities as fairly homogenous –

Table 9.1 *Essential characteristics of post-industrial/postmodern cities*

- Urban life shaped by processes of consumption rather than production
- Fragmentation of economic and social space leading to complexity and multi-nodal structures
- Peripheral nodes of development (edge cities) replace inner urban zones as centres of industrial production and, especially, services
- Central zones reinvented – especially through regeneration – as new centres of consumption
- Conventional distinctions around work, leisure, culture and social class become blurred, but other distinctions (e.g., around wealth and ethnicity) become more sharply drawn and often reinforced through mechanisms such as defence of space
- Urban populations increasingly heterogeneous and tending to form micro social spaces
- Urban landscapes increasingly shaped around theming of space and the promotion of visual and aesthetic media
- Urban landscapes characterised by collages (of signs and symbols) and simulacra

Sources: Adapted from Soja (1989); Davis (1990); Page and Hall (2003); and Selby (2004)

albeit changing – entities, but in practice, of course, cities are often very different places and accommodate tourism is some fundamentally different ways. Fainstein and Gladstone (1999: 25) note that 'the commodity that urban tourism purveys is the quality of the city itself', but that quality differs significantly from place to place. It is therefore important to isolate those key differences between cities that influence how tourism may develop.

As an initial starting point, a number of writers have attempted to develop typologies of cities as a basis for explaining contrasting styles and patterns of urban tourism. Page (1995) attempts to differentiate a quite lengthy typology of cities that includes capital cities, metropolitan centres, large historic cities, industrial cities, cultural cities and resorts. However the lack of consistent definition that surrounds several of these labels and the overlap between at least some categories in the typology is a significant limiting factor (Law, 2002). Cities such as London and Paris, for example, are simultaneously capitals, metropolises and historic, industrial and cultural places.

A more useful approach has been proposed by Fainstein and Judd (1999a). They distinguish a three-fold categorisation based around the following:

- Resort cities – which are urban centres that are created expressly for consumption by visitors. Conventional urban seaside resorts would fall in this category, as does Las Vegas, which is considered in more detail below. The resort city is closely related to Mullins' earlier concept of 'tourist urbanisation' in which – on the basis of a series of case studies of Australia's Gold Coast – several defining features of this form of urban development were proposed. These include distinctive spatial and symbolic attributes (that include well-defined tourism enclaves); rapid population growth based around highly flexible systems of production and consumption; and boosterist approaches to planning and management (Mullins, 1991).
- Tourist-historic cities – which are places that lay claim to a distinctive historic and/or cultural identity that tourists may experience and which forms a primary basis to their attraction. Some tourist-historic cities have been tourist destinations for centuries (such as Venice) whilst others have been transformed into tourist cities through processes of active reconstruction or rediscovery of elements of their urban heritage (such as Boston, USA). An important characteristic of the true tourist-historic city is that since tourist sites and uses tend to be built into the architecture and cultural fabric of the city, tourist space is much more integral to the overall urban structure and tourists become inter-mixed with residents and local workers in ways that are much less typical of the demarcated spaces in resorts.

● Converted cities – which have consciously rebuilt their infrastructures and – most importantly – their identities for the purpose of attracting tourists as a means of supporting new urban economic growth. These places are typically former centres of traditional manufacturing and distribution, and rather like resorts, tourist spaces in the converted city often develop as quite isolated enclaves set within a wider urban environment that may remain comparatively unattractive and sometimes hostile to outsiders. The Inner Harbour redevelopment at Baltimore is an example that is commonly quoted. Occasionally, however, a more seamless integration of tourism into converted cities is achieved, Judd (1999) suggesting, as an example, San Francisco (Plate 9.1) which is widely accessible to tourists and a rare example of a major American city that somehow contrives to function at a human scale.

However, whilst Fainstein and Judd's typology is valuable in making some important distinctions around how tourism may develop in cities, it is clear that many individual cities will often blend elements of each category within their overall make-up rather than simply conforming to a specific model. For this reason it is often more helpful to recognise that most tourist cities are actually comprised of distinctive sub-spaces or functional areas and that the balance between these functions is generally central to defining the nature of the city as a tourist destination. Figure 9.1 represents a development of a model originally

Plate 9.1 The downtown district of San Francisco

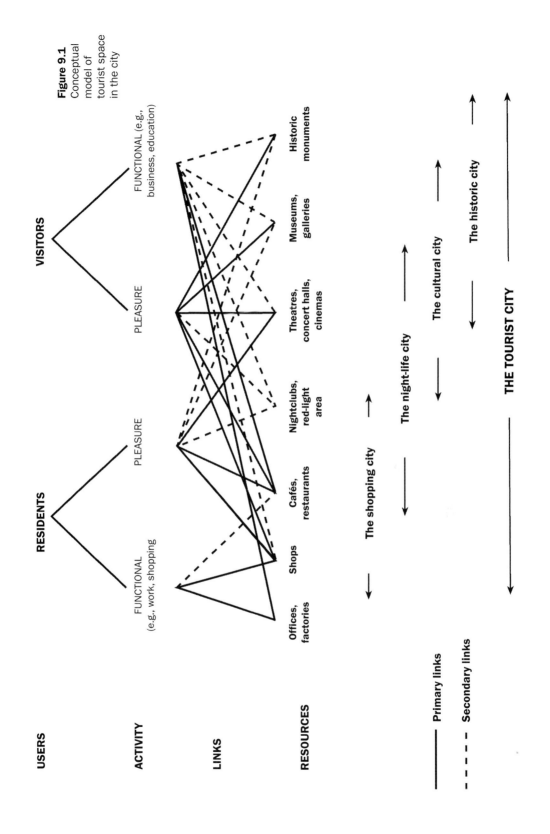

Figure 9.1
Conceptual model of tourist space in the city

USERS

RESIDENTS

VISITORS

ACTIVITY

FUNCTIONAL (e.g., work, shopping)

PLEASURE

PLEASURE

FUNCTIONAL (e.g., business, education)

LINKS

RESOURCES

Offices, factories

Shops

Cafés, restaurants

Nightclubs, red-light area

Theatres, concert halls, cinemas

Museums, galleries

Historic monuments

The shopping city

The night-life city

The cultural city

The historic city

THE TOURIST CITY

——— Primary links

– – – Secondary links

proposed by Burtenshaw et al. (1991) and attempts to show how the functional and leisure demands of both residents and tourists interact to define distinctive 'cities' (or zones) within the overall city space. We can therefore start to understand the tourist city as comprising interconnected sets of functions and associated spaces that reflect different needs and interests of visitors, such as historic and cultural heritage, entertainment, nightlife and shopping, although, importantly, these zones of activity are generally well-defined and are distinct from other parts of the city in which tourists seldom – if ever – penetrate (Shaw and Williams, 2002).

By viewing tourism and urban space as a composite construct, we raise two important questions: what is the nature of demand for urban tourism and how is demand reflected in the supply and organisation of facilities and attractions?

The demand for urban tourism

Cities are probably unique as tourism places in respect to the range of different categories of tourist demand that they attract and accommodate. Law (2002: 55) summarises the main market segments in urban tourism as:

- business travellers;
- conference and exhibition delegates;
- short-break holidaymakers;
- day trippers;
- visiting friends and relatives (VFR);
- long-stay holidaymakers using the city as a gateway or as a short-visit stop on a tour;
- cruise ship passengers (in port cities).

Business travel is an especially important component of urban tourism. On a global scale, business, conference and exhibition travel in 2006 accounted for an estimated 131 million international trips – which represents 16 per cent of the world travel market (WTO, 2007). Domestic business travel markets are – in total – considerably larger. Recent data for the UK suggests that the business tourism industry generates over £20 billion annually in direct expenditures and accounts for over 7 million overseas visitors. Annual attendance figures for UK trade exhibitions exceed 10 million people and more than 80 million travel to attend conferences and meetings (Business Tourism Partnership, 2003, 2005). Similar patterns are evident elsewhere. For example, Law (2002) suggests that in some American cities, the conference and exhibition trade accounts for upwards of 40 per cent of staying visitors, whilst the largest convention in Las Vegas (an annual computer fair) attracts more than 200,000 visitors (Parker, 1999).

The growth of business travel is a direct consequence of processes of globalisation and new organisational structures around production. The 'co-ordination of production, supervision of local managers, design of new facilities, meetings with consultants, purchasing of supplies, product servicing and marketing – all require visits from company officials, technicians, or sales personnel' (Fainstein and Judd, 1999b: 2). This demand for business tourism is associated with a number of key characteristics that are highly beneficial to urban places:

- it is a high quality, high yield sector that is associated with above-average levels of expenditure;
- it is a year-round activity;
- it complements the leisure tourism sectors by supporting much of the infrastructure that underpins other forms of tourism as well as local leisure patterns;

- it is often a key component in tourism-led urban regeneration and because of the high quality of service demanded by most business travellers it creates additional demand for local employment;
- it often stimulates return visits to the same destinations for leisure purposes (Business Tourism Partnership, 2003).

Shaw and Williams (2002) suggest that business travel is generally the dominant sector in urban tourism and will normally bring the highest levels of per capita expenditure. However, in most cities, day trippers – although commonly disregarded in tourism studies and seldom enumerated with any degree of accuracy – will always be the most numerous group. In England in 2005, for example, an estimated 674 million tourist day visits were made to inland towns and cities for activities such as eating out, leisure shopping, entertainment and VFR (Natural England, 2006). Short-break forms of urban tourism have also emerged as a key area of demand. These forms of travel capitalise not only on the growing number of attractions that contemporary cities offer, but in key destination areas such as Europe, short-break city travel has also benefited from enhanced levels of connectivity between major cities via new high-speed rail links and budget airline services, as well as the competitive rates for hotel accommodation at weekends when high-paying business travellers are much less in evidence. In 2001 UK citizens took some 5.2 million short breaks abroad, with city breaks to destinations such as Paris, Amsterdam, Barcelona and Dublin attracting the largest market share (Mintel, 2002).

In practice, of course, much of the demand for urban tourism is multi-purpose (and urban destinations are characteristically multi-functional), so to suggest that people visit for a single, specific reason is usually misleading. Business travellers, for example, will often take in local entertainment and may shop and sightsee as well. By consolidating the various market segments, some sense of the overall scale of demand for urban tourism can be derived. Data on the numbers of urban tourists at specific destinations are notoriously difficult to create, locate and compare with consistency. However Table 9.2 presents an attempt to derive estimated annual levels of visiting to a selection of urban destinations from around the world, and although a breakdown of domestic and foreign visitors is not provided for all destinations, the data still help to illustrate the importance of the urban tourism sector.

Table 9.2 *Estimated levels of tourist visiting to a selection of major cities, 2004–06*

City	Foreign Visitors (millions)	Domestic Visitors (millions)	Total (millions)
New York	7.3	36.5	43.8
Las Vegas	–	–	36.0
Paris	14.3	16.7	31.0
London	15.2	11.4	26.6
Sydney	2.5	7.8	10.3
Singapore	–	–	9.7
Melbourne	1.9	7.6	9.5
Amsterdam *	8.2	–	–
Toronto	–	–	4.0
Vienna	7.6	–	–

Source: compiled, online, from national and city tourist board websites – accessed November 2007
* Figure quoted for Amsterdam is bednights, all other figures are headcount

The supply of urban tourism

According to Law (2002) the attraction base to a city is a key factor in stimulating demand for urban tourism and the spatial arrangement of attractions is also a primary variable that helps to define tourist space and create local geographies of tourism. However, the supply of urban tourism is not grounded just in the incidence of attractions, since these do not sit in isolation but contribute to a more broadly defined tourism 'product'. The urban tourism product will vary from place to place but will normally comprise a blend of tangible facilities, goods and services (such as accommodation, entertainment and cultural facilities) together with intangible elements (such as a sense of place and a place identity). There is, therefore, an important distinction to be drawn between the presence of attractions and the *attractiveness* of a city since the latter quality is not necessarily dependent upon the existence of the former. A city may be an attractive destination by virtue of the qualities of – say – its built environment or its local cultures, rather than through the possession of famous landmarks or 'must-see' tourist sites. It is also important to recognise that urban tourism attractions are not fixed entities and although some are purpose-built to serve that function, others that may not have been designed as attractions per se will acquire this role through shifts in public interest and taste.

This relationship between tourism and urban places has been conceptualised by Jansen-Verbeke (1986). She argues that the essential elements that comprise the basis for urban tourism may be grouped under three headings:

- primary elements that comprise place-specific attractions and facilities (labelled 'activity places') and the broader environments in which the activity places are located (labelled 'leisure settings');
- secondary elements (such as accommodation and retailing);
- tertiary elements (such as parking, information and signage).

However, whilst this framework has a real value in helping to map the basis to urban tourism, the diversity of tourist demand and the differing motives for which visitors come to cities also tends to make this typology problematic. Put simply, elements that Jansen-Verbeke proposes as 'secondary' (such as retail space) will constitute a primary element for some visitors and even some categories that might be assumed to be relatively fixed (such as accommodation) may not necessarily be so. It is, perhaps, an extreme example, but there is no doubt that the truly fantastic, themed hotels of Las Vegas constitute a primary attraction at this resort and are systematically visited and consumed as *attractions* by many of the city's visitors (Plate 9.2). Figure 9.2 therefore presents a reworking of Jansen-Verbeke's original model which strives still to capture the essential message about the nature of urban tourist attractions and a primary–secondary relationship, whilst simultaneously suggesting that the distinction within categories is flexible and contingent upon the particular intentions of the visitor.

The conceptual frameworks reviewed above suggest that we may view tourist space in cities as being essentially organised around several broad areas of tourist interest that in turn draw on particular types of facilities, attractions and places. These will include:

- cultures and heritage;
- entertainment and night-life;
- retailing;
- accommodation.

Culture and heritage is, according to Page and Hall (2003), a strong attraction in urban tourism, although the full range of cultural and heritage attractions sometimes defies concise definition. It embraces 'high' cultural forms such as are encountered in major galleries and

Plate 9.2 Hotel development in a fantasy city: 'New York, New York' hotel and casino on Las Vegas Boulevard

museum collections or in the aesthetics of civic design and monumental architecture, but it is also captured in the 'popular' cultures of local gastronomy, craft industries, festivals, street music and the architecture of the vernacular. Equally, tourist interest in heritage may be revealed in visits to major historical sites and monuments (such as the Tower of London) or – increasingly – in reconstructed celebrations of the artefacts, places and events that defined vanished ordinary urban lives and are reflected in the new generation of museums of industry and working life or, less directly, in the restored landscapes of factories, mills and docklands. The scale of tourist interest in these facets of urbanism is significant. Bull (1997) states that over two-thirds of all visitors to London identify the historic heritage of the city as a primary attraction and this is emphasised by the prominence of cultural and heritage sites shown in Table 9.3. This lists the major free and paid attractions in London in 2005 and it may be noted that only four attractions fall outside the scope of culture and heritage (Visit Britain, 2006).

In spatial terms, cultural and heritage attractions tend to draw visitors to core areas of cities. This is partly a consequence of natural chronologies of urban development that will normally place genuinely old buildings and structures (such as castles or cathedrals) close to the historic points of origin of the settlement, but it also reflects the fact that important civic buildings (such as major galleries and museums) are often given prominent, central locations as an indicator of civic pride and to ensure high levels of accessibility. Figure 9.3 locates the top

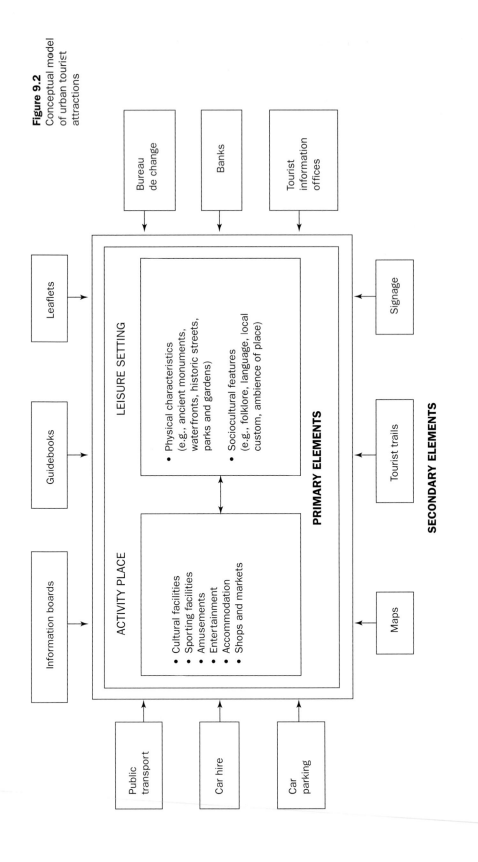

Figure 9.2
Conceptual model of urban tourist attractions

Leaflets

Guidebooks

Information boards

Public transport

Car hire

Car parking

Bureau de change

Banks

Tourist information offices

Signage

Tourist trails

Maps

LEISURE SETTING

ACTIVITY PLACE

- Physical characteristics (e.g., ancient monuments, waterfronts, historic streets, parks and gardens)
- Sociocultural features (e.g., folklore, language, local custom, ambience of place)

- Cultural facilities
- Sporting facilities
- Amusements
- Entertainment
- Accommodation
- Shops and markets

PRIMARY ELEMENTS

SECONDARY ELEMENTS

Table 9.3 *Visitor levels at major paid and free attractions in London, 2005*

Paid attractions	Visitors	Free attractions	Visitors
London Eye	3,250,000	British Museum	4,536,064
Tower of London	1,931,093	National Gallery	4,020,020
Kew Gardens	1,354,928	Tate Modern	3,902,017
Westminster Abbey	1,027,835	Natural History Museum	3,078,346
London Zoo	841,586	Science Museum	2,019,940
St Paul's Cathedral	729,393	Victoria & Albert Museum	1,920,200
Hampton Court Palace	449,957	Tate Britain	1,738,520
Tower Bridge	350,000	National Portrait Gallery	1,539,766
Cabinet War Rooms	311,481	Somerset House	1,200,000
Shakespeare's Globe	269,506	British Library Exhibitions	1,113,114

Source: Visit Britain (2006)

◆ > 130,000 visitors per annum

Figure 9.3 Principal cultural and heritage attractions in London

museums, galleries and historic buildings that are visited by tourists in London and illustrates clearly the importance of the core of the city as a focus for cultural and heritage tourism.

The availability of entertainment and nightlife is also one of the main motivations for people to travel as tourists to urban centres and which, importantly, appeals to a broad spectrum of visitors. For example, one of the factors that has promoted Las Vegas to a position of pre-eminence in the US business convention market is the attraction to delegates of the entertainment and nightlife that is a specialty of the resort (Parker, 1999). Within this category of attractions will be included theatres, cinemas and shows, concert halls, casinos, clubs, bars and restaurants. But since these sites also exert a strong local appeal, they become important points of intersection between visitors and residents and although there is always a temptation to view the entertainment and nightlife of a city as being locally derived and embedded in the leisure behaviours of residents, the demand of tourists is often essential to maintaining the viability of this sector. A study of the theatre industry in London by Hughes (1998) noted that over 30 per cent of overseas visitors to London go to the theatre at some point in their visit (emphasising its attraction) and that around 66 per cent of audiences in London's West End theatres were not residents of the city (emphasising the role of visitors in maintaining demand). It is also the case that the growth in theatre attendances in London (which has risen from just over 10 million in 1986 to exceed 12 million in 2006 (SLT, 2007) has been closely associated with the expansion of tourism.

The entertainment and nightlife sector illustrates well an important synergy between local recreation and tourism, but it also demonstrates the tendency for tourist functions that are actively developed as attractions to cluster within distinct zones and/or on particular streets. The development of London's West End is an excellent example, with its clusters of theatres and cinemas in areas such Leicester Square, Shaftesbury Avenue, Haymarket and Aldwych (Figure 9.4). Similar patterns are evident in other major cities, such as the development of theatres on New York's Broadway or the 'red-light' activity of Amsterdam's Warmoestraat. Such patterns develop partly for reasons of accessibility and the functional relationships between businesses operating in the same sectors (which are then reinforced by the behaviour patterns of users who are drawn to what become known areas), but are also commonly shaped by regulatory controls on activities that may frequently trigger anti-social forms of behaviour (Roberts, 2006).

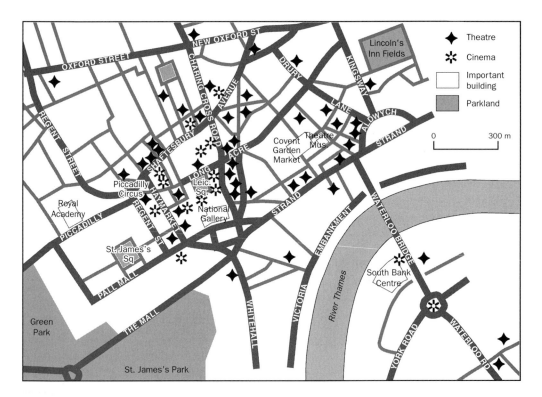

Figure 9.4 Distribution of cinemas and theatres in London's 'West End'

A third area of tourist activity that helps to define and shape the tourist city is retailing. Law (2002) suggests that although tourism and retailing are not immediately associated in most people's minds, the relationship is actually intimate and important. Shopping may not constitute a primary motive for tourist travel (excepting in the day visitor sector where studies suggest that up to a quarter of visits are made for the purpose of shopping (Natural England, 2006)), but it is generally one of the most important activities for tourists at their destination (whether judged by time or expenditure). For this reason tourist space is frequently infused with retail space, and in the most popular sites, retail outlets aimed at tourists will dominate the primary tourist routes. For example, a study of local space around Notre Dame Cathedral in Paris identified a near-continuous line of souvenir, food and other tourist retail units on the main approach to the Cathedral (Pearce, 1998). More widely, the quality of retail provision in cities has been recognised as helping to define the wider attraction of the city as a destination and as we will see in the final section of this chapter, the emergence of new synergies between tourism and retailing, and a general move towards the overt design of new shopping environments around leisure tastes (e.g., in mall development and themed shopping areas) has strengthened this relationship.

The structure of the tourist city is also partly defined by its accommodation sector. The availability of accommodation underpins the urban tourist industry by enabling people to stay in the destination and is critical in sectors such as business tourism where major hotels not only provide rooms for travellers, but may often also provide facilities (such as meeting space) for the actual transaction of business. At the same time, the quality of the accommodation sector also contributes in some fundamental ways to tourist perceptions of the city itself.

Urban tourism is generally dominated by what is termed the 'serviced' accommodation sector which includes hotels (which predominate), guest- or bed and breakfast houses and serviced apartments. In recent years this sector has witnessed some important changes with, in particular, an increased dominance of the urban market by international companies (such as Intercontinental/Holiday Inn, Thistle, Hilton and Marriott) that manage increasingly large hotels. A recent analysis of the London accommodation sector showed a significant contraction in the number of bedspaces in small, private hotels and guest-houses, but a substantial increase in the market-share held by large budget hotel groups such as Premier Travel Inn and Travelodge, alongside smaller (though still significant) growth in capacity in luxury 4- and 5-star hotels. Seventy-three per cent of London hotels contained at least fifty rooms (Visit London, 2007).

As with other key sectors, tourist accommodation maps urban space in some distinctive ways, with a marked tendency towards clustering at preferred locations. Traditional patterns favour quite close proximity between accommodation – especially in hotels – and the attractions that tourists wish to visit. This leads to important areas of hotel development in and around central business districts (with their retailing and entertainment zones) and the historic centres of cities. However, the wider reworking of urban space around zones of regeneration and new transport hubs – especially airports – has encouraged a more dispersed pattern of accommodation to develop with clusters of hotels appearing in regeneration zones (such as waterfronts) and in proximity to transport gateways. This trend is illustrated in Figure 9.5 which maps the distribution of hotel bedspaces across the London boroughs in 2006. This shows that the majority of London hotels are clustered in the three central boroughs of Westminster, Camden, and Kensington and Chelsea (which form a primary zone that has been established for many years), but with a secondary zone in the Borough of Hillingdon (which includes Heathrow Airport) and some emerging tertiary zones in Southwark, Tower Hamlets and Newham which are related to the development of Canary Wharf, London City airport and the sites for the 2012 Olympic Games.

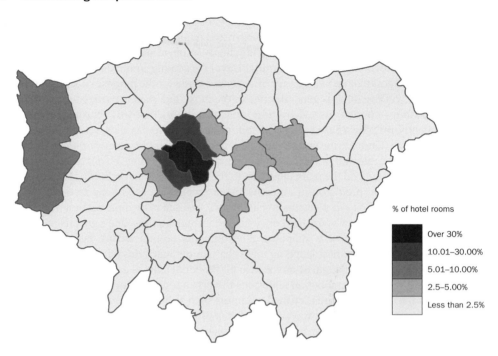

% of hotel rooms

- Over 30%
- 10.01–30.00%
- 5.01–10.00%
- 2.5–5.00%
- Less than 2.5%

Figure 9.5 Distribution of hotel bedspaces in London boroughs

Tourism and the new urban order

Whereas tourism to cities was, until very recently, generally super-imposed upon existing spatial, economic and social patterns, in the post-industrial, postmodern city it has emerged as much more central to the organisation of the new urban spaces and their associated structures. This final section therefore explores some of the key intersections between the changing urban context (as discussed at the start of this chapter) and the emerging patterns of urban tourism (as set out in the preceding section). For convenience the discussion is organised around seven major themes or ideas that strive to capture the fundamental nature of the change as it affects urban tourism. Although each theme is considered separately, it is important to recognise that the processes that are described are, in practice, intimately interrelated.

New spatial opportunities

The reworking of urban space under postmodernity creates new opportunities for tourism. The spatial reorganisation of postmodern cities is commonly shaped by a number of key processes, including:

- increased prominence of consumption as a 'driver' in the urban economy and in which leisure and tourism are critical arenas for the conspicuous display of consumption;
- fragmentation of social and economic spaces that creates diversity and often renders new levels of intrinsic interest to the postmodern urban landscape;
- redevelopment of urban cores around new functional relationships which include a strong emphasis upon leisure and tourism consumption, for example, in the development of cultural and cafe quarters or entertainment districts;

- creation of new centres of development in the outer suburbs and the urban periphery that may contain themed shopping malls with integral leisure spaces, new urban 'villages', transport hubs (such as airports and linked hotel developments), theme parks and sometimes major sporting stadia;
- regeneration (and reinvention) of former industrial zones within inner urban districts which is typically framed around themes of urban heritage and may feature landmark projects such as convention centres, modern museums and galleries and major hotels, as well as areas of gentrification.

Each of these processes enables existing forms of recreation and tourism to develop in novel and exciting ways and new sectors of activity to emerge. Figure 9.6 reproduces a model by

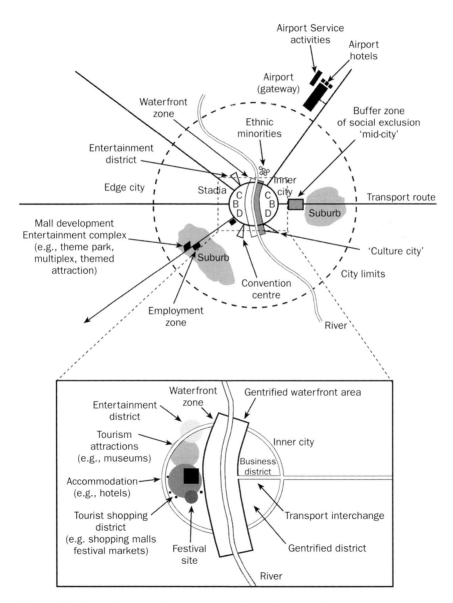

Figure 9.6 Model of tourism development in the postmodern city

Page and Hall (2003) that attempts to show how tourism and leisure maps onto the spaces of the postmodern city and whilst there is perhaps an implicit contradiction in the notion of a 'model' of postmodernity (given that postmodern perspectives explicitly reject totalising approaches), the diagram still has a value in setting out some important spatial patterns and tendencies.

Within the restructuring of urban space, regeneration has emerged as an especially important policy area, not least because it affects such large numbers of former industrial cities. Regeneration is simultaneously a process of physical redevelopment of redundant space to meet new needs, but it is also an intellectual and aesthetic process – a means of rethinking urban space and how the people who live in, or choose to visit the city relate to that space. Wakefield (2007) makes the point that throughout the industrial era, the presence of industry was often a source of civic pride and identity, whereas today the label 'industrial city' more readily denotes a set of negative images. Regeneration is, therefore, very much about reshaping image alongside the physical changes to the urban fabric.

Tourism is an important component of urban regeneration, partly because there is perhaps an implicit assumption on the part of urban developers that by creating regeneration schemes that are capable of attracting tourists, cities will also be capable of attracting other investment as well as permanent residents. It thus becomes a litmus test of what Hagermann (2007) terms a 'liveable city'. Additionally, though, tourism has become a favoured component in urban regeneration because it is widely perceived as creating important new economic linkages. It promotes new firm formation (because of low barriers to entry to the sector); it brings income through tourist expenditures and, as a labour-intensive service industry, it creates new jobs (Robinson, 1999). It is also seen as a means of opening-up areas of the city for new use by both visitors and residents, especially waterfronts and former docklands, which are always prominent zones of regeneration wherever they occur but which in their former state were usually 'no-go' areas for all except those who lived and worked in these zones.

In general these positive attributes are held to outweigh some of the negative aspects that have been associated with tourism-led regeneration. These include:

- doubts over whether actual returns on investment necessarily reach assumed levels (which is a common criticism of major sports stadia development);
- concerns over the quality of jobs and security of employment in tourism;
- issues of social equity (as regeneration schemes often reflect the aspirations of urban elites and disregard the interests of ordinary local people);
- doubts about the extent to which the economic benefits actually diffuse from the enclaves of regeneration into the wider urban environment (McCarthy, 2002; Shaw and Williams, 2002).

However, such concerns have done little to diminish the appetite for regeneration on the part of urban government.

Although the nature of regeneration varies in detail from place to place, recurring themes and tendencies are evident. Typical components of urban regeneration will include physical and environmental improvement; the creation of new zones of employment and housing; enhancement in transport infrastructure; and – in order to attract visitors – the development of new facilities and attractions together with image building and place promotion policies (Page and Hall, 2003). This has led to the creation of what Judd (1999: 39) has described as a remarkably standardised package of key components: 'atrium hotels, festival markets, convention centres, restored historic neighbourhoods, domed stadia, aquariums [and] redeveloped waterfronts'. Figure 9.7 and Plate 9.3 illustrate one example, the Brindley Plaza redevelopment in Birmingham (UK) that deploys virtually all these elements, together with a major concert hall, theatres, retailing, restaurants, public houses (bars) and residential

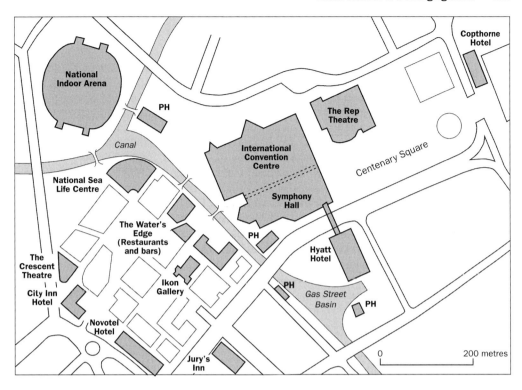

Figure 9.7 Brindley Plaza urban regeneration project, Birmingham, UK

Plate 9.3 Urban regeneration based on leisure and tourism in the inner city: Brindley Plaza, Birmingham, UK

apartments. It is strongly reflective of the importance of arts and cultural activities as catalysts for urban change (Roberts, 2006) and has been a hugely successful project (Williams, 2003).

New urban images

Urban regeneration connects very closely to the wider process of image making and place promotion, both of which have emerged as central planks of entrepreneurial governance in Western cities (Bradley et al., 2002) and reveal strong links to tourism. An important part of this process (and an area of change that intersects directly with urban tourism) is the development of landmark facilities and, particularly, the promotion of hallmark events, both of which form a part of the wider development of the city as a spectacle (Law, 2002; Shaw and Williams, 2002). Hall (2005) suggests that hallmark events can trigger an appeal and an associated impact at a range of geographical scales from the international to the local and may cover a broad range of categories. They may be artistic, cultural, religious, commercial, sporting, or political in nature and may variously be configured as exhibitions, festivals, tournaments or celebrations. However, although diverse in focus they tend to share a number of common characteristics. These include:

- a tendency to be based upon spectacle (i.e., something that is worth seeing and experiencing for its own intrinsic qualities);
- a strong capacity to attract visitors;
- an ability to confer a level of status on the city;
- an ability to raise awareness of the city in the wider world and influence, in a positive direction, the images of the place that are held by outsiders.

Major sporting events such as the Olympic Games, the World Cup (soccer) or the Superbowl (American football) are examples of hallmark events that catch the eye and which – in the case of the Olympics, at least – will trigger the construction of landmark facilities in new stadia. But other sectors are equally capable of generating hallmark events that exert significant impacts on levels of urban tourism. For example, the exhibition of the Chinese Terracotta Army at London's British Museum in 2007/8 was expected to attract more than 800,000 visitors and some 1.7 million people visited the Treasures of Tutankhamun exhibition at the same venue in 1972.

Hallmark events are an important facet of urban tourism not simply because of the numbers of visitors that special events attract, but also because they may support the development of tourism in diverse situations. Hence, in cities with an established tourism industry (such as London) hallmark events (such as the Terracotta Army exhibition and, especially, the forthcoming Olympic Games in 2012) help to reinforce the city's position as a major destination. But as Law (2002) notes, hallmark events may also serve to raise the image of cities that lack a tourist tradition and 'pump-prime' a new industry that might not otherwise become established. For example, in the 1980s a number of British industrial cities (including Liverpool, Glasgow and Stoke-on-Trent) hosted national and international garden festivals as part of a regenerative, image-building strategy (Holden, 1989). In this way, events and their infrastructure become a part of the cultural capital of places.

New urban aesthetics

A third key area of change relates to aesthetics and public taste. The visual and aesthetic character of the postmodern urban landscape promotes the consumption of sights and symbols as tourist activity and leads to an embedding of tourism in daily urban life. In terms of visual and aesthetic qualities, Relph (1987) draws some fundamental points of contrast between, on the one hand, the hard lines, the unity, the sense of order, the functional efficiency and the sheer physical scale of the modernist urban project and, on the other, the disordered, intimate, variegated 'quaintspaces' (with their conscious connections to the vernacular and their collages of signs and symbols) that characterise the postmodern cityscape. The essential point is that the many forms of tourism that are grounded in the enjoyment of places will generally connect much more effectively with the latter than with the former. Selby (2004: 44) comments that the 'built environment of the post-modern city is characterised by a deliberate attempt to refer to the emotions, experiences and sense of place of inhabitants' but in the process, of course, that attempt inevitably creates spaces and places that attract the attention of tourists.

There is, therefore, a reciprocal relationship here. Most forms of tourism are generally dependent upon the creation of spaces and places with high environmental quality and tourist demand thus exerts a wider benefit in helping to make cities attractive as places in which to live permanently. At the same time, the local demands for amenity that reflect the aesthetics and tastes of the local population create demands for space that service not just local needs and interests, but some of those of the visitor too. If, as writers such as Urry (2000) and Franklin (2004) assert, we live in a world in which populations are quintessentially mobile and in which the experiences of travel and tourism become infused into daily lives (in areas such as dress, eating habits, entertainment choices, media consumption or – more fundamentally – the ways in which we shape and direct our various gazes and outlooks) important new synergies between local leisure and tourism serve only to enhance further the appeal of cities as tourist destinations. In other words, the aesthetics of consumption that shape local practice also directly influence the consumption of those same sites and experiences by tourists.

Social and cultural heterogeneity

This tendency is revealed in the way in which the characteristic social and cultural heterogeneity of postmodern cities enhances the appeal of places to tourists. 'Others' readily become central objects of the tourist gaze and ethnicity can quickly become a commodified tourist attraction (Hall, 2005). Ethnic dress, cuisine, music, rituals or customs may all form attractive elements for tourists and will often provide essential local 'colour' in sub-areas of city space. In American cities, 'Chinatowns' and 'Little Italys' have, for example, become commonplace.

However, as with all forms of commodification the process is selective. Hall (2005) observes that ethnicity is often commodified as a heritage product, but this implies that where there is no tradition in the relationship between host communities and ethnic minorities, the value of ethnicity as a tourist attraction may be diminished. Thus, for example, the extended relationship between China and cities on the west coast of the USA such as San Francisco, has ensured that Chinese communities are well-established and form an integral part of the local urban population. 'Chinatown' in San Francisco is, therefore, much visited by tourists and celebrated as a part of the city's heritage, whereas the enclaves of low-paid, contemporary migrants from places such as Vietnam and the Philippines in the same city are widely disregarded.

New synergies

Postmodernity promotes new synergies between tourism, leisure and daily life that blurs distinctions and leads to new forms of ostensibly familiar areas of interaction. Central to several of these emerging synergies is the notion that what were once routine functions of urban life can be reworked in ways that render them simultaneously as entertainment.

Hannigan (1998) discusses several examples of such relationships in detail, including 'shopertainment' and 'eatertainment'.

The combination of shopping with entertainment is not a new idea. Hannigan (1998) notes that large department stores in American cities in the early twentieth century commonly sought to enhance their attraction by providing in-store entertainment – such as orchestras in their tea rooms. Nava's (1997) study of London stores before the First World War also shows that major shops such as Selfridges routinely made reference in their advertising to the ways in which shopping at the store might be considered as a recreation and an essential component of any visit to London. What is different now is that the blending of shopping, leisure and tourism has become much more central to defining the character of urban space, its functionality and the way it is perceived.

This is most evident in the growth of themed shopping malls where trend-setting projects such as the West Edmonton Mall in Canada and Mall of America in the USA have developed a model that, according to Hannigan (1998: 91) 'explicitly and ostentatiously sets out to bring the world of the theme park to the environment of the shopping centre', and which thousands of developments across the urban world have since sought to replicate, albeit normally on a smaller scale. Similarly, the rising popularity of festival markets reveals an alternative approach to blending retailing with leisure and tourism. Pioneered in the USA by a developer named James Rouse, festival markets have become a common feature in reinvented downtown districts and zones of regeneration with a strong tourism dimension. The market brings together small independent retailers and supporting infrastructures such as restaurants and street entertainment, to form an attractive, postmodern version of a very traditional form of retailing. As with themed malls, the success of the early festival markets at sites such as Boston's Faneuil Hall and New York's South Street Seaport, spawned hundreds of imitative schemes elsewhere – such as London's Covent Garden, Sydney's Darling Harbour, Liverpool's Albert Dock and San Francisco's Pier 39 (Plate 9.4).

The concept of 'eatertainment' also collapses boundaries between eating and play, and although this, too, is not an innovation, the scale and extent of the development of synergies between eating and entertainment in the postmodern city is worthy of note. Hannigan (1998: 94) examines the rise of themed restaurants (such as Hard Rock Cafe and Planet Hollywood – which he describes as a 'combination of amusement park, diner, souvenir stand and museum') as exemplars of the trend, but the universal integration of restaurants, cafes and bars into remade downtowns, cultural quarters and zones of regeneration signals the importance of the function to both resident urban populations and the tourists who visit.

Reflexive consumerism

The reflexivity of the postmodern consumer enables urban space to be constructed in multiple ways that favour tourism. Hannigan (1998: 67) characterises the postmodern consumer as 'elusive – a free soul who darts in and out of arenas of consumption which are fluid and non-totalizing . . . constructing individual identity from multiple images and symbols [and] subverting the market rather than being seduced by it'. In other words, as reflexive individuals, many urban tourists are not necessarily the passive recipients of tourist experiences that are made for their consumption by others, but through their agency they are often actively involved in shaping their consumption of tourist places. Individuals carry knowledge, experience and perhaps memories that help to structure and inform their understanding of places and their meanings, but because people understand the world around them in qualitatively different ways, tourism places or products are experienced at different levels and in contrasting ways by different people (Selby, 2004). Thus, as Rojek and Urry (1997) explain, the same sites are subject to multiple readings by different audiences.

Plate 9.4 'Pier 39': a festival market developed from disused wharfs on the waterfront of San Francisco

This has a number of interesting implications for urban tourism. First, the reflexivity of the postmodern tourist means that whilst the urban place promoters and image makers may promulgate a set of preferred sites and experiences for visitors, the tourists themselves may negotiate their own readings of the urban landscape and create individualised geographies of tourism that fashion new and unexpected intersections with local communities. Consequently, tourism practices become progressively more embedded in daily urban life and the presence of tourists becomes an acknowledged and unremarkable fact. Second, the effectiveness of some of the urban regeneration schemes discussed earlier – especially those that showcase former urban industries and transport zones such as waterfronts – is in part dependent upon the broadening of tourist tastes and interests that the postmodern, reflexive tourist is likely to reveal. The reflexivity of the postmodern tourist thus enables many of the contemporary urban tourist spaces to 'work' as attractions. Many contemporary sites of urban tourism are representational, imitative and inauthentic, yet the reflexivity of the subject still permits those sites and experiences to be enjoyed, whilst acknowledging simultaneously their artificiality.

Fantasy cities

Many of these trends come together to create new 'fantasy cities' that directly support urban tourism. The concept of the 'fantasy city' originates in the work of Hannigan (1998) and was initially coined to reflect some key processes in the remaking of downtown areas of American cities as new centres of leisure-based consumption based around themed, entertainment districts and flagship projects such as resort-hotels and convention centres. However, as Page and Hall (2003) observe, with the rapid growth of the consumer society under conditions of postmodernity (and with the increased role of leisure and tourism as cultural capital and as markers of social distinction), the themes and trends that Hannigan describes have become widely diffused across the urban environment and therefore applicable in a range of new leisure and tourism settings.

Six central features define the fantasy city:

- It is 'theme-o-centric' by which it is meant that the key elements (such as attractions, retail provision or entertainment) conform to a scripted theme (or set of themes).
- It is 'branded', such that the place and its products are actively defined and sold.
- It provides round-the-clock activity by conscious blending of a range of functions and attractions that deploy complementary diurnal ranges.
- It is 'modular', mixing an array of standard components (such as cinemas, themed restaurants, leisure retailing) in differing configurations to produce contrasting character to the area.
- It is 'solipsistic', which means the fantasy city zone is focused entirely around its own activity and is functionally isolated from surrounding zones.
- It is postmodern in so far as it is normally constructed around technologies of simulation, virtual reality and spectacle that is contained within an overall landscape that deploys eclectic and unreal blending of style, genres and periods (Hannigan, 1998).

Hannigan (1998: 7) also posits the interesting view that the emergence of fantasy cities in the USA is an end-product of the tension between the American middle-class desire for experience and their parallel reluctance to take risks – especially those involving contact with the urban underclass. He describes how American downtowns are being progressively converted into 'glittering, protected playgrounds for middle-class consumers', environments in which public space is actually replaced by private spaces, with the surveillance and policing that is necessary to protect that privacy. This resonates with Davis's (1990) descriptions of the fractured and defended spaces of postmodern Los Angeles and is also reflected in Fainstein and Judd's (1999b: 12) observation that the construction of tourism enclaves (such as fantasy cities) is a typical way of allaying the sense of threat that most postmodern cities contain. Fantasy city, with its themed environments and simulacra, taps into the postmodern preoccupation with acquiring experiences that might otherwise be unattainable (e.g., by virtue of geography, accessibility, cost or historical disappearance (Hannigan, 1998)), but without the risks that actual travel to real places would entail.

In most urban places, fantasy cities exist – where they exist at all – in micro-level enclaves. However, there is one exception where it may be reasonably argued that the entire logic of the city is shaped by fantasy – Las Vegas. To draw this discussion of tourism and the new urban order to a close, therefore, Case Study 9.1 presents a summary analysis of the development and changing character of Las Vegas as an urban tourism destination and which demonstrates in a particularly well-developed manner, several of the themes that have been presented in the final section of this chapter.

CASE STUDY 9.1

Las Vegas: creating a fantasy tourist city

As an urban centre, Las Vegas is unique. No other city of equivalent size occupies such an outwardly unsustainable location (within a major desert zone and physically isolated from other major urban centres) and no other city of this scale demonstrates such an intimate relationship with tourism and leisure as the 'neon metropolis'. It is presently America's fastest growing city that almost doubled its population in a single decade (rising from 740,000 in 1990 to 1,390,000 in 2000) and with an estimated 36 million visitors annually it is the leading destination on the North American continent and one of the top urban tourist centres in the global market. With more than 125,000 bedspaces (and nine of the ten largest hotels in the world), no other city matches Las Vegas in its tourist capacity and none rivals the truly spectacular designs that are revealed in the latest generation of resort hotels on the city's famous 'Strip' (see Plate 9. 2). These hotels, in particular, have created an urban landscape that blends fantasy and pleasure within spaces and places that have taken the art of theming to new levels, producing what Rothman (2002: xi) describes as a 'spectacle of postmodernism, a combination of space and form . . . that owes nothing to its surroundings and leaves meaning in the eye of the beholder . . . a world [that is] fantastic and unreal, yet simultaneously tangible and available for purchase'. It is the ultimate expression of the ascendancy of consumption over production – a city that 'produces no tangible goods of any significance, yet generates billions of dollars annually in revenue' (Rothman, 2002: xi).

The development of Las Vegas from its origins (in 1905) as a small railroad town, through its formative years as a sleazy gambling resort backed by organised crime (between 1931 and the late 1960s), to its final emergence as a fantasy city founded on corporate investment and with a global reach, has been fully described in fascinating detail elsewhere (see, *inter alia*: Gottdiener et al., 1999; Parker, 1999; Rothman, 2002; Douglass and Raento, 2004). The purpose of this case study is, therefore, to highlight some of the ways in which Las Vegas illustrates many of the wider intersections between urbanism and tourism that are discussed above and, in particular, the emergence of 'fantasy' urban spaces.

Las Vegas illustrates with great clarity the general tendency for urban tourism to focus in well-defined zones of the city, whilst leaving other areas essentially untouched. Apart from a residual area of older hotels and casinos on the fringe of the downtown (mostly around Fremont Street) the dominant tourist zone is a linear development of large-scale resort hotels and related attractions that is quite separate from the downtown area and extends for some three miles along Las Vegas Boulevard (known locally as 'The Strip') towards McCarran International airport. Beyond these narrow confines there are only a handful of attractions within the city limits (mostly museums, galleries and other themed attractions) that are capable of drawing significant numbers of visitors and, consequently, tourist space in Las Vegas is closely prescribed.

As already intimated, Las Vegas is notable for the prevalence of themed environments and the art of the spectacle. For the most part the chosen themes (which are primarily represented in the resort hotels and their interior spaces – see Table 9.4) are either simulations of other times and places or, more typically, simulacra, that blend time and space in impossible combinations and which depend upon staged settings and props that are, for the most part, entirely artificial. Las Vegas is a landscape of signs and signifiers but with certain notable exceptions (such as the collections of original works by artists such as Picasso that adorn the Bellagio, or the touring exhibitions from the Guggenheim that are a regular

continued

Table 9.4 Themes at major Las Vegas resort hotels

Hotel	Theme
Aladdin	Exotic Arabia
Bellagio	Italian Riviera/Lake Como
Caesar's Palace	Imperial Rome
Circus Circus	Circus entertainment
Excalibur	Arthurian legend
Luxor	Egypt of the pharoahs
MGM Grand	Hollywood and the movies
Mandalay Bay	Tropical South-East Asia
Monte Carlo	French Riviera
New York New York	Manhattan street life and architecture
Paris	Parisian life and architecture
Treasure Island	Pirates and buccaneers
Venetian	Venetian canals and architecture

Source: Author's survey (2003)

feature at the Venetian) it is a landscape without substance. But the monumental scale on which these deceptions are arranged inevitably creates a spectacular urban landscape and the spectacle of Las Vegas is arguably one of its greatest attractions.

The most recent phases of development are also strongly illustrative of several of the new synergies between urbanism and tourism. The resort grew (and still sustains its position) on its reputation as a premier centre of gambling and entertainment, but the resort hotels have progressively refined their attractions to cater for non-gambling groups (including families), as well as recognising the appeal (and hence the value to their 'product') of high-quality retail environments and excellent restaurants. Elaborate, themed shopping malls and international-standard restaurants are now a feature of all the major resort hotels. According to Douglass and Raento (2004), the Caesar's Forum mall at Caesar's Palace now contains the world's greatest concentration of luxury retail outlets and nearly a quarter of the country's top-rated restaurants are located on the Strip.

Las Vegas demonstrates a remarkable capacity for reinvention and, in the process, continually asserts the fundamental importance of image. The process of reinvention is evident at both a city-scale and, more typically, within the primary components that define the tourism product. In its initial phases of development, Las Vegas quickly attracted a reputation as a tawdry and tasteless 'sin city', an image that would ultimately have limited the capacity for growth had the problem not been addressed. However, with the progressive development of a corporate investment culture in the 1980s, the resort has consciously reinvented itself as a destination that appeals not just to gamblers, but to a much broader base of tourists that includes families who come simply to be entertained, to shop, to eat and to enjoy the spectacle (Rothman, 2002). Whilst the casinos (which have generally been re-branded as places for *gaming* – with its connotations of fun, rather than *gambling* – with its connotations of risk) still generate almost half of the resort revenue, the larger share comes from sectors that include retailing, entertainment and hospitality. To maintain their competitive position, therefore, the major resort hotels have become engaged on a seemingly endless process of investment and reinvestment in ever-more spectacular and lavish attractions that help to define their image: in world-class entertainment and shows, major sports events (especially boxing), theme parks rides, themed retail malls, dramatic re-enactments (including simulated sea battles), circus acts, wildlife, aquaria, IMAX cinemas and simulations of natural phenomena (such as the regular eruption of the 'volcano' at The Mirage).

Interestingly, although Las Vegas is an ultra-modern city that has no meaningful origin prior to 1905, the relentless drive to reinvention has also prompted some of the forms of regeneration that would not be out of place in cities with a much longer tradition. In particular, the original resort centre around Fremont Street lies on the fringes of a downtown that has become progressively more unfashionable and unattractive, as the primary resort function has migrated to the central and southern reaches of the Strip. To try to regenerate this area, a public–private partnership invested some $70 million during the mid-1990s in programmes of pedestrianisation, environmental improvement and the installation of a canopy over part of Fremont Street, onto which are projected animated images to accompanying pop music in regular, nightly, free 'shows'. These attract reasonable numbers of interested on-lookers, but the project has done little to reverse the decline in the local casino business and the installation of the canopy effectively destroyed one of the great iconic landscapes of urban America (Plate 9. 5). The spectacle is also rather modest when compared to the glitz that typifies the latest phases of development of the Strip.

Las Vegas is a quintessential postmodern city but this quality is not simply a consequence of the depthless, representational qualities of its landmark buildings, the pastiche of styles or the collages of signs and signifiers that bombard the modern *flaneurs* that walk (or ride) the Strip, it also evident in the same kinds of spatial discontinuities that Davis (1990) recognises in Los Angeles. Douglass and Raento (2004) remark on the lack of any organic unity amongst the city's components; the downtown area is a non-descript secondary zone compared to the Strip, whilst the sprawling suburbs (which, at times, have expanded at the

Plate 9.5 Fremont Street, Las Vegas: an iconic American urban landscape now lost through redevelopment

continued

rate of almost 30 homes a day (Gottdiener et al., 1999)) generally comprise decentred and featureless zones of housing. On the Strip itself, the major hotels function almost as micro-states – self-contained entities that compete fiercely with their neighbours to attract the tourist dollar, whilst simultaneously extending private controls over seemingly 'public' space. In Las Vegas, surveillance is an ever-present attribute of tourist space and on some sections of the Strip even the sidewalk (pavement) is owned by the hotels and actively policed by hotel security to ensure only appropriate activity occurs in these spaces.

However, the erosion of public space in favour of the private is just one component in a widening array of difficulties that Las Vegas exhibits. Although, as a city, Las Vegas is almost exclusively a product of the second half of the twentieth century, it appears that the urban developers have not taken too many opportunities to learn lessons from other postmodern cities, such as Los Angeles. Parker (1999) highlights issues around the low wage economy in which most workers in the city's tourism industry are consigned; the relative deterioration of infrastructure to support local populations as the expansion of services struggles to keep pace with the physical growth of suburban Las Vegas; the lack of public amenities such as urban green space; the degradation of the natural desert environment and the excessive consumption of both energy and water by the tourist industry; and traffic congestion (and associated air pollution) of horrendous proportions. For the tourist, whose visit to the city is transient, such issues probably go undetected – masked by the ostentatious promotion of wealth and prosperity that is captured in the main tourism spaces, but the wider assessment of Las Vegas suggests that the creation of fantasy cities, perhaps inevitably, comes at a cost.

Summary

Cities are primary tourist destinations and today most large urban places accommodate a broader range of categories of tourism than any other form of tourist place and present an almost unrivalled range of tourist attractions and supporting infrastructures. As a part of this transformation, tourism has been widely adopted as a key ingredient in the process of remaking cities as post-industrial places shaped around consumption (rather than traditional forms of production), whilst images of cities as places to visit are often fundamental to the wider promotion of these places in global urban systems. Moreover, under processes of postmodern change, the contemporary development of cities is extending significantly the opportunities for tourism in urban places, producing new spaces of tourism and new synergies between tourism and daily urban life that embed tourism in urban experience in new and very influential ways. Tourism has thus become a part of the new urban order by both drawing upon and reinforcing trends and new directions.

Discussion questions

1   Why has the emergence of postmodern cities favoured the development of urban tourism?
2   Critically assess the value of Fainstein and Judd's (1999a) categorisation of tourist cities.
3   What are the primary mechanisms that promote the development of distinctive tourism zones within cities?

4   Why has tourism development become such a conspicuous component in urban regeneration?
5   To what extent has tourism become an embedded aspect of contemporary urban life?
6   Will more urban places develop into fantasy cities in the style of Las Vegas?

Further reading

There are several excellent texts that provide detailed discussions of urban tourism and any of the following will provide a good starting point for further reading:

Judd, D.R. and Fainstein, S.S. (eds) (1999) *The Tourist City*, New Haven: Yale University Press.
Law, C.M. (2002) *Urban Tourism: the Visitor Economy and the Growth of Large Cities*, London: Continuum.
Page, S.J. and Hall, C.M. (2003) *Managing Urban Tourism*, Harlow: Prentice Hall.
Selby, M. (2004) *Understanding Urban Tourism: Image, Culture and Experience*, London: I.B. Taurus.

Additionally, more concise discussions of the general character of urban tourism can be found in:

Hall, C.M. (2005) *Tourism: Rethinking the Social Science of Mobility*, Harlow: Prentice Hall.
Shaw, G. and Williams, A.M. (2002) *Critical Issues in Tourism: A Geographical Perspective*, Oxford: Blackwell.
Williams, S. (2003) *Tourism and Recreation*, Harlow: Prentice Hall.

A very useful essay on the role of tourism in deindustrialising urban centres is provided by:

Robinson, M. (1999) 'Tourism development in de-industrializing centres of the UK: change, culture and conflict', in Robinson, M. and Boniface, P. (eds) *Tourism and Cultural Conflicts*, Wallingford: CAB Publishing, pp. 129–59.

Detailed, and highly readable accounts of the development and significance of Las Vegas are provided by:

Gottdiener, M., Collins, C.C. and Dickens, D.R. (1999) *Las Vegas: The Social Production of an All-American City*, Oxford: Blackwell.
Rothman, H. (2002) *Neon Metropolis: How Las Vegas Started the Twenty-First Century*, London: Routledge.

The concept of the fantasy city is explored in depth in:

Hannigan, J. (1998) *Fantasy City: Pleasure and Profit in the Postmodern Metropolis*, London: Routledge.

10 The past as a foreign country: heritage attractions in contemporary tourism

The notion of the past as a 'foreign country' is derived from the title of David Lowenthal's magisterial examination of the relationships between societies and their history (Lowenthal, 1985). Lowenthal's purview extends well beyond the way in which people use the consumption of history to shape leisure practices, but the metaphor is especially apposite in examining the role of heritage attractions as contemporary spaces of tourism. As we have seen throughout this book, tourism is often essentially concerned with the exploration or experience of the 'foreign', but as the title of Lowenthal's book reminds us, foreignness has a temporal as well as the more familiar spatial dimension. Hence the spaces that we explore as tourists are often those that represent versions of the past rather than the present, but which are still places in which we may derive and enjoy many of the experiences that we acquire through contemporary foreign travel – for example, the sense of the exotic, or of the familiar that is also subtly different.

This chapter is arranged into five linked discussions that review, in sequence: the concept of heritage and heritage tourism; the contemporary significance of heritage; the evolving character of heritage; the heritage tourism market; and, finally, the problematic relationship between heritage and authenticity.

The concept of heritage and heritage tourism

It is helpful to begin by attempting to define 'heritage'. Prentice (1994: 11) notes that in the literal sense, heritage is an inheritance or legacy that is passed from one generation to the next, but he also notes that in the context of heritage tourism it has acquired a much looser usage. Graham et al. (2000) concur, noting that whilst the term was once used only to describe a legal inheritance an individual received in a will, it has now been expanded to include almost any form of intergenerational exchange or relationship. However, underpinning the concept is the notion that heritage possesses a real and/or symbolic value and this encourages Timothy and Boyd (2003: 2) define heritage as those 'elements of the past that society wishes to keep'. In developing this definition, the same authors propose an interesting conceptual framework in which they argue that the sites, objects and artefacts that form the basis to heritage exist initially as part of a world of physical and social facts – in other words a 'phenomenal' environment – but only become part of a 'behavioural' environment when those sites and artefacts are perceived by society to have a value or a utilitarian function. In other words, heritage is actually a socio-cultural construct.

This approach has some significant implications. In particular it emphasises the fact that the identification of heritage is a selective process in which complex value judgements are exercised in order to filter those elements of heritage that are to be retained from those deemed no longer important. Graham et al. (2000) argue that heritage is a product of the way in which we choose to use the past and as in Johnson's conceptualisation of circuits of culture (Johnson, 1986), heritage may be viewed as a product of the interplay between processes of production, regulation and consumption in which heritage emerges as a means of cultural representation. Heritage must therefore be seen as a socially produced, negotiated entity, but because the meanings that society will attach to heritage places on objects will change from one cultural period to the next, heritage is seldom to be seen as a fixed entity. These are critical characteristics that should not be overlooked in trying to understand how tourism relates to heritage.

Although there is an implicit relationship between heritage and history (in which heritage might be seen as a means of consumption of the different readings of the past that history provides), the widening range of environments or contexts in which heritage is now identified certainly dilutes that relationship. Contrasting attempts at differentiating heritage have, for example, proposed basic distinctions between natural, built or cultural forms of heritage (Poria et al., 2003) or between tangible immovable heritage (such as landscapes and buildings), tangible movable heritage (such as museum objects), and intangible heritage (such as values and beliefs) (Timothy and Boyd, 2003). It is therefore important to recognise that heritage resources are not simply confined to a relatively narrow set of places or artefacts that have a conspicuous historic significance or character, but may be applied to a much wider range of settings or practices where the historic dimension, whilst inevitably present, is not necessarily overt – for example, the heritage that is expressed in the use of traditional forms of dress or cuisine. Figure 10.1 represents this idea in a simple, diagrammatic form.

The diversity of heritage resources also means that heritage tourism naturally intersects with other forms of tourism. Richards (1996) demonstrates the many close relations between heritage and cultural tourism, but similar intersections exist with urban tourism (given the prominence of built heritage and the incidence of collections of heritage artefacts in towns and cities), rural tourism (given the increasing interest in nostalgic reconstructions of a rural 'other'), as well as important connections with ecotourism (in relation to the heritage values attached to wildlife and the wider environments of protected areas, such as national parks or other special landscapes).

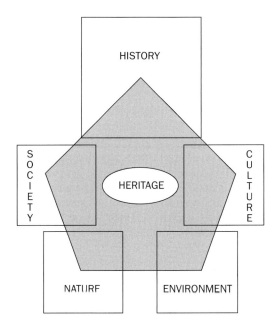

Figure 10.1 Heritage relationships

These intersections between heritage and other common forms of tourism raise interesting questions relating to what may define the core of heritage tourism. Poria et al. (2003) argue that at any heritage site, several categories of tourist will be encountered. Some are tourists who are unaware of the heritage attributes of the site whilst others will be aware of the heritage but were motivated to visit by other attributes of the location. In both these situations, it is argued, simple presence at a site does not infer any participation in heritage tourism and these visitors should not be truly considered as 'heritage tourists'. Instead, the authors suggest that the real core of heritage tourism is expressed by visitors who are motivated by the presence of heritage attributes and, especially, tourists for whom the attributes are perceived to be a part of their own heritage and which therefore acquire a special level of meaning and significance. Although this implies a rather narrower view of heritage tourism than is commonly applied, it serves a very valuable purpose in emphasising that heritage tourism should be considered as being as much a product of demand characteristics as it is a reflection of supply. From this position an even more important observation may be derived, namely, that if engagement with heritage is to be considered as a reflexive response that is expressed at the level of the individual (or groups of individuals), heritage tourism becomes an expression of multiple histories and heritage sites will acquire multiple meanings given the contrasting dispositions of the individuals that are drawn to these spaces. Put simply, different people will apply different meanings and significance to the same heritage sites or objects. So not only will social 'readings' of heritage evolve through time, but heritage sites are also simultaneously subject to differing interpretations.

The contemporary significance of heritage

It is perhaps worth emphasising that although heritage tourism is now a conspicuous aspect of contemporary travel, it is not a modern innovation. Prentice (1994) observes that historic sites have been popular destinations for tourists and sightseers for a lot longer than 'heritage' has been a recognised term. This tendency is variously evident in many of the practices of the Grand Tour (see Chapter 3), in the search for the picturesque that characterised emergent forms of rural tourism in late eighteenth century (Andrews, 1989) and in the 'auratic' gazing on places such as country houses that Walsh (1992) explains was fashionable at much the same time. Franklin (2004) also makes the point that the development of heritage is intimately bound up in the emergence of modern nation states, in which national institutions and common practices form identifiable 'heritage' and provide associated opportunities for collective assertion and celebration of national identity by visiting key sites and observing certain spectacles. (In the UK, many of the spaces and rituals of monarchy are illustrative of this tendency and have been established for a long time.) So the origins of modern heritage are often deeply embedded, but what has changed in recent years is the scale of the attraction that is exerted by heritage places and the way in which an ever-widening range of heritage sites have been entered onto the tourist map.

But why has heritage come to be so significant for contemporary societies and central to so many leisure and tourist practices? Much has been written in explanation of the modern fascination with the past (see, *inter alia*: Boniface and Fowler, 1993; Fowler, 1992; Franklin, 2004; Graham et al., 2000; Hewison, 1987; Lowenthal, 1985; Prentice, 1993, 1994; Walsh, 1992; Urry, 1990) but for the purposes of this chapter, these explanatory discussions are summarised under five linked headings.

Human disposition

In their anxiety to advance sophisticated explanations for the often complex patterns of popular engagement with heritage, academics have sometimes been guilty of overlooking a simple fact. As human beings, we possess what is almost certainly a unique trait – that is, knowledge of our past – and we are likewise disposed (albeit in varying degrees) to an instinctive interest in that past. Who we are and where we come from matters to most people. Lowenthal (1985) provides a detailed exposition of the myriad ways in which our pasts inform the present. He writes of our awareness of the past as being essential to our well-being and as integral both to our imaginations and to our sense of identity. A sense of the past provides a reassuring sense of continuity and, when necessary, a means of escape from aspects of the present we deem temporarily unacceptable. The past endows us with the traditions that act as benchmarks or points of anchorage and it fires the imagination.

The argument here is that although we may not consciously recognise these attributes, there is a subconscious acknowledgement that predisposes us to nurture a latent interest in our past, or at least aspects thereof. Hence, in a world in which people have more time and disposable income to travel, exploration of the past through visiting heritage attractions might reasonably be considered as a natural and expected practice and one that will grow as the opportunities to engage with heritage are themselves expanded. Moreover, partly because of this natural disposition, the fascination with the past has become integral to contemporary life. It is reflected in the media, in hobby-based activities that centre on the collection of all kinds of memorabilia, in entertainment and in many sectors of education (Fowler, 1992). We tend, therefore, to live in a world in which past and present are blended in subtle and often complex ways.

Nostalgia and a sense of loss

A second explanation for the contemporary popularity of heritage attractions emphasises the role of nostalgia and what Urry (1990) terms the 'sense of loss'. This is a line of reasoning that has been strongly influenced by Hewison's (1987) critique of heritage in the UK and which suggests that processes of deindustrialisation created a profound sense of dislocation between people and the ways of life to which they were accustomed. In many communities – of both urban and rural character – the decades between 1970 and 1990 were periods of acute change in which many traditional areas of work, the technologies that supported those activities and the social structures that bound people together in working communities, were dismantled and ultimately destroyed by a combination of the impacts of globalisation and the rise of the New Right political agenda. The emergence of new centres of political and economic influence in areas such as East Asia, began to diminish the position of nation states such as the UK as places of political, economic and strategic influence. These processes, it is suggested, help to construct a nostalgic gaze on the past as a golden epoch and that by visiting heritage sites that either capture and reflect periods in which countries such as Britain were truly dominant, or which convey a sense of the traditions of the past (such as former centres of industry), people find a temporary reconnection with a past that they know cannot be retrieved but which may – in a sense – be revisited. The poet A.E. Housman's famous evocation of the 'blue remembered hills' and 'the land of lost content' captures much of the sense of longing that shapes this view of the past.[1] This is, of course, a socially constructed gaze that is highly selective. The unappealing aspects of the past are filtered out, so that the memories or feelings that remain are essentially positive and reassuring, but in certain sectors of heritage tourism it is powerfully influential.

In the modern context the concept of nostalgia is interesting. The term originated as a description of a form of physical illness that was observed amongst groups of explorers, especially in the very uncertain periods of global exploration in the sixteenth and seventeenth centuries. Today we call it 'homesickness' and we know that it is prompted by the fears and discomforts that some people encounter when they are removed from the security of the familiar and the known. Hewison (1987) suggests that the modern tendency to nostalgia becomes prevalent at times of anxiety, discontent or disappointment and so part of the appeal of heritage may be read as a form of resistance to the alienating and impersonal aspects of life in the twenty-first century. Walsh (1992: 116) comments that 'the expansion of heritage during the late 1970s and 1980s was . . . [in part] . . . a response to the perceived need for a past during a period when the rigours of (post)modern life eroded a sense of history and rootedness'. The observable desire in many communities to capture and conserve aspects of the past that may still remain – or even to remake as replicas things that have already been lost – may be seen as a part of that response.

Aesthetics

There is an influential relationship between some aspects of heritage and aesthetics, particularly those of the nineteenth century Romantic movement. Franklin (2004: 180) comments that 'the authors of Romanticism were also, in part, the authors of heritage', whilst Rojek (1993b: 145) observes that in the nineteenth-century Romantic movement we find 'evidence of a strong aesthetic and ideological association with the past as a place of peace and splendour'. That belief still underpins aspects of the nostalgic tourist gaze. The aesthetics of Romanticism also directly shaped what Graham et al. (2000: 14) describe as the 'deification of nature' that led directly to the eventual conservation of what we now recognise as heritage landscapes and townscapes in designations such as national parks – key components in natural heritage in countries such as the UK and the USA.

Part of the explanation for the recent growth of heritage tourism rests upon a popular rediscovery of some of the aesthetics of Romanticism. Franklin (2004) is instructive on this subject, drawing attention to what he terms the development of 'anti-modern consumerism' as a key response to contemporary change that has fuelled the demand for heritage experiences. In the 1950s and especially during the 1960s, modernist idioms dominated social life – in fashion, domestic interiors, consumer goods, housing design, transportation and urban planning. However, as Franklin notes, 'the relentless procession of new things, new developments, new lifestyles and change . . . engendered a feeling of loss and rootlessness' (2004: 183). As a response, moods and tastes have gradually swung backwards, with a new premium being placed on traditional styles and forms of production – to whole foods, real ale and handmade products, or housing that is imitative of vernacular rather than modernist styles and public spaces that are themed around a traditional past. Franklin labels this trend as 'heritage consumerism' and it is now hugely influential.

Identity, resistance and authenticity

The fourth group of explanations for the popularity of heritage relate to what may be broadly termed as 'identity, resistance and authenticity'. Here there is a clear connection to the work of MacCannell (1973, 1989), who (as we have seen in Chapter 6) argues that one of the primary motives for modern tourism is the immersion in the real, authentic lives of others as an antidote to the inauthenticity of the modern lives that the tourists themselves endure. Some sectors of heritage tourism, especially those attractions that present the heritage of recent industrial and agricultural communities (such as the reconstructed industrial townships in the open air museums at Beamish, Dudley and Ironbridge or the rural folk museum at St Fagans, Cardiff (all in the UK)) may be reflective of this tendency. However, whether visitors are actually seeking the authenticity of other lives, or whether they are simply curious about how their near-ancestors may have lived, is a point of debate.

Although the relationship between heritage and authenticity may be problematic (see below), there are more equivocal links with notions of identity. This operates at a number of geographic scales:

- At a national level where, for example, in the UK the Union flag, landmark historic buildings (such as the Palace of Westminster) and the emblems and trappings of monarchy are all heritage elements that are widely used and recognised as symbols of national identity (see Palmer (2000) for a discussion of heritage tourism and English national identity).
- At a regional and local level, where similar processes are widely used to assert the distinctiveness of particular communities or cultural groups (see, *inter alia*: Hale (2001) on heritage representations of the Cornish; Pritchard and Morgan (2001) on the Welsh; or Halewood and Hannam (2001) on the Vikings). For some writers, a strong communal sense of identity is actually fundamental to the formation of heritage and its subsequent development as a tourism resource (Ballesteros and Ramirez, 2007).
- At an individual level, where heritage is also used to assert personal identity. According to Walsh (1992), for example, the rising levels of consumption of heritage is a key part of the acquisition of new cultural capital by the emerging service classes in the post-industrial economy and provides important markers of what Bourdieu (1984) recognises as distinction and taste.

When viewed in this way, heritage as an expression of identity becomes closely connected to forms of resistance. For example, Urry (1990) suggests that heritage sites are often important locally, since by conserving such places local people find an important way to *signify* their locality. Ashworth and Tunbridge (2004: 210) develop the point by noting that there is often an assumption that since 'history is necessarily unique to a specific place and people, its transformation into heritage should produce a unique product reflecting and promoting a unique place or group identity'. The development of heritage attractions can be interpreted therefore, if only in part, as a form of resistance to the homogenising tendencies of globalisation.

New political economies

The notion of heritage as a product or commodity provides a useful link to the final group of explanatory factors and these relate to the perceived importance (or value) of heritage tourism in the post-industrial political economy. The place of heritage has been actively promoted for a number of reasons.

First, the emergence of neo-liberal, 'New Right' politics during the early 1980s in the USA, Britain and some other parts of Europe created a new political climate in which the principles of market economics widely replaced notions of state support and where governments were happy to see historic resources actively exploited for economic gain. Fowler (1992) notes that this placed new expectations and requirements on historical and conservation bodies to promote their sites and raise revenue directly from visitors as a means to fund their work. This created a new imperative to raise awareness and stimulate demand by developing accessible and often commodified heritage experiences.

The use of the past as a strategic element in place promotion and urban regeneration has also become a ubiquitous feature of post-industrial development, especially in urban contexts. Franklin (2004) writes of how, in the UK, the early years of Margaret Thatcher's administration (between 1979 and *circa* 1985) were a period of acute economic restructuring that plunged many towns and cities into deep recession and triggered a frantic scramble to find new forms of enterprise. (Similar processes were to be seen in the USA under the

Reagan administration.) Urban heritage schemes emerged as an almost universal form of new investment, capitalising on the potential of redundant urban infrastructures to form new places of attraction – in redeveloped docks, mills, factories, in preserved steam railways and restored canal networks – and which have brought rising numbers of tourists to some of the most unlikely destinations (such as the industrial cities of northern England or the eastern USA). Heritage became an economic resource to be developed and promoted as a product (Graham et al., 2000).

Brief mention should be made of the impact of environmental politics. Modern environmentalism has its origins in the late nineteenth century and became initially evident in areas such as landscape conservation through the early development of the national parks in the USA and the formation of the National Trust in Britain. For much of the twentieth century, environmentalism seldom occupied the political centre ground, but since perhaps the early 1970s – with the emergence of the 'green' movement in Europe and especially since the Brundtland Commission report of 1987 – new levels of concern around sustainable development and the environmental impacts of human activity have emerged. This has raised public awareness of the vulnerability of the environment in general and refocused concerns around issues such as conservation. Whilst the politics of sustainability and conservation are not necessarily directed towards heritage issues per se, the natural synergies between conservation, sustainable forms of development and the preservation of heritage ensures that the latter is commonly a natural beneficiary of the political and public interest in the other two.

The evolving character of heritage

Part of the reason why heritage has been widely adopted as a component of urban place promotion and regeneration is that the character of heritage has evolved in ways that have broadened its basis and, in theory, extended its appeal to market segments that were formerly excluded. As was noted in an earlier section, heritage is a negotiated reality – a social construction that evolves through time and which is simultaneously capable of representing a plurality of heritages.

Central to this process has been the dissolution of the hegemony of what we may label as 'high' cultural forms and the promotion of popular and alternative cultures as a basis to heritage. Prior to the 1960s, 'high' cultural forms (which broadly centred on the arts, literature, history and music) were for the most part clearly differentiated from popular (or 'low') cultures that centred on areas such as entertainment and many sports. High culture possessed an aura that was reinforced by the academy and through performance by professional elites to audiences with the refined capability to recognise the intrinsic qualities of that which was conveyed. Franklin (2004: 186) writes that 'culture was absolute, invariable and off-limits to all but those in a position to nurture it through specialised and expensive dedication'.

All this changed with the cultural revolution that occurred in Britain, North America and much of western Europe in the 1960s and the subsequent onset of post-industrialism. The 1960s, in particular, saw the powerful emergence of popular culture (especially youth cultures that centred on fashion, the media of television and on popular music) and the tacit and eventually enthusiastic acceptance of new cultural forms by those who still saw themselves as arbiters of taste. The Beatles, for example, who were initially reviled by an Establishment that objected to their long hair and their working-class, Liverpudlian personas, were soon to be awarded national honours and performed for the Queen, whilst the arrest and prosecution (for possession of cannabis) of Mike Jagger of the Rolling Stones prompted a famous editorial in support of the pop star in the bastion of the English establishment, *The Times* newspaper.

An important part of the process of 'democratising' culture was the progressive diversification in the production of cultural artefacts and a multiplication in the number of producers. The number of television and radio stations, newspaper and magazine titles, and the incidence and extent of advertising, for example, all increased significantly between 1960 and 1980, in recognition of the fact that modern society actually contained multiple audiences with distinctive tastes and preferences. As we move clearly into the epoch of post-industrialism and postmodernity, that recognition becomes fundamental to the emergence of new forms of heritage.

Until quite recently, the identification of 'heritage' reflected the ideologies of people or organisations with power and influence. It is commonly remarked that 'history is written by the winners' and hence, part of the geography of heritage is shaped around symbolic sites of power – such as castles, cathedrals, stately homes or the seats of government – places that generally reflect the hegemony of 'high' cultural forms. This tendency, of course, still persists but it has been supplemented and modified by newer, postmodern predisposition to explore alternative histories and cultures; for example, of working people, of minority groups, or of women. In so doing, distinctions between 'high' and 'popular' culture in terms of meanings and significance become much less influential and entirely different geographies of heritage are created around new attractions and locations, as what Timothy and Boyd (2003) describe as 'excluded pasts' are rediscovered.

These transformations are revealed not just in the thematic range and geographical locations of new heritage attractions, but also in the evolving character of many established places of historic heritage. For example, the museum sector (which is one of the keystones in the traditional presentation of what we now label as heritage) has seen some important shifts in style and emphasis over the last twenty or thirty years. Not only has the number of registered museums risen dramatically since 1980, but Urry (1990) describes how there has also been:

* a marked broadening in the range of objects that are now deemed worthy of preservation and presentation;
* a shift towards the concept of the living, working museum as an alternative to the reverential, hushed aura of the conventional museum;
* an extension of the business of presenting objects of interest for the public gaze beyond the confines of museums themselves and into other types of space, such as the Hard Rock cafes with their displays of memorabilia as described by Hannigan (1998) and previously discussed in Chapter 9.

These are developments designed to make these formerly elite spaces accessible to service and middle classes and which naturally serve to extend the scope and nature of heritage attractions.

One other facet of the evolving character of heritage tourism that is worth noting is the rising significance of what Lennon and Foley (2000) have labelled as 'dark tourism'. Dark tourism refers to the attraction to tourists of visits to sites associated with death and human disaster, such as wars, genocide, assassinations, terrorist attacks and major accidents. The concept applies particularly to events that have occurred within living memory and which have attracted significant media attention. Stone and Sharpley (2008) note that this is not a new phenomenon as there is a lengthy tradition of travel and tourism to sites associated with death and disaster. However, the scale of activity and the widening range of locations that are now visited (which include First World War battlefields, military and civilian cemeteries, former prisons, places associated with the deaths of celebrity figures and, perhaps most surprisingly, the highly sensitive and emotionally charged locations of former Nazi death camps) invites consideration.

The large number of sites that are associated with untimely death, disasters and other catastrophes makes any system of classification problematic, but a number of writers (for example, Miles, 2002; Sharpley, 2005) have suggested there are different levels (or shades) of dark tourism. In broad terms these relate to the intensity and proximity of experience, in which proximity is both spatial and temporal. Hence, for example, a visit to the former concentration camp at Auschwitz is likely to be a much more intense (or darker) experience than a visit to the US Holocaust Memorial Museum in Washington, because the former is the site at which people actually died whereas the latter is a place at which these events are commemorated (Miles, 2002). The analysis by Lennon and Foley (2000) of sites associated with the assassination of US President John F. Kennedy reveals some similar contrasts between the more detached memorialising in the Kennedy Museum in the JFK Library in Boston and the more immediate impacts of the 6th Floor Museum in Dallas, which contains the room from which the fatal shots were allegedly fired. Temporal proximity is also important and Lennon and Foley (2000) suggest that it is only with the passage of time that sites associated with tragedy become accessible to the tourists and after a due period for grief, the paying of respects and commemoration has elapsed.

But why would increasing numbers of tourists want to visit sites associated with human suffering? Dark tourism is a complex amalgam of motives, some of which are grounded in quite basic human instincts – such as the morbid curiosities that routinely draw sight-seers to the sites of major accidents (Rojek, 1993a) – but where other educational, commemorative, reverential and even entertainment purposes may also shape behaviour patterns. Lennon and Foley (2000) argue that the rise of dark tourism is a part of the postmodern condition in which the regular infusion by global media into daily life of images of conflict, death and disaster produces an understandable desire on the part of people to validate for themselves the events that they have seen reported, by visiting the sites in question. There is a very real power associated with 'being there' and seeing for oneself the places at which momentous events have transpired, and in a world that is shaped by mobility such opportunities are becoming more commonplace. In many instances, too, such sentiments become closely aligned with a sense of pilgrimage. Pilgrimage is often associated with the deaths of individuals or groups that acquire a religious or ideological significance that transcends the event and provides a meaning for the pilgrims (Lennon and Foley, 2000). These actions form part of the commemoration of death in many societies and are easily transformed into common tourist practices such as visiting the graves (and sometimes the sites of death) of celebrities – such as the tomb of the American rock singer Jim Morrison (who is buried in the Pere Lachaise cemetery in Paris) or the roadside at Cholame, California (where iconic American actor James Dean met his death in a road accident) and Elm Street Dallas (where John F. Kennedy was assassinated).

To illustrate the shifting nature of heritage (as well as some of the wider themes discussed in the first part of this chapter), Case Study 10.1 describes the development of steam railway preservation in Britain.

CASE STUDY 10.1

Preserving an industrial past: steam railway preservation in Britain

In Britain, the preservation of working steam railways and their presentation as heritage attractions provides an excellent example of several key themes in the changing nature of heritage tourism. First of all, steam railway preservation is an expression of an alternative history that is grounded in the lives of working-class people and in the technology and engineering of the Victorian period. In this way, it forms a part of the wider dissolution of 'high' culture as a basis to heritage and instead becomes part of a culture that is authored by ordinary working men and women. Railways are part of Timothy and Boyd's (2003) 'excluded pasts' – histories that until recently have been written out of narratives of the past as unimportant, but which have acquired new levels of relevance in postmodern societies.

The growth of railway preservation is illustrative of Hewison's (1987) thesis around the relationships between some forms of heritage and processes of deindustrialisation. The railway preservation movement in Britain began in the 1950s in Wales when a number of recently disused narrow-gauge railway systems associated with the slate quarrying industry were bought up by enthusiast groups for the purposes of restoration. The Talyllyn and Ffestiniog Railways both date from this period and were pioneering projects in British railway preservation. Then a major rationalisation of the standard-gauge network following a government report prepared by Lord Beeching in 1962 produced thousands of miles of disused railway lines and their associated infrastructures, mostly in rural areas. This process, when combined with the withdrawal of steam locomotive working in Britain by 1968, created a significant opportunity for enthusiast groups (for whom the sense of loss associated with these changes was often quite profound) to acquire disused lines and redundant locomotives at very favourable rates, for subsequent restoration as tourist lines. Local communities also became widely involved in the process, emphasising the place of steam railways as part of a popular culture and often as a signifier of local identity. Most of the major preserved lines that are operating today – such as the Severn Valley (Shropshire/Worcester), the Bluebell (Sussex), the North Yorkshire Moors and the West Somerset – all originated in this period.

Steam railways are also very much a part of the modern concept of working museums that provide opportunities for people to immerse themselves in the authentic lives of others. Most of the working steam railways rely on the services of volunteers who undertake, as leisure, the work that was formerly performed by engine drivers, firemen, signalmen, guards, station masters, machinists, engineers and repairers. The work that these people then undertake, under the gaze of the tourists who visit, becomes a performance of those authentic working lives for consumption by the visitor. The authenticity of the experience is reinforced by the maintenance of the infrastructures of the railways in their original condition (as far as that is possible), whilst stations are normally dressed with 'props' – such as historic posters, old luggage and the use of authentic, replica uniforms by train crews and station staff – to ensure the sense of the past is actively captured.

Preserved railways also illustrate how modern heritage sites clearly draw people with different motivations for engagement. For some it is a nostalgic reconstruction of a vanished form of transport, for many it is simple entertainment, for others – especially older visitors – the steam railway expresses a wider sentiment around notions of a 'traditional' countryside in which the plume of steam and the sound of a whistle from a distant train often formed an evocative part of the landscapes of childhood. In a related fashion, many of the enthusiasts who visit as photographers, actively seek to capture images that take on a timeless quality in which all sense of the present is deliberately excluded through careful composition of the image (Plate 10.1). For many of the volunteers who work the lines, it is certainly a means of escape and for most it is a hobby and therefore part of the wider integration of heritage into routine lifestyles. Railway preservation might, in the minds of some, also contribute to the wider preservation of national identity, given that the railway was one of many innovations that Victorian Britain gave to the World.

continued

Plate 10.1 Recapturing the past through railway preservation: a vintage steam locomotive from 1938 works a train on the Severn Valley Railway in Shropshire, UK, in 2003

Whatever the motives and the reasons that lie behind steam railway preservation and visiting these sites, the activity in Britain has grown to noticeable proportions. According to the UK Heritage Railways Association (2008) there are now 108 operating steam railways or steam centres in Britain, the locations of the most important being shown in Figure 10.2. During 2005, heritage railways were visited by over 6 million people, with major lines carrying well in excess of 100,000 passengers annually (Table 10.1). With this level of visiting, preserved railways are important to local economies, generating a combined annual turnover in excess of £100 million in 2005. Over 1,300 people are employed directly by the railways and more than 13,000 people work as volunteers. This generates significant local expenditures on goods and services in the towns in which the railways are based, that adds significantly to the tourist expenditures that the railways also bring.

Table 10.1 *Annual passenger figures on leading heritage railways in Britain, 2006*

Railway	Passenger numbers (rounded)
Paignton and Dartmouth Railway	350,000
North York Moors Railway	305,000
Severn Valley Railway	246,000
West Somerset Railway	230,000
Swanage Railway	200,000
Mid-Hants Railway	135,000
North Norfolk Railway	125,000
Keighley and Worth Valley Railway	115,000

Sources: various

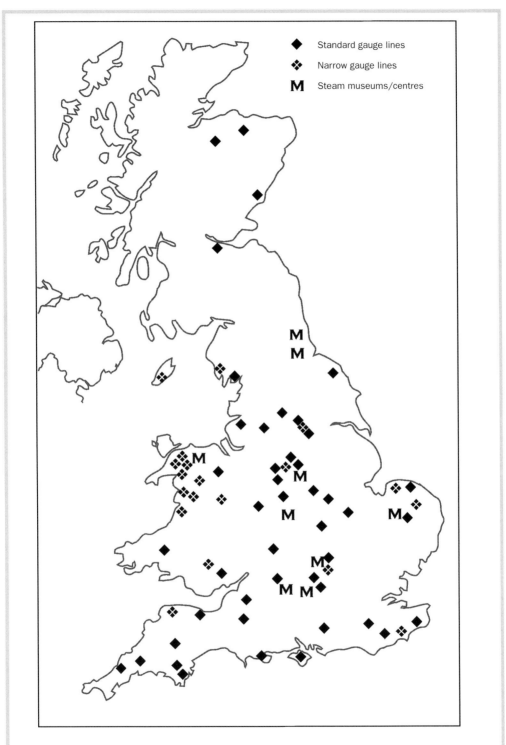

Figure 10.2 Major centres of steam railway preservation in Britain

The heritage tourism market

The nature of heritage attractions is an essential ingredient in defining the 'supply side' of the heritage tourism market (Timothy and Boyd, 2003), but as the preceding discussion demonstrates, this sector has been becoming progressively more diverse as the character of heritage has evolved from the narrow confines of 'high' cultural forms to embrace the much broader realms of 'popular' culture. In this way, heritage attractions that have been recognised for centuries are supplemented by newer attractions that reflect the creation or rediscovery of 'new' heritage.

In an attempt to develop a clearer understanding of the nature and diversity of heritage attractions, Prentice (1994) developed an outline typology. This summarises a wide range of attractions that are grouped under twenty-three headings that described categories such as 'towns and townscapes', 'religious attractions', 'stately and ancestral homes' and 'socio-cultural attractions'. The list of attractions that Prentice provides is comprehensive, but the typology is perhaps limited by the absence of any overarching organising criteria or a clear sense of the connectivity that creates heritage.

To try to address these deficiencies, Figure 10.3 develops a more structured arrangement to deliver the same ends. This figure proposes that heritage is fundamentally the product of the interplay between the environment (in its both natural and non-natural states), people, and the spaces and places that people create through their interactions with their environments and with each other. From these key variables it is possible to define a series of different heritage contexts which are here described as 'scapes' (although not in quite the sense that Urry (2000) deploys the same term). Five such 'scapes' are proposed:

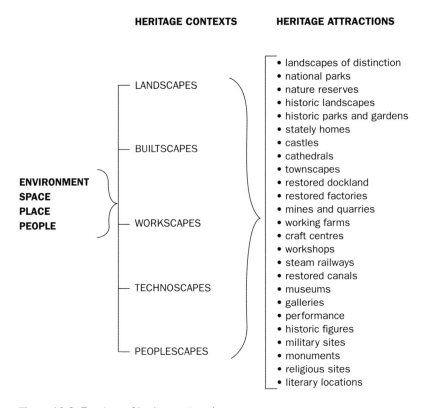

Figure 10.3 Typology of heritage attractions

- 'Landscapes' – which comprise the heritage of the natural environment or, more correctly, areas that although modified by human activity, retain an outwardly natural character.
- 'Builtscapes' – which comprise the heritage of buildings and the built environment as a physical entity.
- 'Workscapes' – which comprise the heritage associated with the world of work.
- 'Technoscapes' – which comprise the heritage associated with technology, science and invention.
- 'Peoplescapes' – which capture the heritage that is expressed by people in their social, cultural or political lives.

The categories are not, of course, necessarily discrete, but the framework then suggests that either directly or – more probably – through a process of interplay between 'scapes', a pattern of heritage attractions may be identified, much as proposed by Prentice (1994). In Figure 10.3 these are 'mapped' (in general terms) against the 'scapes' that most directly shape their character, although the process cannot, of course, be exact.

Timothy and Boyd (2003) suggest that within the range of heritage attractions, particular categories assume a raised level of significance. Their discussion highlights:

- museums and galleries;
- the living culture that is expressed in festivals, ceremonies and performances;
- industrial heritage as a celebration of former productive processes, their societies and technologies;
- archaeological sites, some of which have long been a focus of the tourist gaze at sites (such as the pyramids at Giza), others of which are reflective of more recent discoveries;
- literary sites that are either connected with the real lives of writers – such as Shakespeare's association with Stratford-upon-Avon or Dylan Thomas's home at Laugharne (Herbert, 2001) – and/or provide the settings of their stories – such as Beatrix Potter's Lake District (see Squire, 1993).

To these might be added other key attractions that include religious sites, places associated with historic figures, military sites, monuments and, especially, historic buildings and townscapes. Plate 10.2 shows part of Le Mont St Michel in eastern Brittany, one of the most visited heritage attractions in France and where the appeal is based primarily upon the dramatic setting and the remarkable character of its built environment.

However, in understanding how geographies of heritage tourism are shaped by the pattern of attractions, it is important to remember that heritage is sometimes spatially contingent, that is, it is shaped by its context, and this is variable from place to place. Timothy and Boyd (2003) provide an interesting illustration of this tendency in noting that in the eastern USA, heritage sites are mainly reflective of the colonial era and the period of the War of American Independence, whereas in the western USA, heritage sites reflect the pioneering world of the American frontier, the Spanish and, belatedly, the native Americans (see Plate 10.3).

The heritage tourism market is widely identified as a major growth sector within the overall growth of tourism itself, but as with any large-scale phenomenon, precise measurement is difficult to attain. Timothy and Boyd (2003), for example, provide sample data on the growth of visitors to designated national, natural and cultural heritage sites in the USA that show a rise from 286.5 million visits in 1980 to nearly 430 million in 2000 – an increase of exactly 50 per cent. Since the mid-1970s, Britain has acquired over 1,000 new registered museums, an additional 210,000 listed buildings (i.e., buildings designated as being of architectural importance), over 5,000 new conservation areas and a further 5,400

Plate 10.2 The heritage appeal of historic townscapes: part of Le Mont St Michel, France

Plate 10.3 Alternative heritage: the Spanish mission church at San Xavier del Bac, Arizona

scheduled ancient monuments. Over 2,000 historic buildings and monuments are regularly open to the public, and in 2002, more than 60 million visits were made to historic places in England managed by English Heritage – the primary governmental body responsible for built heritage (English Heritage, 2008).

Table 10.2 shows levels of visiting to the major heritage attractions in England in 2006 and also the percentage change in those levels since 2002. The levels of visiting tend to confirm that heritage forms a significant market segment – especially within the urban environment where traditional heritage attractions such as historic properties, major cathedrals and museums or galleries continue to exert a significant appeal. However, the analysis of change in Table 10.2 shows that the heritage market is also prone to fluctuations. Some of these fluctuations are difficult to anticipate or explain except, perhaps, in the context of the growing range of heritage attractions within the market as a whole, which diverts the gaze of the tourist to new destinations. It is also true, of course, that heritage tourism does not sit in isolation from wider patterns, so falls in visitor levels to specific heritage sites may well reflect the emergence of new attractions that are unrelated to heritage, but nevertheless divert demand. In London, for example, the huge popularity of the London Eye (a giant Ferris wheel providing high-level viewing of the central part of the city), which opened in 2000 and draws some 3.5 million visitors each year, must have exerted some impact on established attractions elsewhere in London.

The majority of the English attractions listed in Table 10.2 are in London and this exemplifies the appeal of traditional heritage sites within a major city with an established reputation in heritage tourism. We do not derive from this data, therefore, any real sense of how widening perspectives on heritage are creating new spaces for heritage tourism. Former industrial cities, for example, have emerged strongly as centres of new forms of heritage. The northern English city of Bradford provides a case in point. Here a study by Davidson

Table 10.2 *Annual visitor levels to major heritage attractions in England, 2006*

Attractions	Visitor numbers	% change 2002–06
Museums and Galleries		
Tate Modern, London	4,915,376	5.4
British Museum, London	4,837,878	5.0
National Gallery, London	4,562,471	10.4
Natural History Museum, London	3,754,496	26.9
Science Museum, London	2,440,253	–10.3
Victoria and Albert Museum, London	2,372,919	7.4
National Portrait Gallery, London	1,601,448	7.9
Tate Britain, London	1,597,359	35.6
National Railway Museum, York	902,149	21.5
The Lowry, Salford *	850,000	4.9
Historic Properties		
Tower of London	2,084,468	7.4
Windsor Castle	986,575	6.0
Stonehenge	879,393	15.8
Roman Baths, Bath	843,693	–0.2
Royal Naval College, London	698,384	112.0
Chatsworth House, Derbyshire	604,400	–2.5
Hampton Court Palace, London	473,013	–10.2
Portsmouth Historic Dockyard	449,933	12.5
Cathedrals		
St Paul's Cathedral, London	1,626,034	108.0
Westminster Abbey, London	1,058,856	–2.8
Canterbury Cathedral	1,047,380	–5.6
York Minster *	895,000	–43.0

Source: Visit Britain (2007)
* estimate

and Maitland (1997) identified a largely domestic tourism market that has expanded dramatically as the area has worked to reconstruct its economy around – in part – a tourism industry that is strongly grounded in forms of heritage associated largely with industrialisation and forms of popular culture. The metropolitan district includes several important heritage attractions, including:

● the National Museum of Film, Photography and Television (now re-branded as the National Media Museum) which draws over 600,000 visitors annually;
● the model nineteenth-century industrial community of Saltaire which, since 2001, has been a designated UNESCO World Heritage Site and includes Salt's Mill, a former woollen mill that showcases the work of the artist David Hockney;
● the Keighley and Worth Valley steam railway, which carried over 110,000 passengers in 2007;
● the village of Haworth, which attracts over 2 million visitors annually and has important literary connections with the Bronte sisters and includes the Bronte Parsonage Museum;
● the landscapes of the moors and dales that provide important components in the wider Yorkshire identity and the images that Yorkshire people hold of the places in which they live.

The diverse nature of these heritage attractions raises interesting questions relating to the differing motives that encourage people to visit heritage sites and to the wider characteristics of heritage visitors. It is generally understood that visitors to what we may label as the more established heritage attractions, such as historic properties, museums, galleries or archaeological sites, tend to display some common characteristics. According to Urry (1994a), such people are overwhelmingly drawn from what he terms the 'service class' of professional, business and white-collar occupations – people who generally exhibit higher levels of education, affluence and mobility, three key attributes that enable the appreciation of heritage sites. For example, 74 per cent of visitors to historic sites managed by English Heritage fall within the top three (ABC1) socio-economic categories, compared with a national average in those categories of 52 per cent (English Heritage, 2008). These types of attractions also draw more heavily on middle-aged and older tourists.

The observed tendency for heritage sites to attract people exhibiting these characteristics has been widely used to make a primary connection between heritage tourism and motives that centre on education and the pursuit of knowledge – and there is ample evidence to show that this is true in many situations. For example, a study by Prentice and Andersen (2007) of visitor motivations to a pre-industrial heritage museum in Denmark identified a desire 'to understand how people used to live' as a primary purpose in shaping the visit. However, interestingly, the motive that was identified as being of almost equal significance was the need 'to find somewhere to visit whilst in the area', in other words, a more commonplace wish to relax and see some interesting sights. Work reported by Chen (1998) and Richards (2001) drew essentially similar conclusions, namely, that heritage tourism is primarily shaped by a pursuit of knowledge, but is often grounded in common leisure motives of relaxation and sightseeing.

However, given the multiple meanings that heritage has acquired, explanation of patterns of tourist visiting to heritage sites needs to reflect a wider range of motives than education and sightseeing. Poria et al. (2006) draw attention to the impacts of several of these wider factors. First, they note the importance of individual needs to understand the self through visiting sites that have connections to personal development or which, more simply, are reflective of one's interests. This is closely related, second, to interests in genealogy. As the incidence of diasporas in the world population multiply, so does the tendency for people to spend leisure time in seeking out their roots and locating ancestral homes – a practice that McCain and Ray (2003) term 'legacy tourism'. Third, heritage tourism is sometimes motivated by a desire to engage with the sacred and to pay homage at sites that act as memorials to key events or influential individuals. Tourists who engage in dark tourism, for example, are – as Poria et al. (2006) point out – engaging in heritage tourism in ways that are not simply explained in relation to recreation and leisure but which draw on some more fundamental motives. Fourth, there are 'must see' locations, places where the iconographic or symbolic importance demands the attention of even the most loosely engaged tourist: the Parthenon, the pyramids or the Taj Mahal, for example.

When these broader motives are placed alongside the emergence of newer forms of heritage attraction that do not conform to the traditional notions of heritage as 'high' culture, does the visitor base becomes broader and more diverse? Although it is outwardly a reasonable expectation that visitors to newer styles of heritage attraction might draw more heavily on sectors of the population that are generally under-represented at traditional heritage attractions, evidence to confirm this expectation is at best fragmentary and often contradictory. Prentice (1993), for example, noted a slight tendency for visitors to heritage sites based on industry and some of the technologies of industry (such as railway or canal museums) to attract higher levels than normal of visiting from lower socio-economic groups – people for whom direct experience of sites of production and transportation often form a part of their 'personal' heritage as working people. A recent survey of museum attendance in Bristol (Bristol City Council, 2005) also produced data to suggest higher numbers of visitors to new industrial museums were drawn from working families or skilled working groups and were also slightly younger than visitors to the older museums and galleries in the city. Data relating to changes in museum attendance between 2001 and 2005 in the English West Midlands region (which includes museums of science and industry in Birmingham and a series of major industrial museums at Ironbridge) have shown rising levels of attendance by visitors from lower socio-economic groups. However, none of these patterns of potential change is – as yet – clearly established and the dominant pattern of visiting to heritage attractions by white, educated, professional groups seems to be as strong in the sectors of new heritage as it is in the old. The development of new types of attraction has been clearly responsible for a significant overall growth in the market, but whether it has been as influential in diversifying the profile of heritage visitors is debatable.

Heritage and authenticity

One of the many lines of argument that are advanced to explain rising levels of public interest in heritage tourism relates to perceptions of heritage destinations as real (or authentic) and which therefore serve as an antidote to some of the patently unreal places that people encounter elsewhere – such as in the 'fantasy cities' that were discussed in Chapter 9. In many heritage contexts, therefore, authenticity is held to be an essential attribute. People desire to see the original objects or to visit the actual places at which famous events transpired, whilst in the burgeoning areas of living museums and places that offer heritage 'experiences', most visitors will carry some expectation of those experiences as being variously realistic, accurate and authentic. However, in heritage tourism the concept of authenticity is deeply problematic.

To try to understand the issues that surround historic authenticity more fully, three key questions are worth addressing. First, is authenticity an attainable attribute in heritage tourism? The pedantic answer to this question is probably 'no'. Herbert (2001), for example, observes that there is no such thing as an authentic past. Heritage and – for that matter history – is created and re-created from surviving memories, records, artefacts and sites and, like culture, is actively invented, remade and periodically reorganised (Chhabra et al., 2003). Memory changes and is historically conditioned as well as being inherently revisionist and selective (Herbert, 2001). Similarly, Fowler (1992) argues that 'living history' is an impossible concept and any attempt to realise it is bound to produce a fraud. We can replicate the look of history but not the meanings, emotions and contemporary experiences of the people of the time. This is often reinforced by the way that heritage is managed. For example, the removal of objects and artefacts from their original cultural milieu and their placement in museums or heritage centres, necessarily alters some of their meanings and symbolic significance.

We also need to remember that the production of authenticity is often dependent upon the nature of its reproduction. For example, when events are played out by indigenous people according to received traditions rather than re-enacted (however realistically) by actors or performers, there is a tendency to ascribe a higher level of authenticity to that performance. But we still may not escape the fact that authenticity is often a relative construct – a subjective quality that is formed by cultural influences and modified through experience. Timothy and Boyd (2003) make the important point that the meanings of heritage sites and objects are seldom attributes of the object or site itself, but are a product of the way the object is presented and the background of the person doing the viewing.

So rather than viewing authenticity as being necessarily an absolute condition, these debates suggest a more nuanced reading in which alternative authenticities at tourist sites may be recognised. This idea has been developed by Brunner (1994) into a four-part typology of heritage that distinguishes:

- original authenticity – in which the actual objects or sites are presented in an essentially unaltered form or context (in so far as that is actually possible);
- authorised authenticity – in which expert verification authenticates the heritage sites or objects;
- perfect reproduction – in which places or objects, whilst not original, are presented in ways that are as complete, flawless and as historically accurate as knowledge permits;
- authentic reproduction – which although based upon elements that are faked or copied, provides the outward appearance of originality and produces a credible representation or a believable experience for the visitor.

A second key question is: how do practices of heritage tourism affect authenticity? Part of the debate around heritage tourism and authenticity turns on the extent to which visitors at heritage sites are truly interested in historical or scientific evidence or the acquisition of deep levels of knowledge, or whether they are mostly after an experience that stimulates appropriate reactions: nostalgia, excitement, curiosity or a sense of awe, but without being entirely authentic. In many of the latter contexts, staged authenticities clearly fit the bill. However, staged authenticities that aim simply to provide 'experience' risk the creation of what Timothy and Boyd (2003) label as 'distorted pasts'. Several forms of distortion are identified, including invented, sanitised and unknown pasts.

Invented pasts relate to a tendency for tourists to seek out places or experiences that are imagined rather than real. This is a common feature of literary forms of heritage tourism in which fiction that is located in real places produces a blurring of the real and the invented, but nevertheless encourages tourists to visit and immerse themselves in settings in which creative works are located or where fictional characters are supposed to have resided. The fact that these are often imagined settings does not necessarily matter to the visitor for, as Herbert (2001) correctly recognises, although literary places have acquired meaning from an imaginary world, they still exert a level of real meaning for the visitor. The Shropshire of A.E. Housman's elegiac collection of poems 'A Shropshire Lad' or, as mentioned earlier, the Lakeland of Beatrix Potter's 'Peter Rabbit' (and friends) are good exemplars of how landscapes of the imagination can become real objects of the heritage tourist's gaze. (Housman, incidentally, was a Worcestershire man who knew the neighbouring county of Shropshire only vaguely – as tourists who travel to the sites described in his poems often discover!)

Invented pasts are often also sanitised or idealised pasts. This is viewed by Timothy and Boyd (2003) as a common shortcoming of many heritage sites, in which the unappealing elements of the original are routinely excised to produce an experience that will be acceptable to the sensibilities of modern leisure visitors. Industrial and folk museums are an interesting case in point as they tend to project an image of these communities as essentially clean, picturesque, fragrant, well-ordered and harmonious. But this often strikes a false chord since as the authors note, 'representations of conflict, anti-social behaviour, death, disease, divorce, orphanages and starvation are notably absent' (Timothy and Boyd, 2003: 251). Waitt (2000), reporting a study of a waterfront heritage development in the shadow of the Sydney Harbour Bridge in Australia (known as The Rocks), reached a similar conclusion. He noted how the commodified heritage experiences at The Rocks were routinely accepted as 'authentic' by a majority of the visitors, yet the Eurocentric narratives of the place completely over-wrote alternative narratives of the indigenous people who had also occupied the site, or the themes of poverty, illness, death and conflict that were very much a part of the real history of the place.

Ultimately, the past must contain elements that are unknown and cannot be recovered, so that any representation of the past is distorted because true authenticity is unattainable. Timothy and Boyd (2003; 253) argue that modern people cannot possibly understand precisely the lives of people in history or know every detail of how they lived, thought, acted, or viewed their world. The past is not a 'known entity', it is an enigma and 'all that can be done is to imagine what it was like'.

This leads to the third key question: are high levels of authenticity actually important in heritage tourism? Urry (1990) argues that although heritage tourism is often distorted – presenting, as it tends to do, largely visual representations that invite the tourist to imagine the real lives and events that are wrapped around the objects or places in question – it still serves a valuable purpose in allowing people to gain a sense of the past in ways that they are otherwise unlikely to achieve. Schouten (1995) argues that most visitors to historic sites are looking for an experience that is grounded in the past but may not be truly reflective of that past. In this respect, true authenticity is often less important than a perceived authenticity that is consistent with a nostalgia for often imagined pasts (Chhabra et al., 2003; Waitt, 2000).

There is, of course, a level at which heritage tourism is essentially entertainment, a spectacle to be consumed by people with time to fill and curiosities to assuage – and this is a context in which authenticity is probably a secondary consideration. But heritage tourism (as is true of most forms of tourism) is also an expression of power relations and the way that heritage is conserved, presented and interpreted ultimately reflects the ideologies of groups that possess power and influence. Where the exercise of such power leads to a partial view of the history from which heritage is created (and in which, for example, the narratives of minority groups are effectively expunged from the record – as is evident in much of the Eurocentric heritage in the USA), then the *in*authenticity of that view matters greatly, not least to those citizens whose heritage is being disregarded.

Summary

In the last thirty years or so, heritage has emerged as one of the primary interests that shape many of the spaces of tourism. Interest in the past is informed by a complex combination of innate curiosity, nostalgia, aesthetics, identity and resistance, as well as by quests for a temporary escape from the perceived intensities of modernity, or for an authenticity that is believed to lie within the past. As demand for heritage tourism has become more embedded, so the range of sites that offer different heritage experiences has extended, eroding the hegemony of 'high' cultural forms of heritage and allowing other forms of popular culture to shape parts of the heritage tourism industry. Through this process new heritage destinations have emerged that have brought tourism to new locations (such as former industrial cities). At the same time, however, the widening range of heritage attractions have raised intriguing questions about whose heritage is being presented and to what extent *any* representations of history can be truly authentic.

Discussion questions

1   In light of the diverse range of locations and environments that now constitute heritage places, how meaningful is it to attempt to identify 'heritage' tourism as a distinctive component in the contemporary tourist industry?
2   Why has heritage tourism become so popular within post-industrial societies?
3   Who are the 'heritage tourists' and why is the profile of heritage visitors so often skewed towards particular groups?

4   How has the developing relationship between heritage and popular cultural forms helped to reshape geographies of heritage tourism?

5   To what extent is authenticity an essential requirement for heritage tourism sites?

Further reading

Although published over twenty years ago, the outstanding discussion of the relationships between societies and their histories remains:

Lowenthal, D. (1985) *The Past is a Foreign Country*, Cambridge: Cambridge University Press.

More recent discussions that explore similar themes are provided in:

Fowler, P.J. (1992) *The Past in Contemporary Society: Then, Now*, London: Routledge.

Graham, B., Ashworth, G.J. and Tunbridge, J.E. (2000) *A Geography of Heritage*, London: Arnold.

Detailed examinations of relationships between heritage and tourism are to be found in:

Boniface, P. and Fowler, P.J. (1993) *Heritage and Tourism in the 'Global Village'*, London: Routledge.

Herbert, D.T. (ed.) (1995) *Heritage, Tourism and Society*, London: Mansell.

Prentice, R. (1993) *Tourism and Heritage Attractions*, London: Routledge.

Timothy, D.J. and Boyd, S.W. (2003) *Heritage Tourism*, Harlow: Prentice Hall.

Useful short discussions of heritage within texts that explore a wider range of tourism issues can be found in:

Franklin, A. (2004) *Tourism: An Introduction*, London: Sage.

Urry, J. (1990) *The Tourist Gaze: Leisure and Travel in Contemporary Societies*, London: Sage.

For a highly readable critique of heritage attractions relating to 'dark' tourism, see:

Lennon, J. and Foley, M. (2000) *Dark Tourism: The Attraction of Death and Disaster*, London: Continuum.

Note

1   A.E. Housman 'A Shropshire Lad' XL

⓫ Tourism, consumption and identity

To conclude this book, the final chapter offers a short excursion into the comparatively new territories in human geography that relate to consumption, identity and the body, and – in particular – explores some of the intersections between tourism and these wider themes. Interest in processes of consumption and their influence has developed strongly in geography over the last twenty years or so, partly through the raised levels of awareness of culturally informed understandings of geographical patterns that have arisen from the so-called 'cultural turn', but also as a reflection of a wider realisation that in the contemporary world, 'an understanding of the processes of consumption is central to debates about the relationship between society and space' (Jackson and Thrift, 1995: 204). If, as it is generally asserted, we have progressed (or are progressing) from a modernist-industrial to a post-industrial/postmodern basis to life – with an associated shift from production to consumption as an organisational logic of economic and social space – the influence of consumption cannot be ignored. Jayne (2006: 1) captures this significance concisely when he writes that 'consumption is understood to be a means and a motor of economic and social change, an active constituent in the construction of space and place, and as playing a vital role in constituting our identities and lifestyles'.

Similarly, a raised awareness of the role of the body as a socially constituted space and a site of embodied forms of consumption has also emerged from the new interest in cultural interpretations of place and space. This area of work focuses attention on the myriad ways through which the body becomes inscribed with social meanings that reflect notions of self-identity (e.g., through dress, adornment, eating habits and personal grooming) and which through the medium of performance (as well as appearance) projects those meanings onto geographical spaces and onto others who share those spaces.

The discussion that follows explores how identities may be shaped around patterns of consumption and through embodied forms of practice that both reinforce and project those identities, and discusses some of the primary ways that these processes are revealed in tourism. Tourism has become an influential area of consumption in post-industrial/postmodern society and an important medium through which we express identity, both through the styles of tourism that we embrace and the performances that we deliver as tourists. Two examples of emerging areas of tourist activity – adventure tourism and wine tourism – are examined to illustrate the wider themes of the chapter.

Acknowledging consumption

What is consumption? In simple terms we may conceive of the process of consumption as relating fundamentally to the acquisition and use of goods and services that meet real or perceived needs. In the narrowest of senses, the act of consumption uses the good or service in such a way that the item is no longer available: the consumption of a meal, or of a mineral resource that is utilised in a production process, for example. But critically, the term has been adopted and adapted to capture a much wider concept in which a whole range of social, cultural and environmental situations are 'consumed' as experiences. Jayne (2006: 5) writes that 'as such, consumption is not just about goods that are manufactured and sold, but increasingly it is about ideas, services and knowledge – places, shopping, eating, fashions, leisure and recreation, sights and sounds can all be "consumed"'. This is a fundamental distinction because it tells us that consumption is not purely confined to an act of purchase and its immediate consequence, but is an on-going process that both prefigures the purchase itself and influences future actions and responses that we make as consumers. Consumption is therefore to be seen as constitutive of daily life, rather than an activity that is self-contained and bounded within fragments of space and time.

For the purposes of this discussion, two primary influences of consumption are worth acknowledging: first, its role in shaping space and the nature of place; and, second, its influence on lifestyle and identity.

In developing an understanding of how space is shaped by consumption, the work of Lefebvre (1991) and Soja (1996) is particularly useful. By drawing on their ideas around spatial practices and the representation of space, we may envisage the nature of space as being defined in relation to three key constructs:

● how space is imagined and represented by groups such as architects, planners, transport engineers and developers;
● how spatial practice is revealed in the production of actual locations and configurations (e.g., in the pattern of homes, streets, shopping precincts and open spaces) and which may be seen as the concretised product of processes of imagination;
● how space is perceived, which relates to the actual lived experiences of people who use spaces and through which space acquires meaning and value.

Consumption intersects with each of these constructs in some important ways. It has underpinned much of the development of sites of spectacle in the modernist city – such as the development of high-class department stores and shopping arcades in cities such as London and Paris from the early decades of the twentieth century (see Nava, 1997). It is similarly evident as a driver for change in the widespread adoption of 'festival' settings of office, retail, entertainment and leisure districts in regenerated, post-industrial cities and in the gentrification of former inner-city zones (Jackson and Thrift, 1995). The allure of consumption infuses contemporary place promotion strategies and is central to creating images of place, in both urban and rural settings.

Each of these examples reveals how 'official' place-makers – the architects, planners and developers – imagine spaces and create them as physical entities that are imbued with a plethora of signs that signify intended purpose and function, but it is also clear that because people view space from contrasting positions and with differing motives, the actual consumption of place through lived experiences serves both to reinforce and to subvert place identities. Shopping malls, for example, have been widely studied as spaces of consumption that blend retail, leisure and entertainment in novel and enticing ways that render the consumption of the place as an integrated experience (Clammer, 1992; Goss, 1993; Langman, 1992; Shields, 1992; Williams, 2003). But it is also clear that the spaces of malls and other contemporary retail environments (such as festival markets) are perceived differently by different users. Hence, consumption of retail environments as leisure spaces by tourists is rather different to the consumption by local people of those same spaces as functional shopping environments and is different again from the consumption by 'mall rats' who colonise malls as social spaces in which to window shop (but seldom buy), or to socialise with others, or to stroll and gaze in the style of the flaneur (Paterson, 2006).

These are important distinctions because they reveal that whilst spaces are imparted by their designers with functions, values or attributes that are intended to produce a particular style of consumption, through the actual processes of consumption, the meanings, values and roles that those spaces acquire for users may be strongly contrasting. Hence, shopping space becomes social space for idle teenagers; spaces of production are consumed as spaces of attraction by tourists; spaces of tourism are places of employment, and so forth. The process of consuming space will often, therefore, alter intended meanings and functions through the divergent ways in which space is apprehended by consumers.

The second area of influence of patterns of consumption relate to our wider lifestyles and the intimate connection between lifestyle and identity. In this area of critical thinking the work of Bourdieu (1984) has been enormously influential. His primary argument is that through acts of consumption, social groups are able to form and maintain distinctive identities. As (post)modern consumers we are confronted with a range of products and services. The ability to choose from this range is predicated on distinctions that we draw between alternative products and by choosing one product in preference to another, we exercise our judgement of taste. The exercise of taste is, so the argument runs, an articulation of a sense of our social class, our background and our identity (Paterson, 2006) and which is shaped by what Bourdieu terms our 'habitus' – that is, our socialisation into a way of life or lifestyle.

In this context Paterson (2006: 41) defines lifestyle as 'a set of positional markers that define a social group and that mark difference from other groups with different lifestyles, through the use and display of consumer goods and cultural goods'. In a related vein, Lury (1996) writes of what she terms 'positional consumption' in which commodities are purchased and used as markers of social position by consumers who are seeking to define their position in relation to others. This draws on a much older argument advanced by Veblen (1994) – which was first set out in 1924 – around what he termed 'social emulation'. This was a practice by which the *nouveau riche* adopted the styles of consumption of higher classes as a means of elevating their social standing, a process evidenced in tourism in the development and changing social tone of early resorts (see Chapters 2 and 3). But however it is conceptualised, the foundational significance of consumption rests on the suggestion that through consumption we affirm our identities: we define ourselves by what we buy and by the meanings that we give to the goods and services that we acquire (Jackson and Thrift, 1995).

The common criticism of Bourdieu's ideas is that they are grounded in assumptions relating to a supposed fixity of class structures and, especially, the values and behaviours of middle-class groups. However, one of the lessons of postmodernity is that class structures are becoming progressively less fixed and the parallel development (under post-Fordist patterns of flexible production) of enhanced levels of consumer choice, enables people to use their patterns of consumption flexibly and sometimes ironically to shape multiple and diverse identities (see Featherstone, 1991). Under these conditions, identity is shaped less by structured class groupings and the self-reinforcing nature of the virtuous circle of lifestyle, consumption and identity that Bourdieu proposes (in which economic capital enables the acquisition of both cultural and social capital that, in turn, reinforce the economic position), but rather by what Paterson (2006: 49) describes as our 'fleeting, capricious, ephemeral' choices as consumers.

This attempt to lift the reading of consumption and identity out of the rather limiting framework of a conventional class structure whose validity is now questionable has a value because it draws our attention to three further attributes of consumption and its relationship with identity that are essential to comprehending these sometimes elusive concepts.

First, the notion that people may adopt multiple and diverse identities reminds us that identity is not a fixed attribute but may be remade through adjustments to patterns of consumption and related lifestyles. This can occur simultaneously – for example, the identities that we adopt in our leisure time may be very different to those that we deploy in a working environment – or change may occur sequentially – most obviously as we progress through our life cycles. Primarily through the acquisition of different levels of economic capital and the progression into a different social milieu, the patterns of consumption and the identities that people may exhibit as – say – middle-aged professionals are very different from those that they revealed as – say – students.

Consumption, through its link to identity, may also be used to signal forms of resistance or adherence to non-conventional lifestyles. Jayne (2006: 5) explains that consumption is related to what he terms a 'matrix of identity positions' (such as constructions of class, gender, sexuality, age and ethnicity). Examples that are widely cited include how overt patterns of consumption have helped to map the territories of 'gay' communities in cities such as Manchester and San Francisco, or the way in which youth culture (which is strongly mediated through patterns of consumption) signals diverse forms of resistance to established values and conventions (see, for example, Chatterton and Hollands, 2003).

We should also acknowledge that the symbolic meanings and values of places and objects of consumption may be altered by the process of consumption. Consumers are not necessarily the passive recipients of products but may rework the nature, uses and applications of the product to give them new meanings and values (Crang, 2005). Du Gay's (1997) adaptation of Johnson's (1986) circuit of culture (Figure 11.1) provides a useful theoretical insight to this process. This shows how rather than pursuing a linear (and uni-directional) pathway from production, through representation (such as advertising), to consumption and the creation of an associated identity for both the product and its user, the process is actually a set of circuits that are continuous and self-regulatory in nature. So when

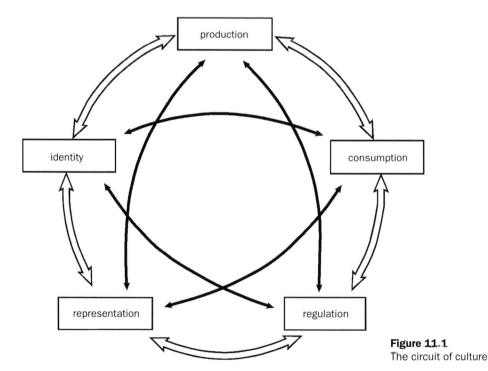

Figure 11.1
The circuit of culture

consumers start to apply their own values to a product through their patterns of use, the identity that the product bestows on the consumer will be modified and it will start to acquire a new set of meanings. This, in turn, informs the production process, encouraging the producer to modify the product and represent it in new ways that are expected to resonate with the changing meanings and values of the consumers. Du Gay (1997) deploys the example of the Sony Walkman to illustrate the theory, but an equally effective example relates to the consumption of mobile phones. These now ubiquitous items started as comparatively exclusive, functional products designed to service the world of business and commerce. However, through the process of consumption and related technical advances, they have been progressively reworked, becoming affordable and acquiring a bewildering array of functions as multi-media communication, recording and entertainment devices. Consequently mobile phones have acquired diverse value positions – from an indispensable business aid to fashion accessory, for people across the social and age spectrums.

Identity and modernity

So far in this discussion the term 'identity' has been used without qualification, but it is a slippery concept that invites closer scrutiny. 'Identity', writes Bauman (1996: 18), 'is a modern invention' and very much a modern preoccupation. Through processes such as industrialisation and the associated reorganisation of capital and labour at an increasingly global scale, modernity has effectively dismantled most of the protective frameworks of the smaller, pre-industrial communities, replacing them with much larger, impersonal forms of organisation. These changes have produced hitherto unseen levels of anonymity and uncertainty in social life but because self-identities are no longer so firmly structured by social hierarchies and traditional authorities, the modern individual faces the challenge of finding new pathways to self-identification whilst simultaneously being presented with a diversity of possible selves (Desforges, 2000). Moreover, the issues that surround the questions of who we are and how we maintain that sense of our own identity have been compounded by the migration of at least some elements of Western societies onto the shifting sands of postmodernity. In postmodern societies that are continuously and actively engaged in remaking themselves, identities need to be adaptable and flexible, rather than fixed and therefore limiting. To quote Bauman (1996: 18) once again, 'the modern "problem of identity" was how to construct an identity and keep it solid and stable, the post-modern "problem of identity" is how to avoid fixation and keep the options open'. Consequently, identities are increasingly fragmented, fractured and multiply-constructed across different areas of discourse and practice (S. Hall, 1996).

These contextual observations raise interesting questions: first surrounding notions of self-identity and how identities are constructed and understood; and, second – and of particular relevance to the themes of this book – how tourism contributes to the formation of identity.

In a landmark study of modernity and self-identity, Giddens (1991) rehearses a number of understandings of self-identity and how it is underpinned by social praxis. A central tenet of Giddens' argument is that identity is a reflexive project of the self in which the individual bears a high degree of responsibility for the outcome. In other words, and as Stuart Hall (1996: 4) explains, identity is a process of *becoming*, rather than *being* (my emphases) in which the choices that we make define who we are and who we become. This process presumes what Giddens terms a 'narrative of the self' which relates to those strands of narrative from our daily lives through which we understand who we are and through which we expect others to understand us. Identity, therefore, becomes something that we make, rather than discover within ourselves and is constituted within processes of representation rather than standing apart from them (S. Hall, 1996).

This key conclusion presumes two further processes. First, Giddens suggests that people need to develop and sustain a reflexive understanding of themselves through coherent, yet continuously revised, biographical narratives that reflect both personal histories and future intentions and which are informed by the multiple choices that may be available to them. (This emphasises the dynamic, reiterative dimensions of contemporary understandings of identity as something that is actively and continuously remade.) Second, Giddens sets out the concept that he terms 'self-actualisation' which relates to the mechanisms by which people envisage and shape lifestyles through which identities may be realised. This implies some ability to exercise control over time in which to engage with activity that shapes identity, but processes of self-actualisation also need to be understood in terms of balancing opportunities against risks. To remake identities 'the individual has to confront novel hazards as a necessary part of breaking away from established patterns of behaviour' (Giddens, 1991: 78).

Giddens (1991) makes one further point that is highly relevant to discussions later in this chapter, which is that an important aspect of the reflexivity of the self is its extension onto the body. He talks of the body as part of an 'action system' rather than merely constituting a passive object. As we will see in a subsequent section, the action systems of the body provide an essential medium through which identity is formed and projected – it is inscribed with an array of encoded 'messages' that others may 'read', and through embodied practices and the performativity of the body, we convey messages that strive to project our self-identities onto others.

This observation serves to cue one other aspect of identity that is relevant. Some writers argue that part of the objective of identifying ourselves in particular ways is to forge relationships with others through the creation of collective identities. Desforges (2000), for example, makes this point and several of the analyses of mountaineering that have informed the discussion of adventure tourism later in this chapter articulate the ways in which participants, through embodied performances and the adoption of styles of clothing and appropriate codes of practice, seek to position themselves as 'belonging' to the fraternity of mountaineers (see, e.g., Beedie, 2003). However, whilst there is clearly some validity in these observations, the practice of reflexive formation of self-identities is more usually based around constructions of difference rather than similarity. Stuart Hall (1996: 4) is quite clear on this point, noting that identities are 'more the product of the marking of difference and exclusion, than they are the signs of an identical, naturally-constituted unity' and that 'identities are constructed through difference'.

Consuming tourism and shaping identity

How does tourism intersect with consumption and the shaping of identity? Tourism is a complex form of consumption in so far as at the heart of the tourist experience, it is often the case that no tangible commodity is being purchased. Many forms of tourist consumption are shaped by the predominantly visual consumption of images and representations through the tourist gaze (Urry, 1990), although as Paterson (2006) notes, the enabling of visual consumption usually entails the direct consumption of a wider range of tangible, supporting goods and services. Tourism exemplifies the idea of 'conspicuous' consumption that acts as a social marker – as intended by Veblen (1994) and Bourdieu (1984), yet whilst the tourist gaze may frequently be attracted by the spectacular sites of consumption, it may equally engage with the mundane in ways that make tourist practices hard to characterise. Links with identity are equally complex. Tourism may help to shape collective identities of the visitors, but often undermines local identities; it is a practice through which we may seek to reaffirm our understanding of ourselves, but it is also provides settings in which those identities may be challenged and remade.

It is clear, therefore, that we may assess the significance of tourism in relation to consumption and identity from multiple positions but for simplicity, three primary areas of significance may be outlined.

First, tourism acquires significance as a form of conspicuous consumption. It is conspicuous in several senses that are simultaneously physical, economic and social. It is a physically conspicuous process because it entails periods of absence from routine settings and engagement with the act of travel. For people in work, such absence is normally formalised through procedures for approving holiday periods and is usually noted informally in the social settings in which people live – through the incidence of homes that are temporarily shut up and where neighbours are perhaps undertaking routine care and security functions. Returning tourists will often reflect embodied attributes (such as a tanned appearance) whilst new possessions in the home that were acquired whilst travelling (such as clothes and souvenirs) may also be used as physical markers of a recent trip.

In economic terms, tourism is frequently an expensive activity (within the overall framework of a household budget), requiring active saving and investment of sums of money to a degree that is not widely replicated in other areas of routine expenditure. Tourism is a luxury commodity for many people and by investing substantial amounts of economic capital in its purchase, tourists clearly expect to derive significant levels of social and cultural capital in return. The holiday – particularly if it has been conducted to foreign and exotic destinations – is therefore a powerful signifier of identity, status and social aspiration and the narration of the experience to friends, neighbours and relatives – through the sending of postcards, the bestowing of gifts, or the showing of photographs and video of the destination – is an integral component in affirming that assumed status.

Second, linked directly to the role of tourism as an arena of conspicuous consumption is its role in shaping identity, whether as an affirmation of an existing one or a pathway to a transformed identity. In a paper published in 1991, Bruner cast doubt upon the notion that tourism provides a vehicle for transformation of the self, arguing that processes of commodification, in particular, often presented tourist experience in ways that were designed to confirm expectations that were formed in advance and which provided little opportunity for travellers to encounter difference. However, more recent work has brought forward opposing perspectives, hence – for example – Desforges (2000: 930) writes that 'tourism practices, and the ways in which they are imagined and enacted, become central to the construction of the self'. Tourism is an obvious arena in which people may confront the novel hazards that are sometimes essential to self-discovery and play out the narratives of self to which Giddens (1991) refers. For example, Noy's (2004: 79) study of young backpackers observes that the trip became a 'moment constructed as formative and transformative in the stories that the youths tell of their self and their identity' and although young people navigating the awkward transition from youth to adulthood are perhaps more susceptible than most to articulating transformative experiences through practices such as travel, other studies report similar evidence from older tourists. Desforge's (2000) study relates an interview with a woman in her sixties who had, through recently adopted patterns of travel to exotic destinations such as Nepal and Peru, constructed an entirely new sense of her self-worth and of herself. In this process, the practice of tourism is critical to validating notions of self-change.

Moreover, the identities that we form as tourists may be variously used to develop a collective identity as well as a means of indicating distinctions that separate us from others. The interviews that Noy (2004) conducted with young Israeli backpackers revealed how participation in this form of travel imparted distinctive cultural capital and admitted them to a sub-culture of travellers who assert a collective identity as backpackers and for whom status within that community is often a direct reflection of the experiences they have gained through this form of tourism.

Third, tourism acquires further significance through its wider relations with contemporary lifestyles and associated patterns and styles of consumption. A central argument in Franklin's (2004) excellent study of contemporary tourism is that tourist practice is now constitutive of, and integral within, most aspects of (post)modern living. It is reflected in the media; in consumer choices in areas such as food and clothing; it is a recurring theme in social exchanges; it helps to structure space and is increasingly central to the postmodern imagination. And imagination is a fundamental part of contemporary consumption. Campbell (1995: 118) proposes that 'the essential activity of consumption is not the actual selection, purchase or use of products, but rather the imaginative pleasure-seeking to which the product lends itself'. That quest for pleasure also translates quite quickly into a desire for novelty. The same author writes that 'modern consumption centres on the consumption of novelty' and that 'modern consumers will desire a novel rather than a familiar product because this enables them to believe that its acquisition and use will supply experiences that they will not have encountered to date' (Campbell, 1995: 118). Tourism is a primary locus for the exercise of the imagination and the experience of novelty.

Tourism and understandings of the body

In the contemporary literature on geographies of consumption and identity, it is now commonplace to find that discourse reflecting new interest in the body and the embodied nature of human experience. Academic interest in the body derived initially from feminist perspectives (see, e.g., Butler, 1993) and the recognition that the body is socially constituted rather than being biologically determined. So whilst gender, for example, is initially located in essential bodily differences between men and women, gendered social practices are the product of routine norms of behaviour and social expectations that define what has been identified as the performative nature of gender (Nash, 2000). But not only is gender a performed role, individually we may also work with the basic physiology with which we are endowed at birth and modify, dress or otherwise adapt our body (e.g., by slimming or body-building) to serve a wide range of personal or collective purposes. In this sense the body is 'not a fixed essence, but is located in a network of political, socio-economic and geographical relations' (Winchester et al., 2003: 157).

This notion of the body as what we might term a 'relational' space is developed in some of the work by Thrift (1996, 1997) that emphasises the significance of the mundane, embodied practices of everyday life that actually shape the conduct of people towards others and themselves in particular settings. In commending what he terms 'non-representational theory' or the 'theory of practices', Thrift attempts to show how the performative dimensions of everyday life provide insights to how such lives are structured, practised and acquire meaning, in ways that representational forms – such as written texts – cannot achieve.

These critical insights are directly relevant to tourism and the heightened level of academic interest in the body in tourism reflects some important shifts in how understandings of tourism are formed. In part, these new critical positions are a reaction against an emphasis that had prevailed for many years in tourism studies on the disembodied subjectivity of the gazing tourist (Urry, 1990) and the outwardly simple role of the bodies of Others as objects of the tourist gaze (Veijola and Jokinen, 1994; Franklin, 2004). Equally, a realisation has emerged that many forms of tourism are fundamentally sensual in nature and that the body takes on a primary role as the space in which those sensuous dimensions of tourism are captured and experienced (Crouch and Desforges, 2003). The body provides the means of connection with the world, thoughts wander and emotions vary, but when you hear, see, smell, sense and taste, you connect with your setting (Veijola and Jokinen, 1994). Another key change has been the recognition of the performative qualities of tourism and that the

266 • Understanding the spaces of tourism

body is a primary site on which a range of more complex motivations, behaviours, beliefs and expectations are inscribed and which are revealed by embodied tourism practices. In this sense, tourism space becomes a stage on which tourists perform, improvise or otherwise adapt a series of roles that may vary from the predetermined to the purely reflexive.

The significance of the body in tourism has therefore been recognised from differing perspectives, but to provide structure to this discussion, three key attributes are examined below. These are, first, the way in which the body relates to identity; second, how embodied tourist practices structure tourist space and ways of understanding that space; and, third, how embodied practice shapes participation or the use of that space.

The body and tourist identity

Although, as we have already seen, identity is strongly aligned to practices of consumption, the body is arguably the primary medium through which we express those identities that our patterns of consumption help to create. The body is a place of consumption but it also reveals much about our view of the self. Through our choice of clothing and bodily accessories or hair styles, through our deportment and bodily dispositions, through the management of our bodies (whether as lithe, fit and tanned or as pale, over-weight and out-of-condition entities), and through the performances that we invoke from our bodies, we make powerful statements about who we believe we are and the values to which we subscribe. As noted earlier, the body is a social construction and it maps power and identity in some influential ways (Winchester et al., 2003). Bodies are also sites of resistance and the social inscription of bodies helps in processes of categorisation and distinction, as well as inclusion and exclusion. It was no accident, for example, that the counter-cultural movements of the Californian 'hippies' of the 1960s and 1970s adopted a range of distinctive bodily practices around dress codes, hair styles, relaxed attitudes to sexuality and social nudity and overt consumption of drugs and alcohol, as ways of signalling resistance and opposition to mainstream cultural values and the politics of the day. As Edensor (2000b) observes, the body is inscribed with cultural meaning and social relations.

Nash (2000) makes the telling point that identity cannot exist outside its performance and performance is intimately linked to the bodily practices through which people make sense of their world and their position within that world. This is highly relevant to understanding the spaces of tourism because not only do the embodied identities that we adopt as tourists tend to distinguish us as tourists – for example, through the deployment of the particular dress codes and body accessories – but also the subsequent locations we inhabit and the embodied performances that we give as tourists maps tourist space and the relations that structure that space. Without the tourist body and its associated identity, many tourist spaces become indistinguishable.

At the same time, we often deploy different styles of performance within the activity of tourism that help to assert a specific identity and distinguish ourselves in relation to others who, whilst sharing the same space, we may not wish to share the same identity. Edensor (2000b) provides an excellent example in analysing the differing styles and embodied practices that surround touristic walking. Here the serious walkers routinely adopt bodily dispositions and accoutrements that are intended to communicate to others a particular set of values and an associated status that marks them out as specialists. Walking boots (which are highly symbolic items), walking poles, particular brands of waterproof clothing and the use of backpacks transmit a form of cultural capital to the onlooker that becomes a discernible mark of status and membership of a fraternity.

However, tourism is not just another context in which we may project onto others an embodied identity and an assumed status. Tourist places are also sites at which we may explore or alter that identity. The individual may use tourism to discover, reaffirm or change their identity and many forms of tourism allow for active renegotiation of identity through body practices (Crouch and Desforges, 2003). Inglis (2000) argues that vacations are a time to try out alternative views of the self, or perhaps to reconnect with the self that you really are, or believe yourself to be. He writes that 'the self one finds one can be on vacation feels so good, so fresh and restorative, that it is taken to be the true self one has to be (Inglis, 2000: 135). Despite the fact that tourist experiences may also be bad, unrefreshing and exhausting, the notion of self-discovery still resonates powerfully across the spaces of tourism. At the time of writing, a TV promotional campaign in the UK for holidays to Ireland exploited directly these sentiments of self-realisation through tourism, with images of tourists reflexively discovering their 'own' Ireland – and by extension themselves – through embodied activities such as horse riding on the strands of the Dingle Peninsula, Gaelic dancing in the 'authentic' space of an Irish pub, or a first encounter with that noted Irish product, Guinness, in a Dublin bar.

The body and tourist space

Connections between the body and the way that we structure and develop shared understandings of tourist space are not immediately evident but once uncovered are often revealed as fundamental and highly influential. Work by MacNaghten and Urry (2000) and Edensor (2000b), for example, articulates a powerful relationship between the nature of tourist places and embodied responses or emotions that such places promote. In his insightful critique of walking as a recreational or tourist practice, Edensor (2000b) makes the point that the rural spaces in which such activity concentrates are widely constructed as environments in which notions of escape, freedom and natural expression are central to the embodied activity of walking. Such sentiments are part of a wider response that developed from the nineteenth-century Romantic movement and in which the aesthetic appreciation of the countryside spilled over into a popular, embodied engagement with natural landscapes – initially through activities such as hiking and camping, through more exclusive practices such as naturism that shaped a particularly intimate relationship between the body and nature (Bell and Holliday, 2000), and latterly through more adventuresome forms of activity (such as surfing and climbing) that engage other embodied responses such as fear and excitement as central to the activity (Franklin, 2004). Many of these constructions of the 'natural' spaces of the countryside as sensual spaces imbued with opportunity for reflexive, self-discovery and personal regeneration are often positioned in opposition to equally embodied constructions of the 'unnatural' spaces of the urban world as constrained, restrictive, regimented, unreflexive and sensually limiting.

Our knowledge of tourist space is generally constituted through a combination of a cognitive and a bodily understanding that is acquired by moving through space, making contact with people and through embodied social practices (Crouch and Desforges, 2003). Embodied practice is often integral to our development of an understanding of particular spaces and places. Edensor (2000b: 82) comments on how walking helps to underpin understandings of rural space by providing a medium through which 'we express ourselves physically, simultaneously performing and transmitting meaning, sensually apprehending "nature" and sustaining wider ideologies about nature, and the role of the body in nature'. In the process, the embodied actions of walkers become integral to the construction of tourist space, by inscribing paths and signage in the spaces that walkers visit and, more widely, by delineating particular kinds of landscape as suitable for particular kinds of walking. For example, the English Lake District has, since the days of Wordsworth and Coleridge (who were prodigious walkers) been widely appreciated as 'good walking country', a construction that is largely derived through cultural precepts and reinforced by the embodied practices of many of the millions of tourists that the area now receives.

There are also interesting associations between the body and the construction of some tourism sites (such as beaches and their resorts) as liminal spaces (Shields, 1990). The concept of liminality relates to notions of boundaries, and liminal spaces are variously defined as places of transition or that lie on the cultural margins, places of ambiguity in which normal social codes and practices are suspended and which permit alternative codes and behaviours to surface (see Preston-Whyte, 2004). The beach – and the tourist practices that are associated with the beach – reveal strong associations with liminality, but it is a liminality that is essentially defined by embodied practices. For example, Franklin (2004) draws on a detailed account of seaside tourism in the English resort of Blackpool in the 1930s developed by Cross (1990) to emphasise the overtly sexualised nature of several types of resort space and the various roles of the body in setting that particular tone. This was evident in a common quest, amongst younger people at least, for sexual encounters, in varying states of exposure or nakedness on the beach, and the use of bodies and bodily functions as a staple subject of humour in saucy seaside postcards or comedic innuendo and lewd jokes at the popular variety shows in the resort theatres.

The role of the body in defining a sexualised nature of some tourist space reminds us that bodies do not just help to structure and define tourist space, in certain situations the body becomes a tourism space in its own right. This is most clearly evident in sex tourism in which an eroticised body of the Other becomes both a primary object of the gaze and, under some circumstances, a site of direct sexual contact and embodied experience. Inglis (2000) makes the point that sex tourism is a much older tradition than might initially be imagined, but in the post-1960s era the general liberalisation of sexual attitudes and behaviours aligned with the development of a global tourism industry in which tourists may travel easily to destinations in the Caribbean or South-East Asia that have acquired a particular reputation for these practices, the sex tourism market has become a distinctive and locally important sector of activity (see also Oppermann (1999) and Franklin (2004)). These spaces of sex tourism are, of course, strongly gendered and whilst the sex tourists are predominantly male, this is not an exclusive tendency. Pruitt and Lafont (1995), in describing what they term as 'romance tourism', chart the practices of women who travel to destinations such as the Caribbean to pursue sexual contacts.

Embodied tourist practices

Readers will discern in the preceding discussion some recurring references to the role of the body as a nexus for a range of performative practices of tourism and the importance of embodied experiences. These are important themes that connect directly to evolving tourism practices and which, in some contexts, provide the stimuli for the development of new tourism spaces or locations. Franklin (2004), once again, is instructive, noting that rather than pursuing the detached role as an observer of Others, it is increasingly their own bodies that tourists attend to. Holidays revolve around bodily experiences – sunbathing, dancing, drinking and eating – and, as Crouch and Desforges (2003) correctly observe, part of the motivation of tourism is to immerse oneself in the sensual, bodily experiences that have previously been only visualised, perhaps whilst contemplating the alluring images of the guide book or the TV travel show. In this way, the embodied experience often becomes critical to the overall process of consumption, marking – as it often does – a culmination.

However, the significance of embodied forms of tourism is not just a product of the role that tourism plays as an important arena of consumption, it also reflects a rather lengthier (though still evolving) relationship between the body, health and nature. The antecedents of the modern preoccupation with body, health and nature lie in the responses to the Enlightenment and then the Romantic Movement that instilled new beliefs in the quintessential value of natural environments as a source of both spiritual as well as physical well-being. This encouraged the emergence of a popular outdoor movement in which activity such as walking in the open countryside was seen as an essential component in self-improvement (Walker, 1985). This 'culture of nature' (MacNaghten and Urry, 2000) continues today: in the use of natural images in marketing; in the promotion of the natural over the artificial; and in the general valorisation of the natural environment, and it is partly in this context that tourism is being reshaped. The evidence for this process lies in a wider incidence and importance of a direct (rather than simply visual) consumption of nature through embodied tourist experience or practices: in sectors such as adventure tourism; in sport-based tourism centred on activity such as horse riding, climbing, caving, surfing, or hang-gliding; or in areas of consumption such as food or wine tourism. It is to a consideration of some examples of these newer forms of tourism – and the spaces that they are creating – that we turn in the final section of this chapter.

Consumption, embodied practices and tourist space

There is a widening range of activities that illustrate how consumption, identity and embodied practices come together in tourism to shape the way that space recognised, valorised, used and understood by tourists, but for convenience the two areas of adventure and wine tourism are examined in the final section of this chapter as exemplars of wider processes of change.

Adventure tourism

Adventure tourism covers a very broad range of air-, water- and land-based activities and there is no clear consensus on its precise definition. At its core lies a focus on recreational activities that rely upon features of the natural terrain (Buckley, 2007) and it is often activity that centres on unusual, remote, exotic or wilderness destinations (Page et al., 2005). It is generally characterised by high levels of sensory stimulation involving physically challenging, experiential components (Pomfret, 2006) and in most definitions of adventure tourism, the presence of varying degrees of risk and an associated level of uncertainty about the outcome is a defining feature (Weber, 2001). Typical adventure tourism activities will include climbing, caving, trekking, white-water sports, off-road driving or cycling, sky diving, hang gliding and bungee jumping, although, as a subsequent part of the discussion will reveal, precise delimitation of the adventure tourism market is problematic.

What is clear is that adventure tourism activity is an important area of expansion in global tourism markets and a particularly strong component in the tourism product of countries such as Nepal and New Zealand which have acquired a particular distinction of centres of adventure tourism (Bentley and Page, 2001). Millington (2001) suggests that the global market for packaged forms of adventure travel is in the order of 5 million tourists per year and is particularly strong in North America where there is a lengthy tradition of adventurous forms of outdoor recreation. European markets are rather smaller with Mintel (2003c), for example, providing estimates of perhaps half a million adventure tourists visiting European destinations. However, although many sectors of adventure tourism clearly constitute niche markets, the evidence – although sometimes fragmentary – suggests that they are expanding quickly. A study of the adventure tourism sector in Scotland, for example, found that almost 80 per cent of Scottish adventure tourism companies had been formed in the last twenty years (Page et al., 2005), whilst in Nepal, the number of permits issued by the authorities for trekking in the Himalayas rose from about 40,000 in 1987 to over 300,000 in 2000 (Beedie and Hudson, 2003).

The expansion of this sector is reflective of both a shift in popular taste as well as key developments in the organisational basis of adventure tourism. Page et al. (2005) argue that the contemporary appeal of adventure tourism is in part a reflection of new levels of global engagement of people with active sports and recreations in natural settings, combined with a more familiar desire for escape from the complexity and commercialism of postmodern urban society. Beedie and Hudson (2003) underscore the importance of structural changes – particularly the proliferation of marketing structures and promotional activity around adventure tourism, the application of new technologies in adventure settings that have helped to make some adventure activities more accessible to people of lesser experience, and the development of expert systems in many adventure tourism sectors to which tourists are happy to defer control – such as trained mountain guides (Beedie, 2003).

Integral to these processes has been a progressive merger of travel, tourism and outdoor recreations. Adventure tourism is essentially founded on recreational activities that define the heart of the tourist experience and the growth of adventure tourism is shaped by a blending of adventurous forms of recreation that developed as the domains of specialist practitioners, with touristic modes of engagement (such as commodification) that make formerly exclusive practices accessible to a wider clientele. This process leads directly to new ways of representing and using activities and spaces. For example, the development of a traditional pursuit such as hill walking – when set in an exotic location – takes on a new guise as 'trekking', whilst conventional forms of cycling have metamorphosed into the adventure sport of mountain biking through the migration of the activity into off-road settings and the related technological development of the mountain bike as an adaptation of the conventional road machine (see Beedie and Hudson, 2003). Through such mechanisms, new spaces of tourism are opened by the spatial patterns of demand revealed by new activities and through the penetration, by tourists, of the terrain of the specialists (Pomfret, 2006).

An important attribute of the adventure tourism sector is, however, its diversity and this is evident in both the range of pursuits that might constitute a basis for adventure tourism and the motivations and expectations of participants. The issue of diversity has been usefully conceptualised in terms of a spectrum of activities that range across a scale from 'hard' to 'soft' forms of adventure tourism. Here the critical factors that serve to distinguish the forms of adventure tourism will include levels of risk, the technical challenge and the levels of skill and endurance that participants are required to deliver (Beedie and Hudson, 2003). Thus, 'hard' spectrum adventure tourism possesses high levels of perceived and actual risks, poses significant physical and technical challenges, and requires developed skills, whereas 'soft' spectrum activities present much lower levels of actual risk (although there may still be a perceived risk on the part of participants) and are much less demanding in terms of levels of endurance, technical challenge and skill. Typical 'hard' spectrum activities will include mountaineering, white-water rafting, hang gliding and diving, whilst 'soft' spectrum activity will embrace hiking, horse riding, cycling (but not mountain biking) and wildlife safaris as examples.

These parameters in turn help to define the nature of the market for adventure tourism in which 'soft' spectrum activities tend to exert the broadest appeal and are often structured around some of the principles of mass tourism (with packaged tours being sold worldwide to comparatively large groups of mixed ages, abilities and interests), whilst 'hard' spectrum activities exert a more selective appeal as defined within niche markets (see Buckley, 2007). Studies of the market profiles of adventure tourists generally reveal a preponderance of people aged between 35 and 50, with high levels of educational attainment and relatively large disposable incomes. Men are normally more numerous than women in most adventure tourism settings (Millington, 2001; Mintel, 2003c). However, such generalisations are not valid across all sectors, especially within certain 'hard' spectrum activities that attract a much younger clientele that exhibit higher levels of desire for thrill seeking and are often better equipped in physical terms to meet the challenge of some of these extreme sports.

This raises interesting questions relating to the motivations and rewards that tourists expect to gain from adventure experiences. Figure 11.2 develops a simple conceptual framework that attempts to map the factors that shape the processes through which adventure tourism is experienced or consumed. It proposes a circular relationship in which motivation, experience and outcomes provide primary determinants, whilst a range of factors that are labelled as 'externalities' (on the basis that they are outside the framework of the individual, embodied attributes of motivation, experience and outcome) are seen as providing a context that mediates the influence and character of the primary factors. Thus, for example, the consumption of adventure tourism might variously reflect the influence of nature and landscapes, or of promotional and advertising media, or the expectations or inducements that arise from social relations – all of which may exert both push and pull effects.

The model attempts to capture core attributes that are generally held to define adventure tourism. Hence, primary motivations are constructed normatively around the notions of challenge, risk and uncertainty that typically define 'adventure', but recognise also that individuals will commonly be seeking varying levels of novelty, escape, stimulation and, especially, a sense of contrast to their routine, daily experiences. Whilst actually engaged on the adventuresome activity, the model anticipates individuals will be susceptible to raised

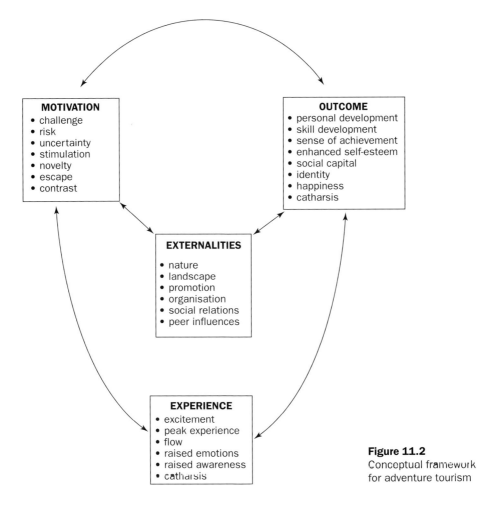

Figure 11.2
Conceptual framework
for adventure tourism

levels of mental and emotional awareness (concentration, fear, excitement, exhilaration) and may also experience what are termed 'flow' and 'peak experience'. The concept of peak experience derives from the work of Maslow (1967). It refers to the forms of elation that participants derive from complete mastery of a situation and is closely aligned with the concept of 'flow'. Csikszentmihalyi (1992: 4) explains that flow is 'the state in which people are so involved in an activity that nothing else seems to matter; the experience is so enjoyable that people will do it for the sheer sake of doing it' (cited in Pomfret, 2006). To 'go with the flow' implies total immersion in the experience, a state in which actions become almost effortless and normal awareness of time and space becomes suspended.

However the activity itself is experienced, there will be outcomes. The model posits an essentially positive set of derived outcomes, in terms of attributes such as catharsis, enhancement of self-esteem through mastery of the challenge, skill development, personal or social development, acquisition of social capital and perhaps a reaffirmation (or even a reworking) of identity. (Of course, a bad experience may trigger countervailing outcomes based around notions of failure, loss of esteem, etc.) But, critically, on the basis of the outcomes, the motivations and expectations that the individual holds with regard to adventure tourism will be modified in light of experience. As Beedie and Hudson (2003) note, as participants gain experience, so their competence increases and the risks and challenges posed by a specific activity (or activity space) diminish. Hence for adventure tourists to acquire a recurring sense of achievement through participation in areas such as extreme sports, the 'bar' needs to be raised successively, by moving to new places and/or to confronting more challenging goals.

There are two further observations that are implicit within this model and which are worth highlighting. First, adventure tourism is essentially a subjective experience. The model proposes that primary motivations lie in attributes such as exposure to challenges, risks and uncertainties, yet these are relative rather than absolute conditions that are constructed at the level of the individual. What is risky and uncertain for one person will be considered as entirely straightforward by another. There is a real sense in which most forms of tourism entail an element of adventure and for many people any encounter with a foreign 'Other' may be seen as challenging, risky or uncertain and will produce heightened levels of emotional awareness and arousal. This is partly why the range of adventure tourism settings is so broad, since for participants in many of the 'soft' spectrum activities, the sense of adventure in – say – camping out in woods or riding a horse for the first time is every bit as challenging as that which faces the mountaineer or the white-water rafter. So 'adventure tourism' cannot be simply demarcated as a discrete and distinctive area of consumption, since the concept of 'adventure' actually infuses many of the spaces of tourism and our understanding of how these spaces work as tourist destinations needs to be reminded regularly of this fact.

Second, adventure tourism is a strongly embodied experience that shapes fundamental relationships to our identities. It will be immediately evident from Figure 11.2 that most of the attributes summarised therein are embodied in their nature – primarily as mental constructs or responses. Moreover, most adventure tourism activities engage the body in direct, kinaesthetic and tactile ways, and that, in itself, is part of the attraction to many exponents. Through the earlier discussion of consumption and identity, we can see that identity is not simply a material thing but is essentially a pattern of behaviour or conduct that we choose to reveal through the performances of our various roles. This tendency is strongly evident in many forms of adventure tourism. Kane and Zink (2004) draw on Stebbins' (1982) concept of serious leisure to make the point. The essence of this argument is that practitioners of serious leisure reveal a level of commitment to a recreational activity that is sufficiently strong as to define their sense of belonging to, and acceptance within, the ethos of a defined culture and which becomes a primary signifier of the person's identity and their stratification in their leisure worlds. For many exponents of adventure tourism, especially towards the 'hard' end of the spectrum, the same characteristics clearly apply. Bourdieu (1984) argues that people gravitate towards the social fields that offer the greatest potential for acquiring cultural capital. Adventure tourism often entails significant investment (in skills, expertise, effort and courage) and for many adventure tourists this investment translates directly into raised levels of cultural capital that assures both status and identity.

Wine tourism

The second example of how consumption and embodied practice helps to construct tourist space relates to the expanding area of wine tourism. Recent work in cultural studies has acknowledged how practices of consumption of food and drink have become fundamental in shaping how we view ourselves as individuals, how we mark our social positions in relation to others, and in so doing, position ourselves within our world. Bell and Valentine (1997: 3) write that 'for most inhabitants of (post)modern Western societies, food has long ceased to be merely sustenance and nutrition. It is packed with social, cultural and symbolic meanings'. Food has become 'woven into the construction of lifestyle and used as a marker of social position' and 'in a world in which self-identity and place-identity are woven through webs of consumption, what we eat (and where and why) signals who we are'.

In many situations, the consumption of food and drink also shapes some of our understanding of space. Regions, for example, are a product of a combination of physical processes (that define natural landscapes) with cultural processes (that shape how land is occupied and used). From this combination of nature and society, powerful place identities may often be derived, some of which may be expressed through associations with particular types of food and drink that constitute regional specialities. As Bell and Valentine (1997) observe, the concept of wine regions (such as Bordeaux, Burgundy or Champagne) provides excellent examples of how the uniqueness of a region may be asserted. Wine is made in particular places where favourable environmental conditions and traditional processes of production have created a distinctive product that is directly associated with the region in question. However, the regional label not only assures the product its identity, it also identifies the spaces of production that for the wine tourist then become defined spaces of tourism (see Figure 11.3).

Tourism holds an ambiguous relationship with food and drink. On the one hand, the opportunity to experience foreign food and drink is a positive attraction for many tourists and the use of cuisine and, in particular, national and regional culinary specialisms as promotional devices is widely practised in destination marketing. But on the other hand, as Cohen and Avieli (2004) reveal, actual practices around the consumption of local food and drink in foreign destinations may be beset with a host of anxieties and insecurities relating to the conditions under which food and drink is prepared, served and consumed, its palatability, and the extent to which local dishes are culturally acceptable to the visitor.

However, because of the regulated patterns of production of wine (e.g., through the standardised practices of matching named grapes to particular styles of wine and especially the systems of appellation that are used as a quality control process), issues of uncertainty and risk in the consumption of wine by tourists is generally removed. Moreover, the development of wine tourism draws directly on a number of positive synergies between tourism and the consumption of wine. Bruwer (2003: 423) comments that, by its very nature, wine production directly complements tourism. 'Wine is a beverage that is associated with relaxation, communing with others, complementary to food consumption, learning about new things and hospitality'. In many contexts wine tourism is organised around wine routes that directly capture touristic notions of travel and discovery, and is generally located in aesthetically pleasing rural settings where the growing requirements of most grapes normally ensures that visitors will be able to enjoy congenial climates (Carmichael, 2005).

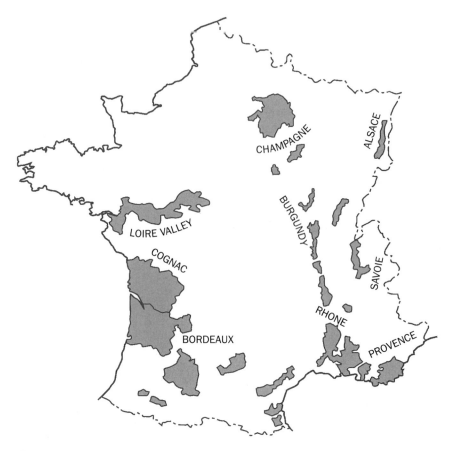

Figure 11.3 Primary wine regions of France

There are several definitions of wine tourism but the one that is perhaps most widely cited is provided by Hall and Macionis (1998: 197) and which describes the activity as the 'visitation to vineyards, wineries, wine festivals and wine shows, for which grape wine tasting and/or experiencing the attributes of the grape wine region are the prime motivating factors for visitors'. This definition is useful because it highlights several of the complexities that have been identified in the wine tourism experience. Getz and Brown (2006), for example, have suggested that from the tourist perspective the wine tourism 'product' comprises three primary elements:

- The wine product – which relates to the actual wines that are tasted or purchased, the wineries and their staff, and special wine events or festivals.
- The destination appeal – which relates to aspects such as climate, scenery, accommodation and information.
- The cultural product – which might comprise the presence of traditional wine villages and their associated societies and cultures, or good local restaurants.

Carmichael (2005), drawing on Jansen-Verbeke's (1986) ideas of activity places and leisure settings, has developed a similar model of wine tourism that proposes that the activity is a blending of specific experiences associated with the visiting of vineyards and wineries (the activity place) with attributes of the wider tourist environment (the regional setting) that enhance and support the core activity (see Figure 11.4).

However, in practice, these key elements will combine in differing ways and the emphasis that different types of tourist will place on these attributes will vary according to levels of interest and essential motivations. Studies of wine tourists and their motivations (see, e.g., Carmichael, 2005; Charters and Ali-Knight, 2002; Getz and Brown, 2006; Sparks, 2007) reveal a segmented market. At one end of the spectrum are located highly motivated and dedicated wine enthusiasts who are using their time as wine tourists to develop further their appreciation and understanding of wine and to augment their private 'cellars' with additional purchases. In contrast, at the other end of the scale are located general tourists who are simply curious or are looking for something interesting to fill part of their time as visitors or sightseers in the wider region. However, as work by Carmichael (2005) and Getz and Brown (2006) reveals, even the dedicated wine enthusiast will commonly combine these wider elements of general tourism with their specific interest in wine (albeit as secondary elements), so that the spaces of wine tourism also need to connect to other tourist space for the tourist product to become fully developed.

With regard to the tourists themselves, studies of visitors to wine tourism locations reveal a market that is dominated by well-educated, affluent professional groups – often aged

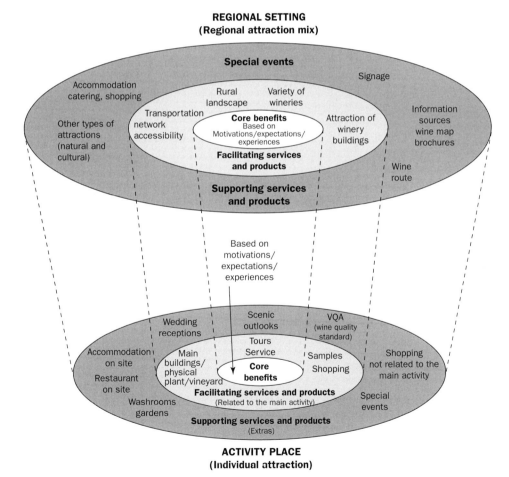

Figure 11.4 Structure of the wine tourism 'product'

between 35 and 50 – people who are well versed (if only subconsciously) in the use of patterns of consumption as a means to achieving social position and status. For such people the time spent acquiring demonstrable appreciation, knowledge and expertise in wine, and in communing with others of like-mind, becomes time well-spent in developing their self-image and their positioning relative to others. So as with so many areas of tourist practice, wine tourism, for the specialist practitioners at least, is an embodied form of consumption (in both a literal and a figurative sense) that contributes to identity.

Summary

Adventure and wine tourism are expanding areas of contemporary tourism that illustrate well both the specific themes of this chapter and several of the broader themes of the book as a whole. Both examples tell us something about the shifting nature of consumption in tourism and, in particular, of trends that are taking some sectors of tourism away from passive and largely visual consumption of places through a general recourse to rest and relaxation, towards much more active, embodied forms of engagement that bring rewards that are perhaps more complex and more diverse than traditional models of tourism allow. Through these newer styles of activity we are able to use tourism practice – and particularly the performance of tourism – to make stronger statements about how we see ourselves and how consumption through tourism – and the identities that we fashion through tourist practice – help to position ourselves within the societies that we occupy. At the same time, adventure and wine tourism are strongly reflective of important wider processes: the commodification and democratisation of experience and, especially, the de-differentiation (or blending) of tourism – not just with leisure and recreation but, more significantly, also with the day-to-day lifestyles that people maintain. Lastly – and perhaps most pertinently from a purely geographical perspective – the development of adventure and wine tourism show us something of how processes of consumption help to map new spaces of tourism and to structure the spatial practices through which those spaces may be understood.

For human geographers, a central theme within the discipline is interpreting and understanding our changing world – a world in which geographic patterns are constantly being reworked by powerful forces of change: population shifts; new patterns of economic production and consumption; evolving social and political structures; new forms of urbanism; and globalisation and the compressions of time and space that are the product of the ongoing revolutions in information technology and telecommunications. This book has attempted to show how tourism has also come to be a major force for change – an integral and indispensable part of the places in which we live, their economies and their societies. When scarcely a corner of the globe remains untouched by the influence of tourism, this is a phenomenon that we can no longer ignore.

Discussion questions

1 To what extent would you concur with the view that tourism provides a quintessential form of postmodern consumption?
2 Why has tourism become central to the contemporary construction of self-identity across major areas of Western society?
3 Using examples from tourism, explain how identity is both exclusive and inclusive.
4 Given recent emphases on tourism as embodied practice, how valuable is Urry's concept of the tourist gaze in understanding tourist practice and the structuring of tourist space?

5 Deploying examples from adventure and/or wine tourism, consider how the increased popularity of these activities has contributed to the extension of tourist space.

Further reading

The literature on consumption is voluminous, but a useful overview is provided by:
Miller, D. (ed.) (1995) *Acknowledging Consumption: A Review of New Studies*, London: Routledge.

A good, concise, recent discussion is given by:
Paterson, M. (2006) *Consumption and Everyday Life*, Abingdon: Routledge.

Other sources that deal with more specific aspects of consumption include:
Bell, D. and Valentine, G. (1997) *Consuming Geographies: We Are Where We Eat*, London: Routledge.
Crang, P. (2005) 'Consumption and its geographies', in Daniels. P. et al. (eds) *An Introduction to Human Geography: Issues for the 21st Century*, Harlow: Prentice Hall, pp. 359–78.
Jayne, M. (2006) *Cities and Consumption*, Abingdon: Routledge.

Identity in the modern world is explored in detail in:
Giddens, A. (1991) *Modernity and Self-identity: Self and Society in the Late Modern Age*, Cambridge: Polity.
Hall, S. and du Gay, P. (1996) *Questions of Cultural Identity*, London: Sage.

Interesting and instructive essays on embodied tourist practice are to be found in:
Crouch, D. and Desforges, L. (2003) 'The sensuous in the tourist encounter: the power of the body in tourist studies', *Tourist Studies*, Vol. 3 (1): 5–22.
Edensor, T. (2000a) 'Walking in the British countryside: reflexivity, embodied practices and ways of escape', *Body and Society*, Vol. 6 (3/4): 81–106.
Franklin, A. (2004) *Tourism: An Introduction*, London: Sage.
Veijola, S. and Jokinen, E. (1994) 'The body in tourism', *Theory, Culture and Society*, Vol. 11 (3): 125–51.

Appendix: a guide to the use of the Internet in tourism geography

Over the period since the first edition of this book was prepared and published, the Internet has emerged as a major source of information and is widely used by students at all levels, but especially by undergraduates, to research information for projects and assignments. It is appropriate, therefore, to offer some guidance to student readers on how the Internet may be used in studying tourism geography. The guidance offered here focuses on how to find reliable information, rather than simply providing a listing of useful website addresses. This is because websites, particularly those belonging to smaller organisations, are prone to frequent changes of address (their URLs) and over the duration that a book is in print, many addresses that might be listed are likely to change. Some sites belonging to global organisations that maintain permanent URLs are identified below as possible starting points for an Internet search for tourism information, but the purpose of this section is really aimed at helping students to understand how to search the Web and to compile personalised lists of sites of their own that suit their specific needs.

First, some words of caution. The Web is an alluring and attractive environment in which to work but it is very important to remember that it is an unregulated environment. Anyone can place information onto a website and unlike – say – peer-reviewed academic papers published in journals, there is no guarantee that information on the Web is necessarily accurate, valid or credible. Even sites that possess an outward appearance of authority – for example Wikipedia – offer no assurance that the information contained therein is reliable. Much of it probably is, but this cannot be assumed, and for academic programmes of work, such websites are best avoided.

The most reliable sites are probably those belonging to governmental organisations and/or associations working at a global scale or across major regions. Such sites are often useful for acquiring information, especially data, on background trends in tourism or other components that perhaps relate to tourism. Examples include:

- The United Nations (www.un.org) which offers a range of online data sets, some of which are specific to tourism, others of which relate to key global indicators or to information on sectors such as population, trade, industrial outputs, food, agriculture and health.
- The European Union (www.europa.eu) provides summaries of EU policies and programmes together with supporting statistical information. At the time of writing there are no specific links to tourism information, but information on related themes such as culture, environment and transport is provided.
- The World Tourism Organization (www.world-tourism.org) offers basic data and commentaries on key indicators such as the growth of international tourism and key market trends. These are often summarised in online publications, excerpts from which

are free of charge (including *World Tourism Barometer* and *Tourism Highlights*), although main reports have to be purchased and are expensive.

● The World Travel and Tourism Council (www.wttc.org) is a business-facing organisation that provides online data and commentary on tourism, especially from an economic perspective.

Below the level of these large-scale organisations, many individual governments will maintain websites where information, statistics and sometimes online reports can be accessed free of charge. In the UK, for example, the government body responsible for the English countryside – Natural England – maintains a website through which hundreds of reports on issues that relate to tourism (including coasts and the sea, designated areas, leisure, countryside planning and management) may be read online or downloaded (www.naturalengland.org.uk/). National level tourism organisations routinely maintain websites, although these tend to be oriented towards marketing and promotional functions – selling destinations to potential tourists – rather than providing the level of factual information that students of tourism geography might require. But occasionally factual information and reports can be accessed so these sites are always worth locating.

Governmental departments, regulatory bodies and national tourist boards represent an obvious starting point for any Internet-based enquiry, but beyond this level exists a myriad of professional bodies, trade associations, commercial and voluntary organisations, all of which have at least a potential to inform our understanding of tourism geography. Locating these types of sources requires effective use of search engines.

There are several Internet search engines that might be used, of which the market leader is probably Google. Google is remarkably quick and efficient but it is not, of course, particularly intuitive. The effectiveness of any search is therefore highly dependent upon the key words that are used to structure a search. It is important to be specific without overloading the search with too many key words. This is more likely to produce manageable and relevant results. So, for example, if you are seeking factual information on tourism in China, search using 'China', 'tourism' and 'statistics' as key words, rather than 'Chinese tourism' which is less specific and therefore more likely to locate sites that do not meet your requirements. It also pays to think about the terminology that you enter into a search engine. For example, you may be after data on tourism, but entering 'data' as a search term might well give you different, and possible inferior results, than if you enter 'statistics'. This is because the types of organisation that maintain numerical information routinely describe that information as 'statistics' rather than 'data'. It is always worth trying alternative key words to structure a search, especially when initial searches prove unproductive.

Most searches using engines such as Google will probably reveal thousands of possible sites but, in general, the most relevant sites will tend to appear on the first two or three pages of the listing and it is rare to find useful links further down the list. There is, of course, a serendipity to using the Web and one can never be certain that a particular search has been exhausted, but as a general rule it is better to alter the search parameters and search again, than to work through sites that are likely to be increasingly removed from what you want. The Bookmark function should be used to store links to sites that are useful and which you may wish to revisit but, to reiterate the caution given above, wherever possible try to confine the storage of sites through Bookmark to sites that are trustworthy.

Searches on engines such as Google will provide links to thousands of sites on which factual information may be located and in many situations sites will also provide reports and commentaries that may be downloaded, alongside facts and figures. However, when the requirement is to locate academic work relating to tourism, Google Scholar may provide a better search facility This is a particularly useful device when used in conjunction with online journal systems such as Science Direct (which most colleges and universities now provide) to access references that search engines such as Google Scholar can help to locate. There is a wealth of information on the Internet to support the study of tourism geography, but it is a resource that needs to be used carefully, with thought, and with due regard for the veracity of the material that it offers.

Glossary

In order to assist readers whose background may not be within the social sciences, some further definitions or elaborations of terms that are used in the text but which may not be generally understood are set out below.

Boosterism a development approach that assumes that development is an intrinsically beneficial process, in which potential negative impacts are often down-played and resources are routinely regarded as objects to be exploited as part of the development.

Brownfield land a description generally applied to land that was formerly used for industrial purposes that has become redundant and which is now available for redevelopment for new use.

Commodification commodities are objects produced for the purpose of being exchanged (i.e., traded) and commodification is the process in which both tangible (e.g., physical goods or services) and intangible (e.g., experiences) elements are combined to produce a product or commodity that may be sold. For example, an East African safari holiday may *commodify* the viewing of wildlife as a packaged tour.

Cultural capital a term that originates in the work of the French social theorist Bourdieu and which refers to the acquisition of social status or position through the adoption of cultural practices that reveal personal (or collective) taste and/or judgement that may distinguish the individual or group as belonging to a particular social class.

De-differentiation a term coined by the sociologists Lash and Urry to describe the breakdown of distinctive spheres of activity (such as work and leisure) or spheres of engagement (such as real and imagined) to produce new patterns in which such distinctions are no longer valid.

Existentialism a philosophy that emphasises the existence of the individual and the capacity of the individual to develop a sense of themselves through processes of self-determination and acts of free will. Many aspects of tourism are held to have an existential dimension through the apparent capacity of people to discover their true selves through tourist practice.

Footloose a descriptive term that is applied to productive processes that can locate virtually anywhere because they have no specific locational ties to either raw materials or markets.

Fordism/Fordist a set of organisational practices that originated in the motor industry under the guidance of Henry Ford and which was based upon principles of mass production of standardised products, generally at low cost. The term has become widely applied to other contexts in which the same principles apply, such as that in tourism, for example, many aspects of mass forms of travel are often described as working on Fordist principles.

Globalisation a process of transformation (in social, cultural, political, economic and even (now) environmental relations) to organisation on a global – rather than national, regional or local scale. Globalisation is also commonly linked to process of increased standardisation in many areas of production and consumption.

Habitus in this text the term is used in the sense in which it has been applied by Bourdieu to refer to the preferred sets of actions, behaviours and interactions by which social groups position themselves within wider social contexts. In this sense the term might be interpreted as a preferred mode of living.

Hegemony this concept originates in the work of Gramsci and refers to the capacity of dominant groups to exercise control over others (by means other than direct control or force), a relationship that is implicitly reinforced by the willingness of the latter to concede power and status to the former and to accept the ideologies and values of the dominant group. In tourism, some aspects of the relationship between tourists and host communities are sometimes conceived in these terms.

Hyper-reality a notion developed by Eco to describe situations in which the distinction between real things and imitations are blurred and/or in which the representation becomes more real than the original on which it is based. In tourism this rather elusive concept is perhaps best represented in themed spaces. In Disneyland, for example, cartoon characters that in their original guise are represented only in artwork, become 'live' figures that stroll through the park.

Market segmentation a process that seeks to differentiate sub-sectors within an overall group of purchasers and develop bespoke products that will appeal particularly to these 'niche' markets.

McDonaldisation a thesis developed by Ritzer to describe the process in which the principles of standardised, low-cost, uniform production that characterise the fast-food industry are extended to other sectors of life. Key attributes of McDonald fast-food systems include efficiency, predictability and control over both production and consumption of the product, attributes that Ritzer proposes are now widely encountered in many fields, including areas such as mass tourism.

Post-Fordism/Post-Fordist a system of productive practices and associated social and cultural systems that are shaped around flexible production that is geared to match products to different market demands. Post-Fordist production often deploys small-scale production and differentiation of products in ways that contrast directly with Fordist principles of mass, standardised production and consumption.

Postmodernism a complex term that has several applications. Originating in the field of architecture, the term originally referred to new styles of building design: often small in scale, eclectic in the mixing of styles, reflective of local traditions and frequently playful. More significantly, perhaps, the term has also been applied to methods of enquiry with a particular emphasis upon developing multiple and alternative readings of the same phenomena. Postmodern critical methods recognise that knowledge is made from differing viewpoints and what are termed 'meta narratives' – that is, supposedly universal explanations (such as Marxism) – are explicitly rejected in postmodern analysis in favour of a plurality of interpretations.

Post-tourists a label applied to participants in what Poon has described as 'new' tourism, which is characterised by attributes such as enhanced levels of choice; wider development of specialist and niche markets; greater emphasis upon individual (and reflexive) engagement with real, natural or authentic experiences; and widespread resistance to standardised forms of mass tourism.

Praxis an habitual action that becomes accepted as custom and practice.

Reductionism an explanatory approach that reduces complex phenomena to their basic constituents as a means of enabling understanding. The obvious weakness of reductionist approaches is the risk of discarding components and/or information that are actually important in the explanation that is sought.

Reflexivity this refers to processes whereby people systematically and critically examine and reflect upon things such as beliefs, values, behaviours and practices in light of changing levels of knowledge, information and experience. Many areas of tourism are subject to reflexive responses by tourists; for example, in tourism marketing and place promotion, the impressions that visitors form of a destination may differ significantly from those that are intended by marketing companies, because of the capacity of people to react reflexively (and differently) to the way that is intended.

Regulation theory this is an approach to understanding the workings of capitalist systems. It proposes that there is a dominant set of principles around which the system is organised and which regulate the system to ensure its continuance. These systems are generally centred upon state or institutional forms of control or governance over matters such as wages or monetary exchange.

Simulacra objects that possess the form or appearance of certain things, without possessing their substance or true qualities. General examples of simulacra will include religious icons; whilst in tourism, examples would include many of the resort hotels of Las Vegas or the latest generation of themcd malls that purport to represent other times or places in settings that are literally staged.

⬤ Bibliography

Agarwal, S. (1994) 'The resort cycle revisited: implications for resorts', in Cooper, C.P. and Lockwood, A. (eds) (1994): 194–208.

—— (1997) 'The resort cycle and seaside tourism: an assessment of its applicability and validity', *Tourism Management*, Vol. 18 (1): 65–73.

—— (2002) 'Restructuring seaside tourism: the resort life cycle', *Annals of Tourism Research*, Vol. 29 (1): 25–55.

Agarwal, S. and Brunt, P. (2006) 'Social exclusion and English seaside resorts', *Tourism Management*, Vol. 27 (4): 654–70.

Agnew, J. (1987) *Place and Politics: the Geographical Mediation of State and Society*, London: George Allen and Unwin.

Aitchison, C. (2001) 'Theorizing Other discourses of tourism, gender and culture: can the subaltern speak (in tourism)?', *Tourist Studies*, Vol. 1 (2): 133–47.

Aitchison, C., Macleod, N. and Shaw, S. (2001) *Leisure and Tourism Landscapes: Social and Cultural Geographies*, London: Routledge.

Albert-Pinole, I. (1993) 'Tourism in Spain', in Pompl, W. and Lavery, P. (eds) (1993): 242–61.

Alipour, H. (1996) 'Tourism development within planning paradigms: the case of Turkey', *Tourism Management*, Vol. 17 (5): 367–77.

Ambrose, T. (2002) 'Cultural tourism opportunities in Morocco', *Locum Destination Review*, Winter: 26–8.

Andrews, M. (1989) *The Search for the Picturesque: Landscape Aesthetics and Tourism in Britain, 1760–1800*, Aldershot: Scolar Press.

Arana, J.E. and Leon, C.J. (2008) 'The impact of terrorism on tourism demand', *Annals of Tourism Research*, Vol. 35 (2): 299–315.

Archer, B.H. (1973) *The Impact of Domestic Tourism*, Bangor Occasional Papers in Economics No. 2, Cardiff: University of Wales Press.

—— (1977) *Tourism Multipliers: the State of the Art*, Bangor Occasional Papers in Economics No. 11, Cardiff: University of Wales Press.

—— (1982) 'The value of multipliers and their policy implications', *Tourism Management*, Vol. 3 (4): 236–41.

—— (1989) 'Tourism and island economies: impact analyses', in Cooper, C.P. (ed.) (1989): 125–34.

—— (1995) 'The importance of tourism for the economy of Bermuda', *Annals of Tourism Research*, Vol. 22 (4): 918–30.

Ashley, C. (2000) *The Impact of Tourism on Rural Livelihoods*, Working Paper No. 128, London: Overseas Development Institute.

Ashworth, G.J. and Voogd, H. (1994) 'Marketing and place promotion', in Gold, J.R. and Ward, S.V. (eds): 39–52.

Ashworth, G.J. and Dietvorst, A.G.J. (eds) (1995) *Tourism and Spatial Transformations*, Wallingford: CAB International.

Ashworth, G.J. and Tunbridge, J.E. (2004) 'Whose tourist-historic city? Localizing the Global and Globalizing the Local', in Lew, A.A. et al. (eds) (2004): 210–22.

Ateljevic, I. (2000) 'Circuits of tourism: stepping beyond the "production/consumption" dichotomy', *Tourism Geographies*, Vol. 2 (4): 369–88.

Ateljevic, I. and Doorne, S. (2003) 'Culture, economy and tourism commodities: social relations of production and consumption', *Tourist Studies*, Vol. 3 (2): 123–41.

Atkinson, J. (1984) *Flexibility, Uncertainty and Manpower Management*, Institute of Manpower Studies Report No. 89, Falmer: University of Sussex.

Bachvarov, M. (1997) 'The end of the model? Tourism in post-Communist Bulgaria', *Tourism Management*, Vol. 18 (1): 43–50.

Baidal, J.A.I. (2003) 'Regional development policies: an assessment of their evolution and effects on the Spanish tourist model', *Tourism Management*, Vol. 24 (6): 655–63.

—— (2004) 'Tourism planning in Spain: evolution and perspectives', *Annals of Tourism Research*, Vol. 31 (2): 313–33.

Balaz, V. (1995) 'Five years of economic transition in Slovak tourism: successes and shortcomings', *Tourism Management*, Vol. 16 (2): 143–59.

Ballesteros, E.R. and Ramirez, M.H. (2007) 'Identity and community: reflections on the development of mining heritage tourism in southern Spain', *Tourism Management*, Vol. 28 (3): 677–87.

Barke, M., Towner, J. and Newton, M.T. (eds) (1996) *Tourism in Spain: Critical Issues*, Wallingford: CAB International.

Baudrillard, J. (1988) *America*, London: Verso.

Baum, T. (1994) 'The development and implementation of national tourism policies', *Tourism Management*, Vol. 15 (3): 185–92.

—— (1996) 'Unskilled work in the hospitality industry: myth or reality?', *International Journal of Hospitality Management*, Vol. 15 (3): 207–9.

Bauman, Z. (1996) 'From pilgrim to tourist – a short history of identity', in Hall, S. and du Gay, P. (eds) (1996): 18–36.

Beard, J. and Ragheb, M.G. (1983) 'Measuring leisure motivation', *Journal of Leisure Research*, Vol. 15 (3): 219–28.

Becheri, E. (1991) 'Rimini and Co.: the end of a legend?', *Tourism Management*, Vol. 12 (3): 229–35.

Becken, S. (2002) 'Analysing international tourist flows to estimate energy use associated with air travel', *Journal of Sustainable Tourism*, Vol. 10 (2): 114–31.

Beedie, P. (2003) 'Mountain guiding and adventure tourism: reflections on the choreography of experience', *Leisure Studies*, Vol. 22 (2): 147–67.

Beedie, P. and Hudson, S. (2003) 'Emergence of mountain-based adventure tourism', *Annals of Tourism Research*, Vol. 30 (3): 625–43.

Beioley, S. (1999) 'Short and sweet – the UK short-break market', *Insights*, Vol. 10 (B): 63–78.

Bell, D. and Valentine, G. (1997) *Consuming Geographies: We Are Where We Eat*, London: Routledge.

Bell, D. and Holliday, R. (2000) 'Naked as nature intended', *Body and Society*, Vol. 6 (3–4): 127–40.

Bentley, T.A. and Page, S.J. (2001) Scoping the nature and extent of adventure tourism accidents', *Annals of Tourism Research*, Vol. 28 (3): 705–26.

Blank, U. (1996) 'Tourism in United States cities', in Law, C.M. (ed.) (1996): 206–32.

Blundell, V. (1993) 'Aboriginal empowerment and souvenir trade in Canada', *Annals of Tourism Research*, Vol. 20 (1): 64–87.

Boniface, B.G. and Cooper, C.P. (1987) *The Geography of Travel and Tourism*, Oxford: Butterworth-Heinemann.

Boniface, P. and Fowler, P.J. (1993) *Heritage and Tourism in the 'Global Village'*, London: Routledge.

Boorstin, D.J. (1961) *The Image: A Guide to Pseudo-Events in America*, New York: Harper and Row.

Borsay, P. (1989) *The English Urban Renaissance: Culture and Society in the English Provincial Town, 1660–1770*, Oxford: Clarendon.

Bourdieu, P. (1984) *Distinctions: A Social Critique of the Judgement of Taste*, London: Routledge and Kegan Paul.

Bradley, A., Hall, T. and Harrison, M. (2002) 'Selling cities: promoting new images for meetings tourism', *Cities*, Vol. 19 (1): 61–70.

Bramwell, B. (2004) 'Partnership, participation and social science research in tourism planning', in Lew, A.A. et al. (eds) (2004): 541–54.

Bramwell, B. and Sharman, A. (1999) 'Collaboration in local tourism policy making', *Annals of Tourism Research*, Vol. 26 (2): 392–415.
—— (2000) 'Approaches to sustainable tourism planning and community participation: the case of the Hope Valley', in Hall, D. and Richards, G. (eds) (2000): 17–35.
Brennan, F. and Allen, G. (2001) 'Community-based ecotourism, social exclusion and the changing political economy of KwaZulu-Natal, South Africa', in Harrison, D. (ed.) (2001): 203–21.
Bristol City Council (2005) *Bristol Museum and Art Gallery Visitor Satisfaction Survey 2004*, Bristol: Bristol City Council.
British Travel Association (BTA) (1969) *Patterns of British Holiday Making, 1951–68*, London: British Travel Association.
British Tourist Authority (BTA) (1995) *Digest of Tourist Statistics No. 18*, London: British Tourist Authority.
—— (2001) *Digest of Tourist Statistics No. 24*, London: British Tourist Authority.
—— (2003) *Visits to Visitor Attractions 2002*, London: British Tourist Authority.
Britton, S. (1982) 'The political economy of tourism in the Third World', *Annals of Tourism Research*, Vol. 9 (3): 331–58.
—— (1989) 'Tourism, dependency and development: a mode of analysis', in Singh, T.V. et al. (eds) (1989): 93–116.
—— (1991) 'Tourism, capital and place: towards a critical geography', *Environment and Planning D: Society and Space*, Vol. 9 (4): 451–78.
Bruner, E.M. (1991) 'Transformations of the self in tourism', *Annals of Tourism Research*, Vol. 18 (2): 238–50.
—— (1994) 'Abraham Lincoln as authentic reproduction: a critique of postmodernism', *American Anthropologist*, Vol. 96: 397–415.
Brunt, P. and Courtney, P. (1999) 'Host perceptions of socio-cultural impacts', *Annals of Tourism Research*, Vol. 26 (3): 493–515.
Bruwer, J. (2003) 'South African wine routes: some perspectives on the wine tourism industry's structural dimensions and wine tourism product', *Tourism Management*, Vol. 24 (4): 423–35.
Bryman, A. (1995) *Disney and His Worlds*, London: Routledge.
Buckingham, D. (2001) 'United Kingdom: Disney dialectics – debating the politics of children's media culture', in Wasko, J. et al. (eds) (2001): 269–96.
Buckley, R. (1999) 'An ecological perspective on carrying capacity', *Annals of Tourism Research*, Vol. 26 (3): 705–8.
—— (2007) 'Adventure tourism products: price, duration, size, skill, remoteness', *Tourism Management*, Vol. 28 (6): 1428–33.
Buhalis, D. (1998) 'Strategic use of information technologies in the tourism industry', *Tourism Management*, Vol. 19 (5): 409–21.
Bull, P. (1997) 'Tourism in London: policy changes and planning problems', *Regional Studies*, Vol. 31 (1): 82–6.
Bunce, M. (1994) *The Countryside Ideal: Anglo-American Images of Landscape*, London: Routledge.
Burns, P. (1999) 'Paradoxes in planning: tourism elitism or brutalism?', *Annals of Tourism Research*, Vol. 26 (2): 329–48.
—— (2004) 'Tourism planning: a third way', *Annals of Tourism Research*, Vol. 31 (1): 24–43.
Burns, P. and Sancho, M.M. (2003) 'Local perceptions of tourism planning: the case of Cuellar, Spain', *Tourism Management*, Vol. 24 (3): 331–9.
Burtenshaw, D., Bateman, M. and Ashworth, G.J. (1991) *The City in Western Europe*, London: David Fulton.
Burton, R. (1991) *Travel Geography*, London: Pitman.
—— (1994) 'Geographical patterns of tourism in Europe', in Cooper, C.P. and Lockwood, A. (eds) (1994): 3–25.
Busby, G. and Rendle, S. (2000) 'The transition from tourism on farms to farm tourism', *Tourism Management,* Vol. 21 (6): 635–42.
Business Tourism Partnership (2003) *Business Tourism Briefing: An Overview of the UK's Business Tourism Industry*, London: BTP.

—— (2005) *Business Tourism Leads the Way*, London: BTP.

Butler, J. (1993) *Bodies That Matter: On the Discursive Limits of 'Sex'*, London: Routledge.

Butler, R. (1980) 'The concept of a tourist area cycle of evolution: implications for management of resources', *The Canadian Geographer*, Vol. 24 (1): 5–12.

—— (1990) 'The influence of the media in shaping international tourism patterns', *Tourism Recreation Research*, Vol. 15 (2): 46–53.

—— (1991) 'Tourism, environment and sustainable development', *Environmental Conservation*, Vol. 18 (3): 201–9.

—— (1994) 'Alternative tourism: the thin end of the wedge', in Smith, V.L. and Eadington, W.R. (eds) (1994): 31–46.

—— (2004) 'Geographical research on tourism, recreation and leisure: origins, eras and directions', *Tourism Geographies*, Vol. 6 (2): 143–62.

Butler, R., Hall, C.M. and Jenkins, J.M. (eds) (1999) *Tourism and Recreation in Rural Areas*, Chichester: John Wiley.

Byrne, E. and McQuillan, M. (1999) *Deconstructing Disney*, London: Pluto.

Campbell, C. (1995) 'The sociology of consumption', in Miller, D. (ed.) (1995): 96–124.

Carmichael, B.A. (2005) 'Understanding the wine tourism experience for winery visitors in the Niagara region, Ontario, Canada', *Tourism Geographies*, Vol. 7 (2): 185–204.

Carr, N. (2002) 'The tourism-leisure behavioural continuum', *Annals of Tourism Research*, Vol. 29 (4): 972–86.

Castells, M. (1996) *The Rise of the Network Society*, Oxford: Blackwell.

—— (1997) *The Power of Identity*, Oxford: Blackwell.

Castree, N. (2003) 'Place: connections and boundaries in an interdependent world', in Holloway, S.L. et al. (eds) (2003): 165–85.

Charters, S. and Ali-Knight, J. (2002) 'Who is the wine tourist?', *Tourism Management*, Vol. 23 (3): 311–19.

Chatterton, P. and Hollands, R. (2003) *Urban Nightscapes: Youth Culture, Pleasure Spaces and Corporate Power*, London: Routledge.

Chen, J.S. (1998) 'Travel motivations of heritage tourism', *Tourism Analysis*, Vol. 2 (3/4): 213–15.

Cheong, S. and Miller, M.L. (2000) 'Power and tourism: a Foucauldian observation', *Annals of Tourism Research*, Vol. 27 (2): 371–90.

Chhabra, D., Healy, R. and Sills, E. (2003) 'Staged authenticity and heritage tourism', *Annals of Tourism Research*, Vol. 30 (3): 702–19.

Child, D. (2000) 'The emergence of "no-frills" airlines in Europe: an example of successful marketing strategy', *Travel and Tourism Analyst*, No. 1: 87–121.

China National Tourism Office (CNTO) (2006) *China Tourism Statistics*. Accessed via www.cnto.org

Choy, D. (1995) 'The quality of tourism employment', *Tourism Management*, Vol. 16 (2): 129–37.

Church, A. (2004) 'Local and regional tourism policy and power', in Lew, A.A. et al. (eds) (2004): 555–68.

Church, A. and Frost, M. (2004) 'Tourism, the global city and the labour market in London', *Tourism Geographies*, Vol. 6 (2): 208–28.

Clammer, J. (1992) 'Aesthetics of the self: shopping and social being in contemporary Japan', in Shields, R. (ed.) (1992): 195–215.

Clarke, J. (1997) 'A framework of approaches to sustainable development', *Journal of Sustainable Tourism*, Vol. 5 (3): 224–33.

Cobbett, W. (1830) *Rural Rides*, London: Penguin. (This edition published in 1967.)

Coccossis, H. (1996) 'Tourism and sustainability: perspectives and implications', in Priestley, G. et al. (eds) (1996): 1–21.

Cohen, E. (1972) 'Towards a sociology of international tourism', *Social Research*, Vol. 39: 164–82.

—— (1979) 'A phenomenology of tourist experiences', *Sociology*, Vol. 13: 179–201.

—— (1988) 'Authenticity and commodification in tourism', *Annals of Tourism Research*, Vol. 15 (2): 371–86.

—— (1993) 'Open-ended prostitution as a skilful game of luck: opportunity, risk and security among tourist-oriented prostitutes in a Bangkok *soi*', in Hitchcock et al. (eds) (1993): 155–78.

Cohen, E. and Cooper, R.L. (1986) 'Language and tourism', *Annals of Tourism Research*, Vol. 13 (4): 533–63.

Cohen, E. and Avieli, N. (2004) 'Food in tourism: attraction and impediment', *Annals of Tourism Research*, Vol. 31 (4): 755–78.

Coles, T. (2004) 'Tourism and leisure; reading geographies, producing knowledges', *Tourism Geographies*, Vol. 6 (2): 135–42.

Compton, J.L. (1979) 'Motivations for pleasure vacation', *Annals of Tourism Research*, Vol. 6 (4): 408–24.

Cooper, C.P. (ed.) (1989) *Progress in Tourism, Recreation and Hospitality Management Vol. 1*, London: Belhaven.

—— (ed.) (1990a) *Progress in Tourism, Recreation and Hospitality Management Vol. 2*, London: Belhaven.

—— (1990b) 'Resorts in decline: the management response', *Tourism Management*, Vol. 11 (1): 63–7.

—— (ed.) (1991) *Progress in Tourism, Recreation and Hospitality Management Vol. 3*, London: Belhaven.

—— (1995) 'Strategic planning for sustainable tourism: the case of offshore islands of the UK', *Journal of Sustainable Tourism*, Vol. 3 (4): 191–207.

—— (1997) 'Parameters and indicators of decline of the British seaside resort', in Shaw, G. and Williams, A.M. (eds) (1997): 79–101.

Cooper, C.P. and Jackson, S. (1989) 'The destination area life cycle: the Isle of Man case study', *Annals of Tourism Research*, Vol. 16 (3): 377–98.

Cooper, C.P. and Lockwood, A. (eds) (1992) *Progress in Tourism, Recreation and Hospitality Management Vol. 4*, Chichester: John Wiley.

—— (1994) *Progress in Tourism, Recreation and Hospitality Management Vol. 5*, Chichester: John Wiley.

Corbin, A. (1995) *The Lure of the Sea*, London: Penguin.

Couch, C. and Farr, S. (2000) 'Museums, galleries, tourism and regeneration: some experiences from Liverpool', *Built Environment*, Vol. 26 (2): 152–63.

Craig-Smith, S.J. and Fagence, M. (1994) 'A critique of tourism planning in the Pacific', in Cooper, C.P. and Lockwood, A. (eds) (1994): 92–110.

Craik, J. (1991) *Resorting to Tourism: Cultural Policies for Tourist Development in Australia*, Sydney: George Allen and Unwin.

Craik, W. (1994) 'The economics of managing fisheries and tourism in the Great Barrier Reef Marine Park', in Munasinghe, M. and McNeely, J. (eds) (1994): 339–48.

Crang, M. (1997) 'Picturing practices: research through the tourist gaze', *Progress in Human Geography*, Vol. 21 (3): 359–74.

—— (1998) *Cultural Geography*, London: Routledge.

—— (2003) 'Cultural geographies of tourism', in Lew, A.A., Hall, C.M. and Williams, A.M. (eds) (2003): 74–84.

Crang, P. (2005) 'Consumption and its geographies', in Daniels. P. et al. (eds) (2005): 359–78.

Crawshaw, C. and Urry, J. (1997) 'Tourism and the photographic eye', in Rojek, C. and Urry, J. (eds) (1997): 176–95.

Cross, G. (ed.) (1990) *Worktowners at Blackpool: Mass Observation and Popular Leisure in the 1930s*, London: Routledge.

Crouch, D. (ed.) (1999) *Leisure/Tourism Geographies*, London: Routledge.

Crouch, D., Aronsson, L. and Wahlstrom, L. (2001) 'Tourist encounters', *Tourist Studies*, Vol. 1 (3): 253–70.

Crouch, D. and Desforges, L. (2003) 'The sensuous in the tourist encounter: the power of the body in tourist studies', *Tourist Studies*, Vol. 3 (1): 5–22.

Csikszentmihalyi, M. (1992) *The Psychology of Happiness*, London: Rider.

Daniels, P., Bradshaw, M., Shaw, D. and Sidaway, J. (eds) (2005) *An Introduction to Human Geography: Issues for the 21st Century*, Harlow: Prentice Hall.

Dann, G.M.S. (1981) 'Tourist motivation: an appraisal', *Annals of Tourism Research*, Vol. 8 (2): 187–219.

—— (1996) 'The people of tourist brochures', in Selwyn, T. (ed.) (1996): 61–81.

Dartmoor National Park Authority (DNPA) (2002) *Dartmoor National Park Local Plan, First Review: 1995–2011*, Newton Abbott: DNPA.

Davidson, R. (1992) *Tourism in Europe*, London: Pitman.

Davidson, R. and Maitland, R. (1997) *Tourism Destinations*, London: Hodder and Stoughton.

Davis, M. (1990) *City of Quartz: Excavating the Future in Los Angeles*, London: Verso.

Davis, S.G. (1996) 'The theme park: global industry and cultural form', *Media, Culture & Society*, Vol. 18 (3): 399–422.

Dear, M. and Flusty, S. (1998) 'Postmodern urbanism', *Annals of the Association of American Geographers*, Vol. 88 (1): 50–72.

Defoe, D. (1724) *A Tour Through the Whole Island of Great Britain*, London: Penguin. (This edition published in 1971.)

de Kadt, E. (1979) *Tourism: Passport to Development?* , Oxford: Oxford University Press.

Demars, S. (1990) 'Romanticism and American national parks', *Journal of Cultural Geography*, Vol. 11 (1): 17–24.

Demetriadi, J. (1997) 'The golden years: English seaside resorts 1950–1974', in Shaw, G. and Williams, A. (eds) (1997): 49–75.

de Oliveira, J.A.P. (2005) 'Tourism as a force for establishing protected areas: the case of Bahia, Brazil', *Journal of Sustainable Tourism*, Vol. 13 (1): 24–49.

Desforges, L. (2000) 'Traveling the world: identity and travel biography', *Annals of Tourism Research*, Vol. 27 (4): 926–45.

Diagne, A.K. (2004) 'Tourism development and its impacts in the Senegalese Petite Cote: a geographical case study in centre-periphery relations', *Tourism Geographies*, Vol. 6 (4): 472–92.

Dieke, P.U.C. (1994) 'The political economy of tourism in the Gambia', *Review of African Political Economy*, No. 62: 611–27.

—— (2002) 'Human resources in tourism development: African perspectives', in Harrison, D. (ed.) (2002): 61–75.

Dilley, R.S. (1986) 'Tourist brochures and tourist images', *The Canadian Geographer*, Vol. 30 (1): 59–65.

Donne, M. (2000) 'The growth and long-term potential of low-cost airlines', *Travel and Tourism Analyst*, No. 4: 1–15.

Douglass, W.A. and Raento, P. (2004) 'The tradition of invention: conceiving Las Vegas', *Annals of Tourism Research*, Vol. 31 (1): 7–23.

Dowling, R.K. (1992) 'Tourism and environmental integration: the journey from idealism to realism', in Cooper, C.P. and Lockwood, A. (eds) (1992): 33–44.

du Gay, P. (ed.) (1997) *Production of Culture/Cultures of Production*, London: Sage/Open University.

du Gay, P., Hall, S., Janes, L., Mackay, H. and Negus, K. (1997) *Doing Cultural Studies: the Story of the Sony Walkman*, London: Sage.

Dunford, M. (1990) 'Theories of regulation', *Society and Space*, Vol. 8: 297–321.

Durie, A.J. (1994) 'The development of the Scottish coastal resorts in the central Lowlands circa 1770–1880: from Gulf Stream to golf stream', *The Local Historian*, Vol. 24 (4): 206–16.

Eadington, W.R. (1999) 'The spread of casinos and their role in tourism development', in Pearce, D.G. and Butler, R.W. (eds) (1999): 127–42

Eco, U. (1986) *Travels in Hyperreality*, London: Picador.

Economic Intelligence Unit (EIU) (1995) 'Thailand', *International Tourism Report*, No. 3: 67–81.

—— (1997) 'Spain', *International Tourism Report* No. 2: 54–77.

Edensor, T. (1998) *Tourists at the Taj: Performance and Meaning at a Symbolic Site*, London: Routledge.

—— (2000a) 'Staging tourism: tourist as performers', *Annals of Tourism Research*, Vol. 27 (2): 322–44.

—— (2000b) 'Walking in the British countryside: reflexivity, embodied practices and ways of escape', *Body and Society*, Vol. 6 (3/4): 81–106.

—— (2001) 'Performing tourism, staging tourism: (re)producing tourist spaces and practices', *Tourist Studies*, Vol. 1 (1): 59–81.

English Heritage (2008) *Profile of Visitors to English Heritage Properties*. Accessed via www.english-heritage.org

English Tourist Board (1991) *The Future of England's Smaller Seaside Resorts*, London: English Tourist Board.

Essex, S., Kent, M. and Newnham, R. (2004) 'Tourism development in Mallorca: is water supply a constraint?', *Journal of Sustainable Tourism*, Vol. 12 (1): 4–28.

European Environment Agency (EEA) (2006) *Bathing Water Quality 2004*. Accessed via http://themes.eea.europa.eu.

Fainstein, S.S. and Gladstone, D. (1999) 'Evaluating urban tourism', in Judd, D.R. and Fainstein, S.S. (eds) (1999): 21–34.

Fainstein, S.S. and Judd, D.R. (1999a) 'Cities as places to play', in Judd. D.R. and Fainstein, S.S. (eds) (1999): 261–72.

—— (1999b) 'Global forces, local strategies and urban tourism', in Judd, D.R. and Fainstein, S.S. (eds) (1999): 1–17.

Falk, P. and Cambell, C. (eds) (1997) *The Shopping Experience*, London: Sage.

Farrell, B.H. and Runyan, D. (1991) 'Ecology and tourism', *Annals of Tourism Research*, Vol. 18 (1): 26–40.

Faulkner, B., Moscardo, G. and Laws, E. (eds) (2000) *Tourism in the Twenty-First Century: Reflections on Experience*, London: Continuum.

Featherstone, M. (1991) *Consumer Culture and Postmodernism*, London: Sage.

Feifer, M. (1985) *Going Places*, London: Macmillan.

Fennell, D. (1999) *Ecotourism: an Introduction*, London: Routledge.

Fisher, D. (2004) 'The demonstration effect revisited', *Annals of Tourism Research*, Vol. 31 (2): 428–46.

Fisher, S. (ed.) (1997) *Recreation and the Sea*, Exeter: Exeter University Press.

Fjellman, S.M. (1992) *Vinyl Leaves: Walt Disney World and America*, Boulder: Westview.

Fowler, P.J. (1992) *The Past in Contemporary Society: Then, Now*, London: Routledge.

Franklin, A. (2004) *Tourism: An Introduction*, London: Sage.

Franklin, A. and Crang, M. (2001) 'The trouble with tourism and travel theory', *Tourist Studies*, Vol. 1 (1): 5–22.

Freethy, R. and Freethy, M. (1997) *The Wakes Resorts*, Bury: Aurora.

Freitag, T.G. (1994) 'Enclave tourism development: for whom the benefits roll?', *Annals of Tourism Research*, Vol. 21 (3): 538–54.

Frenkel, S. and Walton, J. (2000) 'Bavarian Leavenworth and the symbolic economy of a theme town', *The Geographical Review*, Vol. 90 (4): 559–84.

Gale, T. (2005) 'Modernism, post-modernism and the decline of British Seaside resorts as long holiday destinations: a case study of Rhyl, North Wales', *Tourism Geographies*, Vol. 7 (1): 86–112.

Gant, R. and Smith, J. (1992) 'Tourism and national development planning in Tunisia', *Tourism Management*, Vol. 13 (3): 331–6.

Garcia, G.M., Pollard, J. and Rodriguez, R.D. (2003) 'The planning and practice of coastal zone management in southern Spain', *Journal of Sustainable Tourism*, Vol. 11 (2/3): 204–23.

Garin Munoz, T. (2007) 'German demand for tourism in Spain', *Tourism Management*, Vol. 28 (1): 12–22.

Getz, D. (1986) 'Models of tourism planning: towards integration of theory and practice', *Tourism Management*, Vol. 7 (1): 21–32.

Getz, D. and Brown, G. (2006) 'Critical success factors for wine tourism regions: a demand analysis', *Tourism Management*, Vol. 27 (1): 146–58.

Giddens, A. (1991) *Modernity and Self-identity: Self and Society in the Late Modern Age*, Cambridge: Polity.

Gilbert, D.C. (1990) 'Conceptual issues in the meaning of tourism', in Cooper, C.P. (ed.) (1990): 4–27.

Gilbert, E. (1939) 'The growth of inland and seaside health resorts in England', *Scottish Geographical Magazine*, Vol. 55 (1): 16–35.

—— (1975) *Brighton: Old Ocean's Bauble*, Hassocks: Flare Books.

Gillmor, D.A. (1996) 'Evolving air-charter tourism patterns: change in outbound traffic from the Republic of Ireland', *Tourism Management*, Vol. 17 (1): 9–16.

Go, F.M. (1992) 'The role of computer reservation systems in the hospitality industry', *Tourism Management*, Vol. 13 (1): 22–6.

Go, F.M. and Jenkins, C.L. (eds) (1998) *Tourism and Economic Development in Asia and Australasia*, London: Pinter.

Goffman, E. (1959) *The Presentation of Self in Everyday Life*, New York: Doubleday.

Gold, J.R. and Ward, S.V. (eds) (1994) *Place Promotion: The Use of Publicity and Marketing to Sell Towns*, Chichester: John Wiley.

Goodrich, J.N. (2002) 'September 11 2001 attack on America: a record of the immediate impacts and reactions in the USA travel and tourism industry', *Tourism Management*, Vol. 23 (6): 573–80.

Gordon, C. (1998) 'Holiday centres: responding to the consumer', *Insights*, Vol. 9 (B): 1–11.

Gospodini, A. (2001) 'Urban waterfront development in Greek cities', *Cities*, Vol. 18 (5): 285–95.

Goss, J. (1993) 'The "Magic of the Mall": an analysis of form, function and meaning in the contemporary retail built environment', *Annals of the Association of American Geographers*, Vol. 83 (1): 18–47.

—— (1999) 'Once-upon-a-time in the commodity world: an unofficial guide to Mall of America', *Annals of the Association of American Geographers*, Vol. 89 (1): 45–75.

Gottdiener, M., Collins, C.C. and Dickens, D.R. (1999) *Las Vegas: The Social Production of an All-American City*, Oxford: Blackwell.

Graburn, N. (1983a) 'The anthropology of tourism', *Annals of Tourism Research*, Vol. 10 (1): 9–33.

—— (1983b) 'Tourism and prostitution', *Annals of Tourism Research*, Vol. 10 (3): 437–42.

Graham, B., Ashworth, G.J. and Tunbridge, J.E. (2000) *A Geography of Heritage*, London: Arnold.

Gratton, C. and van der Straaten, J. (1994) 'The environmental impact of tourism in Europe', in Cooper, C.P. and Lockwood, A. (eds) (1994): 147–61.

Green, N. (1990) *The Spectacle of Nature: Landscape and Bourgeois Culture in Nineteenth Century France*, Manchester: Manchester University Press.

Gunn, C.A. (1994) *Tourism Planning*, New York: Taylor and Francis.

Hagermann, C. (2007) 'Shaping neighbourhoods and nature: urban political ecologies of urban waterfront transformations in Portland, Oregon', *Cities*, Vol. 24 (4): 285–97.

Hale, A. (2001) 'Representing the Cornish: contesting heritage interpretation in Cornwall', *Tourist Studies*, Vol. 1 (2): 185–96.

Halewood, C. and Hannam, K. (2001) 'Viking heritage tourism: authenticity and commodification', *Annals of Tourism Research*, Vol. 28 (3): 565–80.

Hall, C.M. (1992) 'Sex tourism in South-East Asia', in Harrison, D. (ed.) (1992): 64–74.

—— (1996) 'Gender and economic interests in tourism prostitution: the nature, development and implications of sex tourism in South-East Asia', in Kinnaird, V. and Hall, D. (eds) (1996): 142–63.

—— (2000) *Tourism Planning: Policies, Processes and Relationships*, Harlow: Prentice Hall.

—— (2005) *Tourism: Rethinking the Social Science of Mobility*, Harlow: Prentice Hall.

Hall, C.M. and Jenkins, J.M. (1995) *Tourism and Public Policy*, London: Routledge.

Hall, C.M. and Macionis, N. (1998) 'Wine tourism in Australia and New Zealand', in Butler, R.W., Hall, C.M. and Jenkins, J.M. (eds) (1999): 197–221.

Hall, C.M. and Page, S.J. (2005) *The Geography of Tourism and Recreation: Environment, Place and Space*, London: Routledge.

Hall, C.M., Mitchell, I. and Keelan, N. (1992) 'Maori culture and heritage tourism in New Zealand', *Journal of Cultural Geography*, Vol. 12 (2): 115–27.

Hall, C.M., Sharples, L., Cambourne, B. and Macionis, N. (2000) *Wine Tourism around the World*, Oxford: Butterworth-Heinemann.

Hall, D. and Richards, G. (eds) (2000) *Tourism and Sustainable Community Development*, London: Routledge.

Hall, S. (1996) 'Who needs identity?', in Hall, S. and du Gay, P. (eds) (1996): 1–17.

Hall, S. and du Gay, P. (1996) *Questions of Cultural Identity*, London: Sage.

Hamilton, L., Rohall, D., Brown, B., Hayward, G. and Keim, B. (2003) 'Warming winters and New Hampshire's lost ski areas: an integrated case study', *International Journal of Sociology and Social Policy*, Vol. 23: 52–73.

Hannigan, J. (1998) *Fantasy City: Pleasure and Profit in the Postmodern Metropolis*, London: Routledge.

Hanqin, Q.Z., King, C. and Ap. J. (1999) 'An analysis of tourism policy development in modern China', *Tourism Management*, Vol. 20 (4): 471–85.

Hardy, A.L. and Beeton, R.J.S. (2001) 'Sustainable tourism or maintainable tourism: managing resources for more than average outcomes', *Journal of Sustainable Tourism*, Vol. 9 (3): 168–92.

Harper, D. (2001) 'Comparing tourists' crime victimisation', *Annals of Tourism Research*, Vol. 28 (4): 1053–6.

Harrison, D. (ed.) (1992) *Tourism in Less Developed Countries*, London: Belhaven.

—— (ed.) (2001a) *Tourism and the Less Developed World: Issues and Case Studies*, Wallingford: CAB International.

—— (2001b) 'Less developed countries and tourism: the overall pattern', in Harrison, D. (ed.) (2001): 1–22.

Harvey, D. (1989) *The Condition of Postmodernity*, Oxford: Blackwell.

—— (1996) *Justice, Nature and the Geography of Difference*, Oxford: Blackwell.

—— (2000) *Spaces of Hope*, Berkeley: University of California Press.

Hassan, J. (2003) *The Seaside, Health and Environment in England and Wales since 1800*, Aldershot: Ashgate.

Haworth, J.T. (1986) 'Meaningful activity and psychological models of non-employment', *Leisure Studies*, Vol. 5 (3): 281–98.

Haywood, M.K. (1986) 'Can the tourist area cycle of evolution be made operational?', *Tourism Management*, Vol. 7 (3): 154–67.

Held, D. (ed.) (2000) *A Globalizing World? Culture, Economics, Politics*, London: Routledge.

Herbert, D.T. (ed.) (1995) *Heritage, Tourism and Society*, London: Mansell.

—— (2001) 'Literary places, tourism and the heritage experience', *Annals of Tourism Research*, Vol. 28 (2): 312–33.

Hewison, R. (1987) *The Heritage Industry: Britain in a Climate of Decline*, London: Methuen.

Higgins-Desbiolles, F. (2006) 'More than an "industry": the forgotten power of tourism as a social force', *Tourism Management*, Vol. 27 (6): 1192–1208.

Hiller, H.H. (2000) 'Mega-events, urban boosterism and growth strategies: an analysis of the objectives and legitimations of the Cape Town 2004 Olympic bid', *International Journal of Urban and Regional Research*, Vol. 24 (2): 439–58.

Hitchcock, M., King, V.T. and Parnwell, M.J.G. (eds) (1993) *Tourism in South East Asia*, London: Routledge.

Hodder, R. (2000) *Development Geography*, London: Routledge.

Holden, R. (1989) 'British garden festivals: the first eight years', *Landscape and Urban Planning*, Vol. 18 (1): 17–35.

Hollinshead, K. (1993) 'Encounters in tourism', in Khan, M.A. et al. (eds) (1993): 636–51.

Holloway, S.L., Rice, S.P. and Valentine, G. (eds) (2003) *Key Concepts in Geography*, London: Sage.

Honggen, X. (2006) 'The discourse of power: Deng Xiaoping and tourism development in China', *Tourism Management*, Vol. 27 (5): 803–14.

Hopkins, J. (1999) 'Commodifying the countryside: marketing myths of rurality', in Butler, R. et al. (eds) (1999): 139–56.

Horner, S. and Swarbrook, J. (1996) *Marketing Tourism, Hospitality and Leisure in Europe*, London: International Thomson.

Hudson. J. (1992) *Wakes Weeks: Memories of Mill Town Holidays*, Stroud: Alan Sutton.

Hughes, G. (1992) 'Tourism and the geographical imagination', *Leisure Studies*, Vol. 11 (1): 31–42.

Hughes, H.L. (1998) 'Theatre in London and the inter-relationship with tourism', *Tourism Management*, Vol. 19 (5): 445–52.

Huisman, S. and Moore, K. (1999) 'Natural languages and that of tourism', *Annals of Tourism Research*, Vol. 26 (2): 445–8.

Hunter, C. (1995) 'On the need to reconceptualise sustainable tourism development', *Journal of Sustainable Tourism*, Vol. 3 (3): 155–65.

——(1997) 'Sustainable tourism as an adaptive paradigm', *Annals of Tourism Research*, Vol. 24 (4): 850–67.

Hunter, C. and Green, H. (1995) *Tourism and Environment: A Sustainable Relationship?*, London: Routledge.

Hunter, W.C. (2001) 'Trust between culture: the tourist', *Current Issues in Tourism*, Vol. 4 (1): 42–67.

Huse, M., Gustavsen, T. and Almedal, S. (1998) 'Tourism impact comparisons among Norwegian towns', *Annals of Tourism Research*, Vol. 25 (3): 721–38.

Inglis, F. (2000) *The Delicious History of the Holiday*, London: Routledge.

Inskeep, E. (1991) *Tourism Planning: An Integrated and Sustainable Development Approach*, Chichester; John Wiley.

Instituto Nacional de Estadistica (INE) (2006a) *Spain in Figures 2005*, Madrid: INE.

—— (2006b) *Statistical Yearbook of Spain 2005*, Madrid: INE.

Ioannides, D. and Debbage, K. (1997) 'Post-Fordism and flexibility: the travel industry polyglot', *Tourism Management*, Vol. 18 (4): 229–41.

Iso-Ahola, S. (1982) 'Towards a social psychology of tourism motivation', *Annals of Tourism Research*, Vol. 9 (2): 256–62.

Jackson, E.L. (1991) 'Shopping and leisure: implications of the West Edmonton Mall for leisure and for leisure research', *The Canadian Geographer*, Vol. 35 (3): 280–7.

Jackson, G. and Morpeth, N. (2000) 'Local Agenda 21: reclaiming community ownership in tourism or stalled process?', in Hall, D. and Richards, G. (eds) (2000): 119–34.

Jackson, J. (2006) 'Developing regional tourism in China: the potential for activating business clusters in a socialist market economy', *Tourism Management*, Vol. 27 (4): 695–706.

Jackson, P. and Thrift, N. (1995) 'Geographies of consumption', in Miller, D. (ed.) (1995): 204–37

Jafari, J. (2001) 'The scientification of tourism', in Smith, V.L. and Brent, M. (eds) (2001): 28–41.

Jansen-Verbeke, M. (1986) 'Inner-city tourism: resources, tourists and promoters', *Annals of Tourism Research*, Vol. 13 (2): 79–100.

—— (1995) 'A regional analysis of tourism flows within Europe', *Tourism Management*, Vol. 16 (1): 73–82.

Jayne, M. (2006) *Cities and Consumption*, Abingdon: Routledge.

Jenkins, C.L. (1982) 'The effects of scale in tourism projects in developing countries', *Annals of Tourism Research*, Vol. 9 (2): 229–49.

Jenkins, O.H. (2003) 'Photography and travel brochures: the circle of representation', *Tourism Geographies*, Vol. 5 (3): 305–28.

Johnson, M. (1995) 'Czech and Slovak tourism: patterns, problems and prospects', *Tourism Management*, Vol. 16 (1): 21–8.

Johnson, R. (1986) 'The story so far; and for the transformations', in Punter, D. (ed.) (1986): 277–313.

Jones, T.S.M. (1994) 'Theme parks in Japan', in Cooper, C.P. and Lockwood, A. (eds) (1994): 111–25.

Joseph, C.A. and Kavoori, A.P. (2001) 'Mediated resistance: tourism and the host community', *Annals of Tourism Research*, Vol. 28 (4): 998–1009.

Judd, D.R. (1999) 'Constructing the tourist bubble', in Judd, D.R. and Fainstein, S.S. (eds) (1999): 35–53.

Judd, D.R. and Fainstein, S.S. (eds) (1999) *The Tourist City* , New Haven: Yale University Press.

Kane, M.J. and Zink, R. (2004) 'Package adventure tours: markers in serious leisure careers', *Leisure Studies* , Vol. 23 (4): 329–45.

Khan, H., Seng, C.F. and Cheong, W.K. (1990) 'Tourism multiplier effects on Singapore, *Annals of Tourism Research*, Vol. 17 (3): 408–18.

Khan, M. (1997) 'Tourism development and dependency theory: mass tourism vs ecotourism', *Annals of Tourism Research*, Vol. 24 (4): 988–91.

Khan, M.A., Olsen, M.D. and Var, T. (eds) (1993) *VNR's Encyclopaedia of Hospitality and Tourism*, New York: Van Nostrand Rheinhold.

Kinnaird, V. and Hall, D. (eds) (1994) *Tourism: A Gender Analysis*, Chichester: John Wiley.

Knowles, T. and Garland, M. (1994) 'The strategic importance of CRSs in the airline industry', *Travel and Tourism Analyst*, Vol. 4: 16.

Knowles, T., Diamantis, D. and El-Mourhabi, J.B. (2001) *The Globalization of Tourism and Hospitality: A Strategic Perspective*, London: Continuum.

Konig, U. and Abegg, B. (1997) 'Impacts of climate change on tourism in the Swiss Alps', *Journal of Sustainable Tourism*, Vol. 5 (1): 46–58.

Kun, L., Yiping, L. and Xuegang, F. (2006) 'Gap between tourism planning and implementation: a case of China', *Tourism Management*, Vol. 27 (6): 1171–80.

Lanfant, M-F., Allcock, J.B. and Bruner, E.M. (eds) (1995) *International Tourism: Identity and Change*, London: Sage.

Langman, L. (1992) 'Neon cages: shopping and subjectivity', in Shields, R. (ed.) (1992): 40–82.

Lash, S. (1990) *Sociology of Postmodernism*, London: Routledge.

Lash, S. and Urry, J. (1994) *Economies of Signs and Spaces*, London: Sage.

Law, C.M. (1992) 'Urban tourism and its contribution to urban regeneration', *Urban Studies*, Vol. 29 (3/4): 599–618.

Law, C.M. (ed.) (1996) *Tourism in Major Cities*, London: International Thomson Press.

—— (2000) 'Regenerating the city centre through leisure and tourism', *Built Environment*, Vol. 26 (2): 117–29.

—— (2002) *Urban Tourism: the Visitor Economy and the Growth of Large Cities*, London: Continuum.

Lechner, F.J. and Boli, J. (2000) *The Globalization Reader*, Oxford: Blackwell.

Lefebvre, H. (1991) *The Production of Space*, Oxford: Blackwell.

Leiper, N. (1989) 'Tourism and gambling', *Geo Journal*, Vol. 19 (3): 269–75.

—— (1993) 'Defining tourism and related concepts: tourist, market, industry and tourism system', in Khan, M.A. et al. (eds) (1993): 539–58.

Lennon, J. and Foley, M. (2000) *Dark Tourism: The Attraction of Death and Disaster*, London: Continuum.

Leontidou, L. (1994) 'Gender dimensions of tourism in Greece: employment, sub-cultures and restructuring', in Kinnaird, V. and Hall, D. (eds) (1994): 74–105.

Lew, A.A., Hall, C.M. and Williams, A.M. (eds) (2004) *A Companion to Tourism Geography*, Blackwell: Oxford.

Lew, A.A., Yu, L., Ap, J. and Zhang, G (eds) (2003) *Tourism in China*, Haworth: New York.

Lickorish, L.J. and Jenkins, C.L. (1997) *An Introduction to Tourism*, Oxford: Butterworth-Heinemann.

Light, D. and Andone, D. (1996) 'The changing geography of Romanian tourism', *Geography*, Vol. 81 (3): 193–203.

Lindberg, K. and McCool, S.F. (1998) 'A critique of environmental carrying capacity as a means of managing the effects of tourism development', *Environmental Conservation*, Vol. 25 (4): 291–2.

Lindberg, K., McCool, S.F. and Stankey, G. (1997) 'Rethinking carrying capacity', *Annals of Tourism Research*, Vol. 24 (2): 461–5.

Llewellyn Watson, G. and Kopachevsky, J.P. (1994) 'Interpreting tourism as a commodity', *Annals of Tourism Research*, Vol. 21 (3): 643–60.

Lockhart, D.G. (1993) 'Tourism to Fiji: crumbs off a rich man's table?', *Geography*, Vol. 78 (3): 318–23.

Loverseed, H. (1994) 'Theme parks in north America', *Travel and Tourism Analyst*, No. 4: 51–63.

Lowenthal, D. (1985) *The Past is a Foreign Country*, Cambridge: Cambridge University Press.

Lowerson, J. (1995) *Sport and the English Middle Classes 1870–1914*, Manchester: Manchester University Press.

Lundgren, J.O.J. (1973) *Tourist Impact and Island Entrepreneurship in the Caribbean*, unpublished paper presented to the Annual Conference of Latin American Geographers, cited in Mathieson and Wall (1982).

Lury, C. (1996) *Consumer Culture*, Cambridge: Polity.

Mabogunje, A.L. (1980) *The Development Process: A Spatial Perspective*, London: Hutchinson.

MacCannell, D. (1973) 'Staged authenticity: arrangements of social space in tourist settings', *American Journal of Sociology*, Vol. 79 (3): 589–603.

—— (1989) *The Tourist*, London: Macmillan.

—— (2001) 'Tourist agency', *Tourist Studies*, Vol. 1 (1): 23–37.

MacEwan, A. and MacEwan, M. (1982) *National Parks: Conservation or Cosmetics?* , London: George Allen and Unwin.

MacNaghten, P. and Urry, J. (2000) 'Bodies of nature', *Body and Society*, Vol. 6 (3/4): 1–16.

—— (eds) (2001) *Bodies of Nature*, London: Sage.

McCain, G. and Ray, N.M. (2003) 'Legacy tourism: the search for personal meaning in heritage travel', *Tourism Management*, Vol. 24 (6): 713–17.

McCarthy, J. (2002) 'Entertainment-led regeneration: the case of Detroit', *Cities*, Vol. 19 (2): 105–11.

McCool, S.F. and Lime, D.W. (2001) 'Tourism carrying capacity: tempting fantasy or useful reality?', *Journal of Sustainable Tourism*, Vol. 9 (5): 372–88.

McIntosh, R.W. and Goeldner, C.R. (1986) *Tourism: Principles, Practices, Philosophies*, Chichester: John Wiley.

McIntyre, G. (1993) *Sustainable Tourism Development: a Guide to Local Planners*, Madrid: World Tourism Organization.

McKercher, B. (1993a) 'Some fundamental truths about tourism: understanding tourism's social and environmental impacts', *Journal of Sustainable Tourism*, Vol. 1 (1): 6–16.

—— (1993b) 'The unrecognised threat to tourism: can tourism survive "sustainability"?', *Tourism Management*, Vol. 14 (2): 131–6.

McNally, S. (2001) 'Farm diversification in England and Wales', *Journal of Rural Studies*, Vol. 17 (2): 247–57.

Mak, B. (2003) 'China's tourist transportation: air, land and water', in Lew, A.A. et al. (eds) (2003): 165–93.

Maslow, A. (1954) *Motivation and Personality* , New York: Harper.

—— (1967) 'Lessons from the peak experience', *Journal of Humanistic Psychology* , Vol. 2 (1): 9–18.

Mathieson, A. and Wall, G. (1982) *Tourism: Economic, Physical and Social Impacts*, Harlow: Longmans.

Mbaiwa, J.E. (2005) 'Enclave tourism and its socio-economic impacts in the Okavango Delta, Botswana', *Tourism Management*, Vol. 26 (2): 157–72.

Medina, L.K. (2003) 'Commoditizing culture: tourism and Maya identity', *Annals of Tourism Research*, Vol. 30 (2): 353–68.

Meethan, K. (2001) *Tourism in Global Society*, Basingstoke: Palgrave.

Messerli, H.R. and Bakker, M. (2004) 'China', *Travel and Tourism Intelligence Country Reports*, No. 2: 1–43.

Miles, W.F.S. (2002) 'Auschwitz: museum interpretation and darker tourism', *Annals of Tourism Research*, Vol. 29 (4): 1175–8.

Miller, D. (ed.) (1995) *Acknowledging Consumption: A Review of New Studies*, London: Routledge.

Millington, K. (2001) 'Adventure travel', *Travel and Tourism Analyst*, November, London: Mintel.

Miossec, J.M. (1977) 'A model of the spaces of tourism', *L'Espace Geographique*, Vol. 6 (1): 41–8 (in French).

Milne, S. (1992) 'Tourism and development in South Pacific microstates', *Annals of Tourism Research*, Vol. 19 (2): 191–212.

Milne, S. and Ateljevic, I. (2001) 'Tourism, economic development and the global-local nexus: theory embracing complexity', *Tourism Geographies*, Vol. 3 (4): 369–93.

Mintel (2002) 'Short breaks abroad', *Leisure Intelligence*, June, London: Mintel.

—— (2003a) 'No frills/low cost airlines', *Leisure Intelligence*, February, London: Mintel.

—— (2003b) 'Rural tourism in Europe', *Travel and Tourism Analyst*, August, London: Mintel.

– – (2003c) 'European adventure travel', *Travel and Tourism Analyst*, October, London: Mintel.

Mitchell, R.E. and Reid, D.G. (2001) 'Community integration: island tourism in Peru', *Annals of Tourism Research* , Vol. 28 (1): 113–39.

Molina, A. and Esteban, A. (2006) 'Tourism brochures: usefulness and image', *Annals of Tourism Research*, Vol. 33 (4): 1036–56.

Morgan, N. (2004) 'Problematizing place promotion', in Lew. A.A. et al. (eds) (2004): 173–83.

Moroccan Ministry of Tourism (2005) *2010 Vision and Future* . Accessed via www.tourisme.gov.ma.

Mowforth, M. and Munt, I. (2003) *Tourism and Sustainability: Development and Tourism in the Third World*, London: Routledge.

Mullins, P. (1991) 'Tourism urbanization', *International Journal of Urban and Regional Research*, Vol. 15 (3): 326–42.

Munasinghe, M. and McNeely, J. (eds) (1994) *Protected Area Economics and Policy: Linking Conservation and Sustainable Development*, Washington DC: World Bank.

Muroi, H. and Sasaki, N. (1997) 'Tourism and prostitution in Japan', in Sinclair, M.T. (ed.) (1997): 180–217.

Murphy, P.E. (1985) *Tourism: A Community Approach*, London: Routledge.

—— (1994) 'Tourism and sustainable development', in Theobald (ed.) (1994): 274–90.

Nash, C. (2000) 'Performativity in practice: some recent work in cultural geography', *Progress in Human Geography*, Vol. 24 (4): 653–64.

Natural England (2006) *England Leisure Visits: Report of the 2005 Survey*, Cheltenham: Natural England.

Nava, M. (1997) 'Modernity's disavowal: women, the city and the department store', in Falk, P. and Campbell, C. (eds) (1997): 56–91.

Nepal, S.K. (2005) 'Tourism and remote mountain settlements: spatial and temporal development of tourist infrastructure in the Mt. Everest region, Nepal', *Tourism Geographies*, Vol. 7 (2): 205–27.

Noy, C. (2004) 'This trip really changed me: backpackers' narratives of self-change', *Annals of Tourism Research*, Vol. 31 (1): 78–102.

Okello, M.M. (2005) 'A survey of tourist expectations and economic potential for a proposed wildlife sanctuary in a Maasai group ranch near Amboseli, Kenya', *Journal of Sustainable Tourism*, Vol. 13 (6): 566–89.

Olsen, K. (2001) 'Authenticity as a concept in tourism research: the social organization of the experience of authenticity', *Tourist Studies*, Vol. 2 (2): 159–82.

O'Neill, C. (1994) 'Windermere in the 1920s', *The Local Historian*, Vol. 24 (4): 217–24.

O'Neill, M.A. and Fitz, F. (1996) 'Northern Ireland tourism: what chance now?', *Tourism Management*, Vol. 17 (2): 161–3.

ONS (Office of National Statistics) (2000) *Social Trends*, No. 30, London: The Stationery Office.

—— (2003) *Social Trends*, No. 33, London: The Stationery Office.

—— (2006) *Travel Trends 2004: A Report on the International Passenger Survey*, Basingstoke: Palgrave Macmillan.

Oppermann, M. (1992) 'International tourism and regional development in Malaysia', *Tijdschrift voor Economische en Sociale Geografie*, Vol. 83 (3): 226–33.

—— (1999) 'Sex tourism', *Annals of Tourism Research*, Vol. 26 (2): 251–66.

Oppermann, M. and Chon, K.S. (1997) *Tourism in Developing Countries*, London: International Thomson.

Page, S.J. (1990) 'Sports arena development in the UK: its role in urban regeneration in London Docklands, *Sport Place*, Vol. 4 (1): 3–15.

—— (1995) *Urban Tourism*, London: Routledge.

—— (1999) *Transport and Tourism*, Harlow: Addison Wesley Longman.

Page, S.J. and Dowling, R.K. (2002) *Ecotourism*, Harlow: Prentice Hall.

Page, S.J. and Hall, C.M. (2003) *Managing Urban Tourism*, Harlow: Prentice Hall.

Page, S.J. and Thorne, K.J. (1997) 'Towards sustainable tourism planning in New Zealand: public sector planning responses', *Journal of Sustainable Tourism*, Vol. 5 (1): 59–75.

Page, S.J., Bentley, T.A. and Walker, L. (2005) 'Scoping the nature and extent of adventure tourism operations in Scotland: how safe are they?', *Tourism Management*, Vol. 26 (3): 381–97.

Palmer, C. (2000) 'Heritage tourism and English national identity', in Robinson, M. et al. (eds) (2000): 331–47.

Paradis, T W. (2004) 'Theming, tourism and fantasy city', in Lew, A.A. et al. (eds) (2004): 195–209.

Parker, R.E. (1999) 'Las Vegas: casino gambling and local culture', in Judd, D.R. and Fainstein, S.S. (eds) (1999): 107–23.

Paterson, M. (2006) *Consumption and Everyday Life*, Abingdon: Routledge.

Patmore, J.A. (1983) *Recreation and Resources: Leisure Patterns and Leisure Places*, Oxford: Blackwell.

Pearce, D.G. (1987) *Tourism Today: A Geographical Analysis*, Harlow: Longman.

—— (1989) *Tourism Development*, Harlow: Longman.

—— (1994) 'Alternative tourism: concepts, classifications and questions' in Smith, V.L. and Eadington, W.R. (eds) (1994): 15–30.

—— (1997) 'Tourism and the autonomous communities in Spain', *Annals of Tourism Research*, Vol. 24 (1): 156–77.

—— (1998) 'Tourist districts in Paris: structure and function', *Tourism Management*, Vol. 19 (1): 49–65.

Pearce, D.G. and Butler, R.W. (eds) (1993) *Tourism Research: Critiques and Challenges*, London: Routledge.

—— (1999) *Contemporary Issues in Tourism Development*, London: Routledge.

Pearce, P.L. (1993) 'Fundamentals of tourist motivation', in Pearce, D.G. and Butler, R.W. (eds) (1993): 113–34.

Perry, A. (2006) 'Will predicted climate change compromise the sustainability of Mediterranean tourism?', *Journal of Sustainable Tourism*, Vol. 14 (4): 367–75.

Picard, M. (1993) 'Cultural tourism in Bali: national integration and regional differentiation', in Hitchcock, M. et al. (eds) (1993): 71–98.

—— (1995) 'Cultural tourism in Bali', in Lanfant, M-F. et al. (eds) (1995): 44–66.

Pigram, J. (1983) *Outdoor Recreation and Resource Management*, London: Croom Helm.

—— (1993) 'Planning for tourism in rural areas: bridging the policy implementation gap', in Pearce, D.G. and Butler, R.W. (eds) (1993): 156–74.

Pile, S. and Keith, M. (eds) (1997) *Geographies of Resistance*, London: Routledge.

Pimlott, J.A.R. (1947) *The Englishman's Holiday: A Social History*, London: Faber.

Pitchford, S.R. (1995) 'Ethnic tourism and nationalism in Wales', *Annals of Tourism Research*, Vol. 22 (1): 35–52.

Pizam, A. and Pokela, J. (1985) 'The perceived impact of casino gambling on a host community', *Annals of Tourism Research*, Vol. 12 (2): 147–65.

Pizam, A. and Mansfeld, Y. (eds) (1996) *Tourism, Crime and International Security Issues*, Chichester: John Wiley.

Pizam, A.. Milman, A. and King, B. (1994) 'The perceptions of tourism employees and their families towards tourism: a cross-cultural comparison', *Tourism Management*, Vol. 15 (1): 53–61.

Poirier, R.A. (1995) 'Tourism and development in Tunisia', *Annals of Tourism Research*, Vol. 22 (1): 157–71.

Pollard, J. and Rodriguez, R.D. (1993) 'Tourism and Torremolinos: recession or reaction to environment?', *Tourism Management*, Vol. 14 (4): 247–58.

Pomfret, G. (2006) 'Mountaineering adventure tourists: a conceptual framework for research', *Tourism Management*, Vol. 27 (1): 113–23.

Pompl, W. and Lavery, P. (eds) (1993) *Tourism in Europe: Structures and Developments*, Wallingford: CAB International.

Poole, R. (1983) 'Oldham Wakes', in Walton, J.K. and Walvin, J. (eds) (1983): 72–98.

Poon, A. (1989) 'Competitive strategies for "new tourism"', in Cooper, C.P. (ed.) (1989): 91–102.

Poria, Y., Butler, R. and Airey, D. (2003) 'The core of heritage tourism', *Annals of Tourism Research*, Vol. 30 (1): 238–54.

Poria, Y., Reichel, A. and Biran, A. (2006) 'Heritage site management: motivations and expectations', *Annals of Tourism Research*, Vol. 33 (1): 162–78.

Potter, R., Binns, J., Smith, D and Elliott, J. (eds) (1999) *Geographies of Development*, Harlow: Longman.

Prentice, R. (1993) *Tourism and Heritage Attractions*, London: Routledge.

—— (1994) 'Heritage: a key sector of the "new" tourism', in Cooper, C.P. and Lockwood, A. (eds) (1994): 309–24.

Prentice, R. and Andersen, V. (2007) 'Interpreting heritage essentialisms: familiarity and felt history', *Tourism Management*, Vol. 28 (3): 661–76.

Preston-Whyte, R. (2004) 'The beach as a liminal space', in Lew, A.A. et al. (eds) (2004): 349–59.

Price, M.F. (ed.) (1996) *People and Tourism in Fragile Environments*, Chichester: John Wiley.

Prideaux, B. (1996) 'The tourism crime cycle: a beach destination case study', in Pizam, A. and Mansfeld, Y. (eds) (1996): 59–75.

—— (2000) 'The resort development spectrum: a new approach to modelling resort development', *Tourism Management*, Vol. 21 (3): 225–40.

—— (2004) 'The resort development spectrum: the case of the Gold Coast, Australia, *Tourism Geographies*, Vol. 6 (1): 26–58.

Priestley, G. and Mundet, L. (1998) 'The post-stagnation phase of the resort life cycle', *Annals of Tourism Research*, Vol. 25 (1): 85–111.

Priestley, G., Edwards, J. and Coccossis, H. (eds) (1996) *Sustainable Tourism? European Experiences*, Wallingford: CAB International.

Pritchard, A. and Morgan, N. (2001) 'Culture, identity and tourism representation: marketing Cymru or Wales?', *Tourism Management*, Vol. 22 (2): 167–79.

Pruitt, S. and Lafont, S. (1995) 'For love and money: romance tourism in Jamaica', *Annals of Tourism Research*, Vol. 22 (2): 422–40.

Punter, D. (ed.) (1986) *Introduction to Contemporary Cultural Studies*, London: Longman.

Qian, W. (2003) 'Travel agencies in China at the turn of the Millenium', in Lew, A.A. et al. (eds) (2003): 143–64.

Reid, D.G. and Mannell, R.C. (1994) 'The globalization of the economy and potential new roles of work and leisure', *Loisir et Societe*, Vol. 17 (1): 251–66.

Relph, E. (1976) *Place and Placelessness*, London: Pion.

—— (1987) *The Modern Urban Landscape*, London: Croom Helm.

Reynolds, P.C. and Braithwaite, R. (2001) 'Towards a conceptual framework for wildlife tourism', *Tourism Management*, Vol. 22 (1): 31–42.

Richards, G. (ed.) (1996) *Cultural Tourism in Europe*, Wallingford: CAB International.

—— (ed.) (2000) *Cultural Attractions and European Tourism*, Wallingford: CAB International.

—— (2001) 'The market for cultural attractions', in Richards, G. (ed.) (2000): 31–53.

Riley, M., Ladkin, A. and Szivas, E. (2002) *Tourism Employment: Analysis and Planning*, Clevedon: Channel View.

Ringer, G. (ed.) (1998) *Destinations: Cultural Landscapes of Tourism*, London: Routledge.

Ritzer, G. (1998) *The McDonaldization Thesis*, London: Sage.

Ritzer, G. and Liska, A. (1997) '"MacDisneyization" and "post-tourism": complementary perspectives on contemporary tourism', in Rojek, C. and Urry, J. (eds) (1997): 96–112.

Roberts, M. (2006) 'From "creative city" to "no-go areas" – the expansion of the night-time economy in British towns and city centres', *Cities*, Vol. 23 (5): 331–8.

Roberts, R. (1983) 'The corporation as impresario: the municipal provision of entertainment in Victorian and Edwardian Bournemouth', in Walton, J.K and Walvin, J. (eds) (1983): 138–57.

Robins, K. (1997) 'What in the world is going on?', in du Gay, P. (ed.) (1997): 12–47.

Robinson, M. (1999) 'Tourism development in de-industrializing centres of the UK: change, culture and conflict', in Robinson, M. and Boniface, P. (eds) (1999): 129–59.

Robinson, M. and Boniface, P. (eds) (1999) *Tourism and Cultural Conflicts*, Wallingford: CAB Publishing.

Robinson, M., Evans, N., Long, P., Sharpley, R. and Swarbrooke, J. (eds) (2000) *Tourism and Heritage Relationships: Global, National and Local Perspectives*, Sunderland: Business Education.

Rojek, C. (1993a) 'De-differentiation and leisure', *Loisir et Societe*, Vol. 16 (1): 15–29.

—— (1993b) *Ways of Escape: Modern Transformations in Leisure and Travel*, London: Macmillan.

—— (1997) *Decentring Leisure: Rethinking Leisure Theory*, London: Sage.

Rojek, C. and Urry, J. (1997) 'Transformations in travel theory', in Rojek, C. and Urry, J. (eds) (1997): 1–19.

Rojek, C. and Urry, J. (eds) (1997) *Touring Cultures: Transformations in Travel Theory*, London: Routledge.

Romeril, M. (1985) 'Tourism and the environment: towards a symbiotic relationship', *International Journal of Environmental Studies*, Vol. 25 (4): 215–18.

Rothman, H. (2002) *Neon Metropolis: How Las Vegas Started the Twenty-First Century*, London: Routledge.

Ruhanen, L. (2004) 'Strategic planning for local tourism destinations: an analysis of tourism', *Tourism and Hospitality Planning and Development*, Vol. 1 (3): 1–15.

Ryan, C. (1991) *Recreational Tourism: A Social Science Perspective*, London: Routledge.

—— (1993) 'Tourism and crime: an intrinsic or accidental relationship?', *Tourism Management*, Vol. 14 (3): 173–83.

—— (1997) *The Tourist Experience: A New Introduction*, London: Cassells.

Ryan, C. and Crotts, L. (1997) 'Carving and tourism: a Maori perspective', *Annals of Tourism Research*, Vol. 24 (4): 898–918.

Ryan, C., Page, S. and Aitken, M. (eds) (2005) *Taking Tourism to the Limits: Issues, Concepts and Managerial Perspectives*, Oxford: Elsevier.

Scarborough Borough Council (1994) *Scarborough Local Plan – Tourism*, Scarborough: Scarborough Borough Council.

Scheyvens, R. (2002) *Tourism for Development*, Harlow: Prentice Hall.

Schouten, F.F. J. (1995) 'Heritage as historical reality', in Herbert, D.T. (ed.) (1995): 21–31.

SCPR (Social and Community Planning Research) (1997) *UK Day Visits Survey – Summary Findings*, London: SCPR.

Sears, J.F. (1989) *Sacred Places: American Tourist Attractions in the Nineteenth Century*, Oxford: Oxford University Press.

Selby, M. (2004) *Understanding Urban Tourism: Image, Culture and Experience*, London: I.B. Taurus.

Selwyn, T. (ed.) (1996) *The Tourist Image: Myths and Myth Making in Tourism*, Chichester: John Wiley.

Sharpley, R. (2000) 'Tourism and sustainable development: exploring the theoretical divide', *Journal of Sustainable Tourism*, Vol. 8 (1): 1–19.

—— (2005) 'Travels to the edge of darkness: towards a typology of dark tourism', in Ryan, C. et al. (eds) (2005): 217–28.

Shaw, G. and Williams, A.M. (1994) *Critical Issues in Tourism: A Geographical Perspective*, Oxford: Blackwell (First Edition).

—— (eds) (1997) *The Rise and Fall of British Coastal Resorts*, London: Pinter.

—— (2002) *Critical Issues in Tourism: A Geographical Perspective*, Oxford: Blackwell (second edition).

—— (2004) *Tourism and Tourism Spaces*, London: Sage.

Sheller, M. and Urry, J. (2004) *Tourism Mobilities: Places to Play, Places in Play*, London: Routledge.

Sherlock, K. (2001) 'Revisiting the concept of hosts and guests', *Tourist Studies*, Vol. 1 (3): 271–95.

Shiebler, S.A., Crotts, J.C. and Hollinger, R.C. (1996) 'Florida tourists' vulnerability to crime', in Pizam, A. and Mansfeld, Y. (eds) (1996): 37–58.

Shields, R. (1990) *Places on the Margin: Alternative Geographies of Modernity*, London: Routledge.

—— (ed.) (1992) *Lifestyle Shopping: the Subject of Consumption*, London: Routledge.

Shirato, T. and Webb, J. (2003) *Understanding Globalization*, London: Sage.

Shurmer-Smith, P. and Hannam, K. (1994) *Worlds of Desire: Realms of Power. A Cultural Geography*, London: Edward Arnold.

Sidaway, R. (1995) 'Managing the impacts of recreation by agreeing the Limits of Acceptable Change', in Ashworth, G.J. and Dietvorst, A.G.J. (eds) (1995): 303–16.

Simpson, K. (2001) 'Strategic planning and community involvement as contributors to sustainable tourism development', *Current Issues in Tourism*, Vol. 4 (1): 3–41.

Simpson, P. and Wall, G. (1999) 'Consequences of resort development: a comparative study', *Tourism Management*, Vol. 20 (3): 283–96.

Sinclair, M.T. (ed.) (1997) *Gender, Work and Tourism*, London: Routledge.

Sinclair, M.T. and Stabler, M. (1997) *The Economics of Tourism*, London: Routledge.

Sindiga, I. (1999) 'Alternative tourism and sustainable development in Kenya', *Journal of Sustainable Tourism*, Vol. 7 (2): 108–27.

Singh, T.V., Theuns, H.L. and Go, F.M. (eds) (1989) *Towards Appropriate Tourism: The Case of Developing Countries*, Frankfurt am Mein: Peter Lang Publishing.

Smith, A.J. and Newsome, D. (2002) 'An integrated approach to assessing, managing and monitoring campsite impacts in Warren National Park, Western Australia', *Journal of Sustainable Tourism*, Vol. 10 (4): 343–59.

Smith, C. and Jenner. P. (1989) 'Tourism and the environment', *Travel and Tourism Analyst*, Vol. 5: 68–86.

Smith, M.K. (2003) *Issues in Cultural Tourism Studies*, London: Routledge.

Smith, R.A. (1991) 'Beach resorts: a model of development evolution', *Landscape and Planning*, Vol. 21: 189–210.

Smith, V.L. (1977) *Hosts and Guests: The Anthropology of Tourism*, Philadelphia: University of Pennsylvania Press.

—— (1997) 'The Inuit as hosts: heritage and wilderness tourism in Nunavut', in Price, M.F. (ed.) (1996): 33–50.

Smith, V.L. and Eadington, W.R. (eds) (1994) *Tourism Alternatives: Potentials and Problems in the Development of Tourism*, Chichester: John Wiley.

Smith, V.L. and Brent, M. (eds) (2001) *Hosts and Guests Revisited*, New York: Cognizant Communications.

Soane, J.V.N. (1993) *Fashionable Resort Regions: Their Evolution and Transformation*, Wallingford: CAB International.

Society of London Theatres (SLT) (2007) *Box Office Data Report 2006*. Accesssed via www.officiallondontheatre.co.uk.

Sofield, T.H.B. and Fung, M.S.L. (1998) 'Tourism development and cultural policies in China', *Annals of Tourism Research*, Vol. 25 (2): 362–92.

Soja, E.W. (1989) *Postmodern Geographies: The Reassertion of Space in Critical Social Theory*, London: Verso.

—— (1995) 'Postmodern urbanization: the six restructurings of Los Angeles', in Watson, S. and Gibson, K. (eds) (1995): 125–37.

—— (1996) *Thirdspace: Journeys to Los Angeles and Other Real and Imagined Places*, Oxford: Blackwell.

—— (2000) *Postmetropolis: Critical Studies of Cities and Regions*, Oxford: Blackwell.

Sonmez, S.F. (1998) 'Tourism, terrorism and political instability', *Annals of Tourism Research*, Vol. 25 (2): 416–56.

Sonmez, S.F. and Graefe, A.R. (1998) Influence of terrorism risk on foreign tourism demands', *Annals of Tourism Research*, Vol. 25 (1): 112–44.

Sparks, B. (2007) 'Planning a wine tourism vacation? Factors that help to predict tourist behavioural intentions', *Tourism Management*, Vol. 28 (4): 1180–92.

Squire, S.J. (1993) 'Valuing countryside: reflections on Beatrix Potter tourism', Area, Vol. 24 (1): 5–10.

Stebbins, R. (1982) 'Serious leisure: a conceptual statement', *Pacific Sociological Review*, Vol. 25: 251–72.

Stone, P. and Sharpley, R. (2008) 'Consuming dark tourism: a thanatological perspective', *Annals of Tourism Research*, Vol. 35 (2): 574–95.

Svoronou, E. and Holden, A. (2005) 'Ecotourism as a tool for nature conservation: the role of WWF Greece in the Dadia-Lefkimi-Soufli Forest Reserve in Greece', *Journal of Sustainable Tourism*, Vol. 13 (5): 456–67.

Theobald, W. (ed.) (1994) *Global Tourism; the Next Decade*, Oxford: Butterworth-Heinemann.

Thomas, B. and Townsend, A. (2001) 'New trends in the growth of tourism employment in the UK in the 1990s', *Tourism Economics*, Vol. 7 (3): 295–310.

Thomas, C. (1997) 'See your own country first: the geography of a railway landscape', in Westland, E. (ed.) (1997): 107–28.

Thornton, P. (1997) 'Coastal tourism in Cornwall since 1900', in Fisher, S. (ed.) (1997): 57–83.

Thrift, N.J. (1996) *Spatial Transformations*, London: Sage.

—— (1997) 'The still point: resistance, expressive embodiment and dance', in Pile, S. and Keith, M. (eds) (1997): 124–51.

Tickell, A. and Peck, J.A. (1992) 'Accumulation, regulation and the geographies of post-Fordism: missing links in regulationist research', *Progress in Human Geography*, Vol. 16 (2): 190–218.

Tickell, A. and Peck, J. (1994) 'Accumulation, regulation and the geographies of post-Fordism: missing links in regulation research', *Progress in Human Geography*, Vol. 16 (2): 190–218.

Timothy, D.J. (1999) 'Participatory planning: a view of tourism in Indonesia', *Annals of Tourism Research*, Vol. 26 (2): 371–91.

—— (2001) *Tourism and Political Boundaries*, London: Routledge.

Timothy, D.J. and Boyd, S.W. (2003) *Heritage Tourism*, Harlow: Prentice Hall.

Tobin, G.A. (1974) 'The bicycle boom of the 1890s: the development of private transportation and the birth of the modern tourist', *Journal of Popular Culture*, Vol. 7 (4): 838–48.

Tosun, C. and Jenkins, C.L. (1996) 'Regional planning approaches to tourism development: the case of Turkey', *Tourism Management*, Vol. 17 (7): 519–31.

Towner, J. (1996) *An Historical Geography of Recreation and Tourism in the Western World, 1540–1940*, Chichester: John Wiley.

Travis, J. (1997) 'Continuity and change in English sea-bathing, 1730–1900: a case of swimming with the tide', in Fisher, S. (ed.) (1997): 8–35.

Tsartas, P. (1992) 'Socio-economic impacts of tourism on two Greek isles', *Annals of Tourism Research*, Vol. 19 (3): 516–33.

Tuan, Y-F. (1977) *Space and Place: the Perspective of Experience*, Minneapolis: University of Minnesota.

Turner, L. and Ash, J. (1975) *The Golden Hordes: International Travel and the Pleasure Periphery*, London: Constable.

UK Heritage Railways Association (2008) *UK Heritage Railways: Facts and Figures*. Accessed via www.ukhrail.uel.ac.uk.

Urierly, N. (2005) 'The tourist experience: conceptual developments', *Annals of Tourism Research*, Vol. 32 (1): 199–217.

Urry, J. (1990) *The Tourist Gaze: Leisure and Travel in Contemporary Societies*, London: Sage.

—— (1991) 'The sociology of tourism', in Cooper, C.P. (ed.) (1991): 48–57.

—— (1994a) *Consuming Places*, London: Routledge.

—— (1994b) 'Tourism, travel and the modern subject', in Urry, J. (1994a): 141–51.

—— (1994c) 'The making of the English Lake District', in Urry, J. (1994a): 193–210.

—— (2000) *Sociology Beyond Societies: Mobilities for the 21st Century*, London: Routledge.

United States Department of Commerce (2002) *The United States National Data Book 2002*, Austin: Hoover Business Press.

Veblen, T. (1994) *The Theory of the Leisure Class*, Harmondsworth: Penguin (first published in 1924).

Veijola, S. and Jokinen, E. (1994) 'The body in tourism', *Theory, Culture and Society*, Vol. 11 (3): 125–51.

Visit Britain (2003) *Visits to Visitor Attractions 2002*, London: Visit Britain.

—— (2006) *Visitor Attraction Trends in England 2005*, London: Visit Britain.

—— (2007) *Visitor Attraction Trends in England 2006*, London: Visit Britain.

Visit London (2007) *Accommodating Growth: A Guide to Hotel Development in London*, London: Visit London.

Vukovic, B. (2002) 'Religion, tourism and economics: a convenient symbiosis', *Tourism Recreation Research*, Vol. 27 (2): 59–64.

Wagar, J. (1974) 'Recreational carrying capacity reconsidered', *Journal of Forestry*, Vol. 72 (5): 274–8.

Wahab, S. and Pigram, J.J. (eds) (1997) *Tourism, Development and Growth: the Challenges of Sustainability*, London: Routledge.

Waitt, G. (2000) 'Consuming heritage: perceived historical authenticity', *Annals of Tourism Research*, Vol. 27 (4): 835–62.

Wakefield, S. (2007) 'Great expectations: waterfront redevelopment and the Hamilton Harbour Waterfront Trail', *Cities*, Vol. 24 (4): 298–310.

Walker, H. (1985) 'The popularisation of the Outdoor Movement, 1900–1940', *British Journal of Sports History*, Vol. 2 (2): 140–53.

Wall, G. (1997) 'Is ecotourism sustainable?', *Environmental Management*, Vol. 21 (4): 483–91.

Wall, G. and Mathieson, A. (2006) *Tourism: Change, Impacts and Opportunities*, Harlow: Prentice Hall.

Wallace, G. and Pierce, S. (1996) 'An evaluation of ecotourism in the Amazon, Brazil', *Annals of Tourism Research*, Vol. 23 (4): 843–73.

Walsh, K. (1992) *The Representation of the Past: Museums and Heritage in the Postmodern World*, London: Routledge.

Walton, J.K. (1981) 'The demand for working class seaside holidays in Victorian England', *Economic History Review*, Vol. 34 (2): 249–65.

—— (1983a) *The English Seaside Resort: A Social History*, Leicester: Leicester University Press.

—— (1983b) 'Municipal government and the holiday industry in Blackpool, 1976–1914', in Walton, J.K. and Walvin, J. (eds) (1983): 160–85.

—— (1994) 'The re-making of a popular resort: Blackpool Tower and the boom of the 1890s', *The Local Historian*, Vol. 24 (4): 194–205.

—— (1997a) 'The seaside resorts of western Europe, 1750–1939', in Fisher, S. (ed.) (1997): 36–56.

—— (1997b) 'The seaside resorts of England and Wales, 1900–1950: growth, diffusion and the emergence of new forms of coastal tourism', in Shaw, G. and Williams, A.M. (eds) (1997): 21–48.

—— (2000) *The British Seaside: Holidays and Resorts in the Twentieth Century*, Manchester: Manchester University Press.

Walton, J.K. and Walvin, J. (eds) (1983) *Leisure in Britain 1780–1939*, Manchester: Manchester University Press.

Walton, J.K. and Smith, J. (1996) 'The first century of beach tourism in Spain: San Sebastian and the Playas del Norte from the 1830s to the 1930s', in Barke, M. et al. (eds) (1996): 35–60.

Walvin, J. (1978) *Beside the Seaside: A Social History of the Popular Seaside*, London: Allen Lane.

Wang, N. (1999) 'Rethinking authenticity in tourism experience', *Annals of Tourism Research*, Vol. 26 (2): 349–70.

Ward, C. and Hardy, D. (1986) *Goodnight Campers! The History of the British Holiday Camp*, Mansell: London.

Ward, S.V. and Gold, J. R. (1994) 'Introduction', in Gold, J.R. and Ward, S.V. (eds) (1994): 1–17.

Wasko, J. (2001) 'Is it a Small World, after all?', in Wasko, J. et al. (eds) (2001): 3–28.

Wasko, J., Phillips, M. and Meehan, E.R. (eds) (2001) *Dazzled by Disney? The Global Disney Audiences Project*, London: Leicester University Press.

Watson, S. and Gibson, K. (eds) (1995) *Postmodern Cities and Spaces*, Oxford: Blackwell.

Weaver, D. (1998) 'Peripheries of the periphery: tourism in Tobago and Barbuda', *Annals of Tourism Research*, Vol. 25 (2): 292–313.

—— (2000) 'Sustainable tourism: is it sustainable?' in Faulkner, B. et al. (eds) (2000): 300–11.

Weber, K. (2001) 'Outdoor adventure tourism: a review of research approaches', *Annals of Tourism Research*, Vol. 28 (2): 360–77.

Westland, E. (ed.) (1997) *Cornwall: The Cultural Construction of Place*, Penzance: Patten Press.

Wheeller, B. (1994) 'Ecotourism: a ruse by any other name', in Cooper, C.P. and Lockwood, A. (eds) (1994): 3–11.

White, P.E. (1974) *The Social Impact of Tourism on Host Communities: A Study of Language Change in Switzerland*, School of Geography Research Paper No. 9, Oxford: Oxford University.

Wilkinson, P.F. (1987) 'Tourism in small island nations: a fragile dependence', *Leisure Studies*, Vol. 6 (2): 127–46.

Williams, A.M. (1996) 'Mass tourism and international tour companies', in Barke, M. et al. (eds) (1996): 119–36.

Williams, A.M. and Shaw, G. (1995) 'Tourism and regional development: polarization and new forms of production in the United Kingdom', *Tijdschrift voor Economische en Sociale Geografie*, Vol. 86 (1): 50–63.

—— (eds) (1998) *Tourism and Economic Development: European Experiences*, Chichester: John Wiley.

Williams, S. (1998) *Tourism Geography*, London: Routledge.

—— (2003) *Tourism and Recreation*, Harlow: Prentice Hall.

—— (2004a) 'General introduction' in Williams. S. (ed.) (2004b): 1–21.

—— (2004b) (ed.) *Tourism: Critical Concepts in the Social Sciences Vol. 1*, London: Routledge.

Winchester, H.P.M., Kong, L. and Dunn, K. (2003) *Landscapes: Ways of Imagining the World*, Harlow: Prentice Hall.

World Commission on Environment and Development (WCED) (1987) *Our Common Future*, Oxford: Oxford University Press.

World Tourism Organization (1993) *National and Regional Tourism Planning*, London: Routledge.

—— (1994) *Tourism to the Year 2000: Recommendations on Tourist Statistics*, Madrid: World Tourism Organization/United Nations.

—— (1995) *Compendium of Tourism Statistics*, Madrid: World Tourism Organization.

—— (2001) *Tourism 2020 Vision*, Madrid: World Tourism Organization.

—— (2005a) *Tourism Highlights 2005*, Madrid: World Tourism Organization.

—— (2005b) *World Tourism Barometer*, Vol. 3, No. 2, Madrid: World Tourism Organization.

—— (2005c) *World Tourism Barometer*, Vol. 3, No. 3, Madrid: World Tourism Organization.

—— (2006) *World Tourism Barometer*, Vol. 4, No. 1, Madrid: World Tourism Organization.

—— (2007) *Tourism Highlights 2007*, Madrid: World Tourism Organization.

—— (2008) *World Tourism Barometer*, Vol. 6, No. 1, Madrid: World Tourism Organization.

Xiaolun, W. (2003) 'China in the eyes of Western travellers, 1860–1900', in Lew, A.A. et al. (eds) (2003): 35–50.

Xiaoping, S. (2003) 'Short and long-haul international tourism to China', in Lew, A.A. et al. (eds) (2003): 237–61.

Yingzhi, G., Samuel, K.S., Timothy, D. and Kuo-Ching, W. (2006) 'Tourism and reconciliation between mainland China and Taiwan', *Tourism Management*, Vol. 27 (5): 997–1005.

Yu, L., Ap. J., Zhang, G. and Lew, A.A. (2003) 'World trade and China's tourism: opportunities, challenges and strategies', in Lew, A.A. et al. (eds) (2003): 297–307.

Yuksel, A. and Akgul, O. (2007) 'Postcards as affective image makers: an idle agent in destination marketing', *Tourism Management*, Vol. 28 (3): 714–25.

Zhang, G. (1997) 'China's domestic tourism: impetus, development and trends', *Tourism Management*, Vol. 18 (8): 565–71.

—— (2003) 'China's tourism since 1978: policies, experiences and lessons learned', in Lew, A.A. et al. (eds) (2003): 13–34.

Zhang, G. and Lew, A.A. (2003) 'China's tourism boom', in Lew, A.A. et al. (eds) (2003): 3–11.

Zhang, P. (2007) *Culture and Ideology at an Invented Place*, Newcastle: Cambridge Scholars Publishing.

Zube, E.H. and Galante, J. (1994) 'Marketing landscapes of the Four Corner states', in Gold, J.R. and Ward, S.V. (1994) (eds): 213–32.

Index

Colwyn Bay 96
commodification 135, 136, 137, 145, 196, 264, 270, 281; of culture 146–9, 227; and heritage 241, 255; and power relations 139, of rural experience 48
communities: moral drift in 149–51; and planning 160, 174–9, and social values 149–51
computer reservation systems 65–6
Coney Island (USA) 33
consumption 258–62, 268; and aesthetics 227; and cities 210, 222; and daily life 259; and demonstration effect 138; and fantasy cities 230; geographies of 21; and heritage 237, 241; and identity 22, 137, 260, 266; and lifestyle 260; of places 185, 259; and reflexivity 228–9; and resistance 261–2; and tourism 263–5; in urban economies 48
Cook, Thomas 36, 47, 54
crime 149; and tourism 150
'cultural capital' 137, 241, 273, 281
cultural identity 140, 147–8; and religion 151
'cultural turn' 20, 258
culture: and heritage tourism 237; popular 242–3; as travel motivation 134; see also tourism impacts, tourist encounter

Dallas (USA) 244
'dark tourism' 243–4, 253
day visits 6, 215; see also excursions
'de-differentiation' 6, 188, 204, 281
defence of space 210
'demonstration effect' 135, 138–9, 141, 142, 149
dependency 102, 139
development: concept of 81–2; see also tourism development
Dieppe 30
Disney (Disneyland) 18, 136, 195–6, 200–1, 201, 202
Doberan (Germany)
Doxey's 'irridex' 143–4
drifters 14, 28
Dublin 215, 267
Dudley (UK) 240

Eastbourne 41, 92
Eastern Europe: development of tourism in 59, 66
'eatertainment' 228
eco-tourism 20, 112, 130–2, 237
Eftling (Netherlands) 200
employment: seasonality of 105; in tourism 104–6, 151, 224
empowerment 151–2; see also power relations
enclaves 87–90, 142, 197
environmental change: and carrying capacity 123–6; and community consultation

128–30; and impact assessment 127–8; limits of acceptable change 126–7; management of 122–30; relationship with tourism 109–10
environmental impact assessment 127–8
environmental impacts: and biodiversity 116; and erosion 117–18; and pollution 118–19; and resources 119–21; and seasonality 113; and tourism 112–21; and visual change 121; on wildlife 116
environmentalism 242
erosion 117–18
Eurodisney 202, 204
European Regional Development Fund 172
European Union 66, 84, 121, 165, 172, 173
excursions 36, 47, 215
existential experience 131, 281
explorers 14, 15, 28, 113, 141

fantasy cities 229–34, 254
fashionability of travel 30–1, 34, 36, 37, 67, 100, 183–4, 238
Fes (Morocco) 168
festival markets 228, 229, 259
Florence 53
'flow' 272
Fordism/ist 16, 21, 83, 209, 281
French Riviera 54

galleries 210, 216, 219, 223, 231, 251–2, 254
gambling 149, 231, 232; effects of 150
garden festivals 226
Glasgow 226
global distribution systems 66
global information systems: development of 65–6
globalisation 21, 48, 51–3, 83–4, 185, 241, 281; and business travel 214–15; and cities 209–10; and heritage 239; through theme parks 202
global warming 119
golf 40, 96, 119, 169
Grand Tour 53–4, 208, 238
gross domestic product (GDP) 102

'habitus' 189, 282
hallmark events 226
Harrogate 91
Haworth (UK) 252
health spas 29
heritage: and aesthetics 240; attractions 248–9; and authenticity 240–1, 254–6; character of 242–7; and commodification 241, 255; concept of 236–8; and consumption 241; definition of 236; and genealogy 253; and heritage tourism 237–8; and human disposition 239; and identity 239, 240–1; and new political economies 241–2; and